CLASS REPTILIA

ENCYCLOPAEDIA OF LIFE SCIENCE SERIES

CLASS REPTILIA

BARON CUVIER

ANMOL PUBLICATIONS PVT. LTD.
NEW DELHI-110 002 (INDIA)

ANMOL PUBLICATIONS PVT. LTD.
4374/4B, Ansari Road, Daryaganj
New Delhi - 110 002

Class Reptilia
First Edition, 1998
ISBN 81-261-0112-1

PRINTED IN INDIA

Published by J.L. Kumar for Anmol Publications Pvt. Ltd., New Delhi and Printed at Tarun Offset Press, Delhi.

LIST OF PLATES IN THE NINTH VOLUME.

LIST OF PLATES IN THE NINTH VOLUME.

Class REPTILIA.

Reptiles have the heart disposed in such a manner, as that, on each contraction, it sends into the lungs only a portion of the blood which it has received from the various parts of the body, and the rest of that fluid returns to the several parts, without having passed through the lungs, and undergone the action of respiration.

From this it results, that the oxygen acts less on the blood than in the mammifera. If the quantity of respiration in the latter animals in which the whole of the blood passes through the lungs, before returning to the parts, be expressed by unity, the quantity of respiration in the reptiles must be expressed by a fraction of unity so much the smaller, as the portion of the blood sent to the lungs on each contraction of the heart, is less.

As respiration communicates to the blood its heat, and to the fibres their nervous irritability, so we find that reptiles have cold blood, and that their muscular power is less, upon the whole, than that of quadrupeds, and *à fortiori* than that of birds. Accordingly, they do not often perform any movements, but those of creeping, and of swimming; and though many of them leap, and run fast enough, on some occasions, their general habits are lazy, their digestion excessively slow, their sensations obtuse, and in cold and temperate climates, they pass almost the entire winter in a state of lethargy. Their brain, proportionally smaller, is not so necessary to the exercise of their animal and vital faculties, as it is in the first two classes of the animal kingdom. Their sensations appear less referable to a common centre. They continue to live and exhibit voluntary motion, after having lost the brain, and even after decapitation, and that for a very considerable time. The connexion with the nervous system is also much less necessary to the contraction of their fibres, and their flesh, after having been separated from the rest of the body, preserves its irritability much longer than in the preceding classes. Their heart will beat for several hours after it has been plucked out, and its loss does not hinder the body from moving for a long time. In many of them, it has been observed, that the cerebellum is remarkably small, which perfectly accords with their little propensity to motion.

The smallness of the pulmonary vessels permits reptiles to suspend their respiration without arresting the course of the blood; accordingly, they dive more easily, and for a longer time, than mammifera or birds: the cellules of their lungs being less numerous, because they have fewer vessels to lodge on their parietes, are much wider, and those organs have sometimes the form of simple sacs, which are scarcely cellular.

Reptiles are provided with a trachea and larynx, though the faculty of an audible voice is not accorded to them all.

Not possessing warm blood, they have no occasion for teguments capable of retaining the heat, and they are covered with scales, or simply with a naked skin.

The females have a double ovary, and two oviducts. The males of many genera have a forked or double organ of generation. In the last order—that of the batracians—they have none.

No reptile sits upon its eggs. In many genera of batracians, the eggs are not fecundated until after they have been laid; and hence they have only a membranous envelope. The young of this last-mentioned order, on issuing from the egg, have the form and the gills of fishes; and some genera preserve these organs, even after the development of their lungs. In many of the reptiles which lay eggs, especially in the colubri, the young one is already formed, and considerably advanced in the egg at the moment when the mother lays it; and it is the same with the species, which may, at pleasure, be rendered viviparous by retarding their laying.*

The quantity of respiration in reptiles is not fixed, like that of mammifera and birds, but varies with the proportion which the diameter of the pulmonary artery bears to that of the aorta. Thus tortoises, and lizards, respire considerably more than frogs, &c. From this proceed differences of energy and sensibility, much greater than can exist between one mammiferous animal and another, or one bird and another.

Accordingly, the reptiles exhibit forms, movements, and properties, much more various than the two preceding classes; and it is more especially in their production that nature seems to have sported in the formation of fantastic shapes, and to have modified, in all possible ways, the general plan which she has followed for vertebrated animals, and especially for the viviparous classes.

The comparison of their quantity of respiration, and their organs of motion, has, however, given a foundation for M. Brogniart, to divide them into four orders, *viz.*: —

I. The CHELONIANS, or TORTOISES (CHELONIA), in which the heart has two auricles, and the body supported on four legs, is enveloped by two

* For example, colubri when deprived of water, according to the experiment of M. Geoffroy.

plates, or shields, formed by the ribs and the sternum;—

II. The Saurians, or Lizards (Sauria), in which the heart has two auricles, and the body supported on four, or on two feet, is covered with scales;—

III. The Ophidians, or Serpents (Ophidia), in which the heart has two auricles, and the body is always destitute of limbs;--and

IV. The Batracians (Batrachia), in which the heart has but one auricle, the body is naked, and the majority of them pass, with age, from the form of a fish respiring by gills, to that of a quadruped respiring by lungs. Some, however, never lose their gills, and some have never more than two feet.*

* Other writers, as Merrem, make a different arrangement of the Sauria and Ophidia. They make a separate order of the crocodiles, and unite, to the rest of the Sauria, the first family of the Ophidians (*Anguis*), a distribution founded on some peculiarities of organization in the crocodiles, and on a certain resemblance between the angues and the lizards. We have thought it sufficient merely to hint at these relations, which are all internal, while we preserve a division more easy of application.

THE FIRST ORDER OF REPTILES.

The Chelonia,

Better known under the name of tortoises, have the heart composed of two auricles and one ventricle, with two unequal chambers which communicate with each other. The blood of the body enters into the right auricle, that of the lungs into the left: but both are mingled more or less in passing through the ventricle.

These animals are distinguishable at the first glance, by the double buckler in which the body is enclosed, and which leaves only the head, neck, tail, and four feet to pass out.

The upper buckler, which is named *carapace*, or back plate, is formed by the ribs, eight pairs in number, widened and united by dentelated sutures between them, and having plates adhering to the muscular portions of the dorsal vertebræ, so that all these parts are without motion. The lower buckler, termed *plastron*, or breast-plate, is formed of pieces which represent the sternum, and which are usually nine in number. A frame-work, composed of osseous pieces, in which there has been supposed to be some analogy to the sternal or cartilaginous portion of the ribs, surrounds the carapace with a cincture, and unites all the ribs which compose it. In one subgenus, this frame-work always remains cartilaginous. The vertebræ of the neck and tail alone are moveable.

These two osseous envelopes being covered immediately by the skin or scales, the omoplate, and all the muscles of the arm, and of the neck, instead of being attached to the ribs and the spine, as in other animals, are attached underneath. It is the same with the bones of the pelvis, and all the muscles of the thigh, which might lead us to term the tortoise, in this point of view, an *inverted* animal.

The vertebral extremity of the omoplate is articulated with the carapace; and the opposite extremity, which may be considered as analogous to the clavicle, is articulated to the plastron, so that the two shoulders form a ring, into which the œsophagus and the trachea pass.

A third osseous branch, greater than the other two, is directed downwards, and hindwards, and represents, as in the birds, the coracoïd process; but its posterior extremity remains free.

The lungs are very much expanded, and in the same cavity as the other viscera.* The thorax being immoveable in the greater number of species, it is by the play of the mouth that the tortoise respires, holding the jaws well closed, and lowering

* Remark, that in all the reptiles, in which the lungs penetrate into the abdomen (and the crocodile is the only one in which this does not take place), they are enveloped, as well as the intestines, in a fold of the peritoneum, which separates them from the abdominal cavity.

and raising alternately the hyoid bone. The first movement lets the air enter by the nostrils; and the tongue then closing their interior aperture, the second motion forces this air to penetrate into the lungs.

The tortoises have no teeth: their jaws are clothed with horn, like those of birds, except in the chelydes, where they are only furnished with skin. Their tympanic cavity, and palatine arches, are fixed to the cranium, and immoveable. Their tongue is short and bristly, with fleshy threads. Their stomach is simple, and strong. Their intestines of moderate length, and without cœcum, and they have a very large bladder.

The male has a simple and considerable penis. The female produces eggs covered with a hard shell. The male may often be recognised by its external appearance, because the plastron is concave.

The tortoises possess amazing tenacity of life. Some have been known to move without a head for many weeks. Very little nourishment is necessary for them; and they can pass entire months, and even years, without eating.

The Chelonians, all united by Linnæus in the genus

TORTOISE: TESTUDO. L.

Have been divided into five subgenera, principally, according to the forms and teguments of their carapaces, and their feet.

Land Tortoises. Testudo. Brongn.*

Have the carapace convex, sustained by an osseous frame-work, altogether solid, and synostosized by the greatest part of its lateral edges to the plastron. The limbs seem as if truncated, and the toes are very short, and very closely united as far as the claws. The limbs, as well as the head, can be withdrawn entirely within the shields. The fore feet have five claws, the hinder four—all thick and conical. Many species subsist on vegetable substances.

The Common Tortoise. Testudo Græca, L. Schœpf. pl. viii. ix.

Is the most common species in Europe. It inhabits Greece, Italy, and Sardinia, and, as it would seem, the entire circuit of the Mediterranean. It is distinguished by its broad carapace, uniformly convex; by its elevated scales, granulated in the centre, striated at the edge, and with large marbled spots of black and yellow; also by the hinder edge, which has, in the middle, a prominence, a little recurved over the tail. It rarely attains a foot in length; subsists on leaves, fruits, insects, and worms; excavates a hole to pass the winter in; couples in spring, and lays four or five eggs, resembling those of pigeons.

Among the foreign species, there are many be-

* Merrem has changed this name into Chersine.

longing to the East Indies, of an enormous size, and three feet or more in length. One of them has been, too, exclusively named,—

The Indian Tortoise. *Test. Indica,* Vosm. Schœpf. Tort. pl. xxii.

Its carapace is compressed in front, and the anterior edge is raised above the head. Its colour is a deep brown.

There are also several species remarkable for the handsome distribution of their colours, as—

The Geometrical Tortoise. *Test. Geometrica,* L. Lacep. I. ix. Schœpf. x.

A small tortoise, whose black carapace has each of its scales regularly adorned with yellow lines, in rays parting from a disk of the same colour. And—

The Radiated Tortoise. *T. Radiata,* Shaw. Gem. Zool. III. pl. ii, and Daudin II. xxvi.

A New Holland species, almost as well and regularly lineated as the geometrical, but which attains a much larger size.*

* Add: *T. stellata,* Schœpf. xxv.;—*T. angulata,* Schweig;—*T. areolata,* Schw., xxiii;—*T. denticulata,* Sch. xxviii. i.:—*T. cafra,* Schw.;—*T. signata,* Schw.;—*T. marginata,* Schw. xii., 1, 2;—*T. carbonaria,* Spix, xvi.; *T. Hercules,* id. xiv.;—*T. cagado,* id. xvii.;—*T. tabulata,* Sch. xiii.; —T. *sculpta,* Spix, xv.;—*T. Nigra,* Quoy et Gaim. Voy. de Freyc., Zool., xxxvii;—*T. depressa,* Cuv.;—*T. biguttata,* id.;—*T. Carolina,* Leconte, &c.

Some species, as the *Pyxis*, Bell. have the anterior part of the buckler moveable, like the box-tortoises; and others, *Kinixys*, Id., can move the posterior part of the carapace.*

THE FRESH-WATER TORTOISES. EMYS. Brongn.†

Have no other constant characters to distinguish them from the preceding than more separated toes, terminated by longer claws, and the intervals occupied by membranes, though in this respect there are gradations. Like the last subdivision, they have five claws before, and four behind. The form of their feet suits their aquatic habits. The majority live on insects, small fish, &c. Their envelope is generally more flattened than that of the land tortoises.

The Speckled Tortoise. Testudo Europæa. Schn. *Orbicularis,* Linn. Schœpf. pl. i.‡

Is the most extended species. It has been observed in all the south and east of Europe as far as Prussia.

* See Mr. Bell's papers in The Linnæan Trans., Vol. XV., Second Part, p. 392. In two of these *Kinixys*, which we have seen alive, the edges of the juncture of the carapace were unequally worn; and as it were carious, so much so, that one might easily believe that there was something morbid in this conformation.

† From εμὺς. (Tortoise.)

‡ It is the same as *la verte et jaune,* Lacep. pl. 6; and his *Ronde,* pl. 5. On this species, the excellent monograph of M. Bojanus, Vilna, 1819, in folio, ought to be consulted.

Its carapace is oval, not very convex, pretty smooth, blackish, and altogether sown with yellowish points disposed in rays. It is about two inches long. Its flesh is an article of food, and for that purpose it is reared on bread and young herbs; it also eats insects, slugs, small fish, &c. Marsigli says, that its eggs take a year before the young are excluded.

The Painted Tortoise. Test. Picta. Schœpf., pl. 4.

Is one of the handsomest species. It is smooth, brown, and each of its scales is surrounded with a yellow band, very broad at the anterior edge. It is found in North America, along streams, on rocks, and trunks of trees, from which it lets itself drop into the water as soon as it is approached.*

We must remark, among the fresh water tortoises,

* Add: *Em. lutaria,* Lacep., pl. 4;—*Em. Adansonii,* Schweig; *Em. Senegalensis,* Dumer; —*Em. subrufa,* Lacep., xiii.; —*Em. contracta,* Schweig;—*Em. punctata,* Schœp., 5;—*Em. reticulata,* Leconte;—*Em. rubriventris,* id.;—*Em. serrata,* Daud. ii, xxi.;—*Em. concinna, Lec.* or *geometrica,* Lesueur;—*Em. pseudogeographica,* Lesueur; *Em. scripta,* Schœpf,. iii., 4;—*Em. scabra,* id., iii.;—*Em. cinerea,* id., ii., 3.—*Em. centrata,* Daud., or *terrapen,* Lin. Schœpf., xv.;—*Em concentrica,* Lec.;—*Em. Odorata,* id.;—*Em. fusca,* Lesueur;—*Em. leprosa,* Schw.;—*Em. nasuta,* id.;—*Em. dorsata,* Schœp.;—*Em. pulchella,* Schœp., xxvi., or *insculpta,* Lec.;—*Em. lutescens,* Schw.;—*Em. expansa,* id.;—*Em. macquaria,* Cuv.

M. Fitzinger separates under the name of Chelodina, and Mr. Bell under that of Hydraspis, the species with more elongated neck, such as, *Em. longicollis,* Shaw; *Em. planiceps,* Schœp, or *canaliculata,* Spix, viii.;—*Em platicephala,* Merrem;—*Em. depressa,* Spix, iii., 2;—*Em. carunculata,* Aug. St. Hil.;—*Em. tritentacula,* id.

The Box Tortoise.*

In which the plastron, or breast-plate, is divided into two lids, by a moveable articulation, and which can entirely close their carapace, when the head and limbs are withdrawn into it.

Some have the anterior lid alone moveable.†

In others, the two lids move equally.‡

There are, on the contrary, tortoises of the fresh water, whose long tail and voluminous limbs cannot enter completely within their bucklers. In this respect they approach the subsequent sub-genera, and especially the chelydes, and consequently merit a distinct notice.§

Such is,

The Snake Tortoise, *T. Serpentina*, L. Schœpf. pl. vi.

Which may be recognised by its tail, which is almost as long as its carapace, and bristling with sharp and denticulated crests, and by its scales, which are raised

* It is of this subdivision that Merrem has made his genus Terrapene; Spix his genus Kinosternon; and Fleming his genus Cistuda. The European species, and others, have also something of this mobility, which renders it difficult to define the genus.

† *Test. subnigra.* I. vii. 2;—*T. Clausa*, Schœpf, vii.

‡ The box-tortoise of Amboyna. Dand. II. 309; *Test. tricarinata*, Schœpf. II.; *Test. Pensilvanica.* I. d, xxiv.

§ M. Fitzinger has made of this subdivision his genus Chelydra, and Fleming his genus Chelonura.

in a pyramidical form. It inhabits the warm parts of North America, destroys many fish and water-fowl, removes occasionally pretty far from rivers, and weighs sometimes above twenty pounds.

The SEA-TORTOISES, OR TURTLES. CHELONIA. Brongn.*

Have their envelopes too small to receive their head, and especially their feet, which are extremely elongated (particularly the fore-feet), flatted into fins, and all the toes are closely united and enveloped in the same membrane. The first two toes of each foot alone have pointed claws, which fall not unfrequently at a certain age. The pieces of their plastron do not form a continuous plate, but are variously denticulated, and leave great intervals, which are occupied only by cartilage. The ribs are narrow, and separated one from the other at their external part. Nevertheless, the round of the carapace is entirely occupied by a circle of pieces correspondant to the sternal ribs. The temporal foss is covered above by a vault, formed by the parietals, and other bones; so that the whole head is provided with a continuous osseous casque. The œsophagus is armed throughout, inside, with cartilaginous and sharp points, directed towards the stomach.

* *Chelonia* from χελωνη. Merrem has preferred the barbarous name of CARETTA.

The Green Turtle, Testudo Mydas, Lin. *T. Viridis.* Schn. Lacep, i. 1.

Is distinguished by its greenish scales, thirteen in number, which do not cover each other like tiles; and those of the middle range are pretty nearly in regular hexagons.

It is about six or seven feet long, and weighs as much as seven or eight hundred pounds. Its flesh furnishes an agreeable and salutary aliment to sailors in all the latitudes of the torrid zone. It feeds in large flocks on the alga at the bottom of the sea, and approaches the mouths of rivers to respire. Its eggs, which it deposits on the sand in the sun, are very numerous, and are excellent eating. Its scales are not employed.

An approximating species (*Chel. Maculosa* Nobis) has the middle plates twice as long as they are broad, and of a fawn-colour, marked with large black spots; and another (*Chel. Lachrymata* Nobis), with plates like the preceding, has the last one elevated into a boss, and broad irregular black stripes on the fawn-colours. Their scales are employed with utility.

Imbricated Turtle. Testudo Imbricata. L. Lac. I. ii. Schœpf. xvii. A.

Smaller than the green turtle, with a more elongated muzzle, and denticulated jaws. It has thirteen fawn-coloured and brown scales, which cover each other like tiles. The flesh of this species is disagreeable and unwholesome; but the eggs are very delicate, and it furnishes the finest tortoise-shell which is employed in the art. It is found in the seas of warm climates.

There are also two species approximating to this, *Chel. Virgata* nobis—Bruce, Abyss. pl. 42—which has the plates less elevated; those of the middle equal, but with sharper lateral angles, and black and radiated streaks on its scales :—and *Chel. radiata,* Schœpf. xvi. B., which does not differ from the preceding but in having the last of its middle plates broader. It is probably only a variety.

Loggerhead Turtle. *Test. Caretta.* Gen. Schœpf. pl. xvi.

Is more or less brown or red, and has fifteen scales, the middle ones of which are raised in ridges, especially towards their extremity. The point of the upper bill is crooked, and the fore-feet are longer and more narrow than in the preceding species, and preserve, too, more marked claws. It inhabits many seas, and even the Mediterranean, and lives on shell fish. Its flesh is bad, and the scales in no estimation; but it furnishes a good oil for burning.

Merrem has recently distinguished, under the name of SPHARGIS, the Chelonians whose shell is not scaly, but covered with a sort of leather only.*

* Fleming names them CORIUDO; M. Lesueur, DERMOCHELIS.

Such is one great species of the Mediterranean—

The Coriaceous Turtle. *Testudo Coriacea.* L. Lacep. I. iii. Schœpf. xxviii.

Its oval carapace, and pointed behind, presents three longitudinal ridges projecting through the hide.*

THE CHELYDES. CHELYS. Dumeril.†

Resemble the emydes in the feet and claws. The envelope is much too small to receive the head and feet, which have considerable volume. The nose is prolonged into a small trunk; but the most marked of their characters is the mouth, which is cleft crosswise, not being armed with a horny beak, like the other Chelonians, but resembling that of certain batracians, particularly the *pipa*.

Fimbriated Tortoise. *Testudo Fimbriata.* Gm. Bruguières Journ. d'Hist. Nat. I. xiii. Schœpf. xxi.

The carapace bristling, with pyramidal eminences; the body edged all round with an unequally indented fringe: it is found in Guiana.

* Add, *Dermochelis adantica,* Lesueur.

† Merrem has preferred for this genus the barbarous name of MATAMATA.

The Soft Tortoises. Trionyx. Geoff.

Have no scales, but only a soft skin to envelope their carapace and plastron, neither of which is completely supported by the bones, the ribs not reaching the edges of the carapace, and not being united together but in a portion of their length, the part analagous to the sternal ribs being formed of a simple cartilage, and the sternal pieces partly indented, as in the sea-tortoises, not filling all the lower face. It is easy to perceive, after death, through the dried skin, that the surface of the ribs is very rugged and uneven. The feet, as in the fresh-water tortoises, are palmate, without being elongated; but three of their toes only are provided with claws. The horn of their beak is clothed externally with fleshy lips, and their nose is prolonged into a small trunk. The tail is short, and the anus pierced under its extremity. They live in the fresh water, and the flexible edges of their envelope assist them in swimming.

The Tyrsé, or *Soft Tortoise of the Nile*. *Testudo Triunguis*. Forsk. and Gm. *Trionyx Ægyptiacus*. Geoff. Ann. du Mus. xiv. 1.

Sometimes three feet long; green, spotted with black, and the carapace not very convex. It devours the little crocodiles at the moment when they

1. **EMYS.** *concinna.*

2. **CHELONIA.** *virgata.*

3. **TESTUDO.** *depreſsa.*

4. **TRYONIX.** *gangeticus.*

escape from the egg, and renders more service in this way to Egypt than the ichneumon.*

The Fierce Tortoise. Testudo Ferox. Gm. Penn. Trans. Phil. LXI. x. 1—3. Cop. Lacep. I. vii. Schœpf. xix.

Inhabits the rivers of Carolina, Georgia, Florida, and Guiana; remains in ambuscade under the reeds, &c., and seizes birds, reptiles, &c.; devours the young alligators, but becomes the prey of the old ones. Its flesh is good eating.†

* Sonnini, Voyage en Egypt, t. ii. p. 333.

† Add, *Trionyx Javanicus*, Geoff. Ann. du Mus. xiv.;—*Tr. Carinatus*, id.;—*Tr. Stellatus*, id.; *Tr. Euphraticus*, Olivier, Voy. en Turq. &c. pl. 42;—*Tr. Gangeticus*, Duvaucel;—*Tr. Granosus*, Leach; or *Test. Granosa*, Schœpf. xxx. A. and B.

N. B. The tortoise of Bartram appears to be the *testudo ferox*, to which the artist by mistake has given two claws too many on each foot.

SUPPLEMENT ON REPTILES IN GENERAL.

In the history of nature, exhaustless as its subject, and varied as the prodigious multitude of productions which it examines, there are parts capable of occupying, for a series of years, with ever-growing interest, the true lover of science, though presenting to the vulgar mind no images but those of terror and disgust. They relate to animals, from which the majority of mankind start with involuntary abhorrence, and which in almost all nations, and all ages, have been dreaded for their malignancy, or despised for their stupidity. In the popular superstitions of different lands, the reptile races have almost invariably been clothed in revolting attributes, and even the worship which has sometimes been paid to them, was a religion, not of gratitude, but of fear. The God of Day was armed by the Grecian mythology with his unerring shafts to pierce the enormous Python; the terrific Acheloüs was strangled, by the son of Jove, in spite of his contorted foldings. The Garden of the Hesperides, and the Golden Fleece, were protected by furious dragons. The serpents of the dripping head of Medusa were sown by Perseus on the arid Lybian sands. The Gorgons, the Furies, Discord and Envy, are armed with snakes by the poets, as an appropriate emblem of their ministry of vengeance and torture.

But of what consequence, in the estimation of the sage, are the vain opinions and absurd prejudices of mankind? To him are equally indifferent the dream of the poet and the prepossession of the clown. Like a new Cadmus, he becomes the vanquisher of monsters, assisted by the ægis of science and the wand of discovery. He finds that the power of nature is manifested with as much glory in these vile objects of universal animadversion as in the more favourite races, which excite our admiration, or awaken our cupidity. Her energies are equally exerted, and her manifold resources equally developed in their production. The philosopher calmly proceeds to their examination and study—to their enumeration and classification. He sees nothing in the class of reptiles but animals singular in their forms, curious in the diversity of colours by which they are embellished, marvellous in the metamorphoses of some species, and in the extraordinary habits of almost all. Scarcely one-sixth of all the individuals of this entire class is venomous. Many among its species furnish wholesome and abundant aliment, restorative medicines, and productions useful in the arts. Some, even the fiercest, such as crocodiles, have been tamed, and will suffer infants to sport upon their backs. So true it is that the superiority and domination of man extend over beings of every class. He can derive from the most ignoble, or the most odious, useful supplies for the necessities of his existence, or new ideas for the extension of his intelligence. Such are among the especial privileges granted to man by the Author of his being, which elevate him so considerably above all other animals, and mark the dignified character of his destiny upon earth.

Few beings, indeed, are more worthy of the attention of the thinking observer than the proscribed and persecuted animals to whose history the course of our labours now conducts us. If the graphic and eloquent descriptions, suggested to the historians of nature by the two preceding classes of the animal kingdom, have power to instruct and delight us, with no less pleasure and profit may we accompany them in their researches on the present, and penetrate into the sombre retreats of the reptile races in the bosom of the earth, behind the broken masses of the rock, or under the scattered débris of gigantic vegetables. We may pursue their evolutions over the tranquil surface of lakes, of streams, and rivers; mark the tortuous folds by which they attach themselves to the branches; and unveil the mechanism by which they creep, climb, walk, run, leap, and even fly.

Reptiles consist of oviparous quadrupeds and serpents. To the first, the name of reptile is as suitable as to the last: for though they have feet, they make little use of them, except in creeping, and their belly almost always touches the ground. Tortoises, lizards, frogs, toads, and salamanders, afford sufficient proof of this. Though the three last-mentioned genera live in the water, and swim there with facility, they also live on land very well. For this reason some naturalists have considered them as true amphibia. But, in fact, for an animal to be amphibious, in the strictest acceptation of the term, it is necessary that it should possess the power of respiring under the water like fishes, and on the earth like man—none, therefore, of these animals are true amphibia, except, perhaps, the siren and the proteus, which possess both lungs in the chest, and external gills. Frogs, toads, and salamanders, when in the tadpole state, are provided only with gills, which respire the water, and, accordingly, in this tadpole state, they cannot live out of the liquid element. When they become perfect animals the gills disappear, and they breathe by lungs, consequently, they are then obliged to respire the air, and would perish by suffocation under water, were they forced to remain submerged for too long a period of time.

It is the more necessary to insist on all this, as the term *amphibious*, in popular language, like most other terms of Natural History, is grievously abused—we have heard it applied to the hippopotamus, the seal, and even to the beaver.

There are fewer links of connexion in reptiles than in birds to viviparous quadrupeds, which may be considered as the *élite* of the animal kingdom, notwithstanding that the majority of reptiles have four feet. It is rather with birds and fishes that they seem to exhibit an alliance of manners and habits. There is an inter-resemblance in these classes in many points of internal organization, and even in the external forms of some species; but it is more especially striking in the reptiles and the fish.

Vertebrated animals, with cold blood, may, in fact, be considered as almost forming another world. They preserve some analogy, it is true, with the superior classes, in the bony skeleton, in the general arrangement of the brain, of the senses, and of the principal viscera; but the heart, both in reptiles and fishes, has but one ventricle, or cavity. The vesicular lungs of the reptiles, instead of receiving, as in mammifera and birds, the entire blood to be impregnated with the vital air, receives but a small streamlet of the venous blood, which is even oxygenated but feebly, for these animals breathe but very slowly through this pulmonary viscus, the tissue of which is so very lax. From this it results, that the blood, scarcely warmed and vivified by combination with the vital air, excites but languidly the entire organization; accordingly, we find the reptiles nearly cold to the touch like inanimate bodies: for this reason, they are observed to seek and court atmospheric heat, or the warm sunshine; and the cold of winter reduces them to a state of torpidity. They seem, for the most part, to vegetate rather than live, to be insensible of a wound, and even scarcely to discover any considerable degree of anguish when cut in pieces. Their organization very speedily renews many parts, such as the tail or toes, when they have been removed. As these animals have but very little cerebellum in proportion to their size, and a brain composed of but six small tubercles, their existence is not so absolutely concentrated in their head as ours. It seems rather to be attached to their spinal marrow, and to be more generally disseminated throughout their body. A tortoise has been known to live for eighteen days after the brain was removed, still walking about, but groping its way, for its eyes were closed, and the power of vision lost in consequence of the cutting of the optic nerves. A salamander has lived several months although decapitated by means of a ligature fastened tightly round the neck. The heart of a viper, when plucked out, will beat and contract on being pricked for the space of forty hours. From all this it appears that these animals have not such a centralized life as that of a quadruped or a bird, which would instantly perish from similar amputations. This pertinacious irritability in frogs and serpents renders them very proper subjects of galvanic and electrical experiments. Electricity is found to exercise a most powerful influence on them. Reptiles are exceedingly sensible to storms, and to an electric state of the atmosphere, of which they appear to forsee all the changes, as appears by the croaking of frogs, &c.

This want of concentration of vitality in the brain has, in the reptiles, as its natural accompaniment, a marked diminution of intelligence; and though, as we hinted before, some of them can be tamed, it is next to impossible to teach them any thing.

The system of respiration in reptiles is the principal character which separates them from all other animals, and exercises the most powerful influence over all the parts of their organization. In organized bodies there are certain general modes of conformation which necessitate a multitude of particular conformations. Thus, for example, the animal whose stomach is found to digest flesh, must be furnished with teeth proper for tearing it, robust muscles for vanquishing his prey, agile limbs for overtaking it, &c. In like manner, the external organs of every being are all in relation to the wants of the internal organs; the latter must, therefore, be investigated, if we want to ascertain the cause which determines the conformation of the former.

Moreover, there are in each class of animals, and of plants, certain parts which may be said to give the impulse to the whole organized machine, and certain organs which assume an ascendancy over the other organs, in consequence of the extent and the energy of their functions. Thus, the pulmonary apparatus is predominant in birds; in man, the brain and the nervous system; and in the carnivora, the muscular system, because the said organs in these different animals are more developed and more active than all the rest. In reptiles and in fishes, it is the muscular contractility which claims pre-eminence, above all other properties, for its influence and duration.

But if it is essential to recognize the predominating organs in any class of animals, it is equally important to study those which are feeble and comparatively inactive, because they produce effects inverse of those of the former in the living economy. The knowledge, in fact, of the one, necessarily determines the others, and weakness thus becomes as influential as strength on the animated machine. In truth, it is to the imperfect state of the respiratory apparatus that the entire constitution of reptiles must be referred.

These animals all respire the air because they have lungs. But this organ is vesicular, and the blood vessels which arrive at it, are only branches of the vena cava and the aorta, so that those vessels form no considerable system, and transport but a small quantity of blood to the lungs, instead of a mass of this fluid almost equal to that of the rest of the body, as in warm-blooded animals. For this same reason the heart of reptiles possesses but a single ventricle, which suffices to make their blood circulate, independently of respiration. The latter may remain suspended for some time without interrupting the course of life, and the circulation of the fluids. This is witnessed in frogs, salamanders, and marine tortoises, which dive under water, or bury themselves in mud for entire days. The colder the atmosphere is, the longer these animals can subsist under water, without having occasion to respire the air, and without perishing, for they are then in a state of semi-torpor. But in warm weather, the respiration of the atmospheric air becomes necessary, and the reptiles enjoy a more active life.

In consequence of the imperfect respiration of reptiles, the vital air combines with the portion of blood which is exposed to it, in small quantity only, and disengages very little heat from the sanguine fluid; for the degree of heat is always in proportion to the extent of the respiratory function. Accordingly, these animals are all cold, and the temperature of their blood exceeds, by a very few degrees, that of the atmosphere; while the birds, as we have formerly observed, which respire so vigorously, are always exceedingly hot, and almost burning; in fact, they are naturally, and habitually, in a state which would be pronounced fever in a mammiferous animal.

If the reptiles are naturally cold, the rigour of winter may be presupposed to exercise a powerful influence over them. This, accordingly, is the fact. They all grow torpid during the cold season, and do not awaken until spring expands its genial warmth. In this state of torpor they are immoveable and almost frozen. Their blood circulates very slowly. Their sensibility and life are suspended; they remain plunged in a profound sleep, and lose scarcely any thing by transpiration, because they are covered by a thick, coriaceous, and almost impermeable skin.

It is also in consequence of this natural coldness of reptiles, that we must account for their almost total disappearance from the polar regions, and the more rigorous latitudes of the north, while on the contrary they are excessively multiplied in the ardent climates of the tropics, where the heat of the earth supplies the deficiencies of their respiration. This feebleness of respiration presents to our notice another result equally remarkable in the reptiles. As it causes a sort of stagnation, a slowness, an almost continual insensibility in the life of these animals, it follows that this life is less rapidly worn out and exhausted. We see, in general, that life is longer in proportion as it is less active, unless its course be abridged, or its thread be cut, by maladies or unforeseen accidents. We have, all of us, a determinable sum of existence, as it were, which we may expend more or less quickly. The reptile, which lives, if we may so express ourselves, but little at a time, and remains torpid for a part of the year, should naturally exist a good while. This we find to be the fact. It has even been asserted that the crocodile continues to grow during almost its entire life, which is a certain mark of longevity: for as long as the growth is going on, the animal is young, and its cessation is the mark of approaching age. The serpent, among the Egyptians, the Greeks, and the Mexicans, was the emblem of eternity, or of time, in consequence of its long life. It even seems every year to assume a rejuvenescence in casting its former skin, like the earth, which in winter throws aside its faded livery only to invest itself with richer verdure on the return of spring. Were not the reptiles exposed to be destroyed by their adversaries, in consequence of their tardy movements, and general want of the means of defence, they would become too numerous, as to a long life they unite an exceeding fecundity.

There is one very singular property in the reptile races, which has been noticed in the text. This is the power of reproducing certain parts, such as the tail, feet, &c., when they have been lost. This fact is particularly demonstrated in salamanders and lizards, and was known as long ago as the time of Aristotle. They are also, as has before been hinted, remarkable for their extreme tenacity of life, and the long duration of their fibrous irritability after life is extinguished.

The weakness of respiration diminishes the activity of the nutritive system in reptiles, because the one is always in relation with the other. Accordingly, these animals eat but little, and digest slowly.

The small quantity which reptiles eat, is another reason for the slowness of their growth, and the length of their existence; and the same character is also connected with the inactivity of their senses. Their organs of sensation seem scarcely developed. Their touch is very obtuse, in consequence of the density and hardness of their skin. The sense of taste cannot be otherwise than dull, because the torgue is either cartilaginous, or covered with a thick and viscous humour. The smallness of the organs of smelling, indicate the weakness of that sense. That of hearing appears to be less imperfect, though its organ in reptiles is destitute of many useful parts, such as the cochlea, the conch, and the meatus externus. Even the tympanum is usually covered with skin, scales, or muscles. Sight is the most perfect sense in reptiles. They have, for the most part, very large eyes, a contractile pupil, like that of cats (especially the geckos, which appear to see clearly by night), and a nictitating membrane, the same as in birds. This indicates a great sensibility in this organ in these two classes of animals, and the necessity under which they labour, of having the intensity of the light moderated in its action on their eyes. Nevertheless, the CÆCILIA, a genus of serpents approaching the batracians, have excessively small eyes concealed under the skin. The brain of reptiles is remarkably small, and does not even completely fill the cavity of the cranium, though that is far from being capacious.

We shall now consider the nature of those localities in which the reptile races most generally abound.

Though shaded and humid tracts of land, and slimy

marshes, nourish in our European climates the majority of those reptiles which breed among us, it is necessary to study this class of beings in their natural empire, in the deep morasses, the lakes, the rivers, the wastes of uncultivated vegetation, where every thing concurs to their development and multiplication, under the burning suns of tropic and equatorial regions. We shall avail ourselves here of an eloquent description relative to this subject, given by Buffon, in his history of the *Screamer*, or *Kamichi*, a bird already described in our account of the order Grallæ.

"We have already painted," says this delightful writer, "the arid deserts of Arabia Petrea, those naked solitudes where man has never yet breathed beneath the shade; where the earth, destitute of verdure, presents no subsistence to quadruped, to bird, or to insect; where all appears dead, because nothing can be born, and because the element necessary to the developement of the germs of every being, whether animal or vegetable, far from impregnating the earth with streams of living water, or fertilizing it by penetrating showers, does not even moisten its surface with a simple dew. To this picture of absolute drought, in a land too ancient, let us oppose that of the mighty morasses and inundated savannahs of the New Continent. Here we shall behold the excess of that of which we there lamented the deficiency. Rivers of immense breadth, such as the Amazons, La Plata, and the Orinoco, rolling onward their huge and foaming billows, seem to menace the earth with invasion, and attempt to overwhelm it altogether. From them originate stagnant waters, which spreading far and near, cover the alluvial slime which they have deposited. These vast marshes exhaling their vapours in fetid fogs, would soon communicate the infection of the earth to the air, did not their exhalations quickly fall again in precipitous rains, or become rapidly dispersed by storms and hurricanes. These regions, alternately inundated and left dry, where earth and water appear to contend for undefined possessions; the masses of aquatic herbage, which are cast on the undecided confines of those belligerent elements, are peopled only by unclean animals, which swarm in those revolting haunts, the very cloaca of nature, where every thing retraces the image of those monstrous births which mythological poetry has fabled to have sprung from the slime of the ancient earth, on the recession of the waters of the deluge. Enormous serpents trailing their slow length along, impress the miry soil with broad and indented furrows; crocodiles, toads, lizards, and a thousand other reptiles, knead with broad feet the yielding clay. Millions of insects, developed by the humid heat, rise in dense masses from their muddy cradle, obscuring the atmosphere with their clouds, and deafening the ear with their hummings. All this impure population of creeping, or of winged vermin, attracts in numerous cohorts the birds of prey, whose piercing cries, mingled with the hoarse croakings of their reptile victims, breaking the awful silence of those frightful deserts, seem to interdict the approach of every other living being. Regions like these, impracticable and unformed, recal our imagination to chaotic ages, ere the elements were yet separated, when earth and water made but one common mass, and the living tribes had not yet received their allocation in the different districts of nature."

It is amid these masses of aquatic vegetation that the alligator lies concealed in ambush for his unsuspecting prey. It is in these impure marshes that the tortoise buries itself —that serpents devour each other—that large and hideous toads withdraw themselves from the light of day, and secrete their venom. Painful it is, without doubt, and disgusting to the mind, to dwell on images like these; but strange marvels are hidden in these gloomy recesses, where the most wonderful operations of nature are proceeding in the very bosom of putrefaction ;—where those fearful legions of amphibious monsters, and those incalculable multitudes of pestiferous insects are engendered, developed, and reproduced.

The reptile, cast by nature into this intermediate domain, between the waters and the land, and as it were in the battle-field of those two elements, is neither a perfect quadruped, like the mammiferous animal which treads the solid earth, nor a true fish, like the multitudinous population of the seas. It is a sort of divided being, one of those ambiguous productions of nature, now appearing as a quadruped, and now as a fish.

This inconstancy of the medium inhabited by reptiles, is the cause of the fantastic variety of their conformation, and the strange singularity of their habitudes. It was necessary that they should be fashioned to those perpetual elementary changes, and enabled equally to subsist in the water, on the earth and in the air. It is as necessary that those species, which seem, as it were, to have been deformed and disgraced by the hand of nature, which denuded of defence, or even deprived of limbs, trail themselves along with effort and with difficulty, should be protected by their prudence—should glide in obscurity—should retreat from the light of heaven and the persecutions of their foes. The slow-paced tortoise, retiring under his osseous roof, awaits without apprehension the various shocks to which he is exposed. The more agile lizard shoots into some hole or cavern, at the hazard of abandoning his tail, the loss of which he can easily repair ; but the serpent, unable, for want of limbs, to avoid his enemies, would in vain exert himself to escape. Nature has, therefore, provided some of the slowest and most feeble species of this kind with a terrible weapon, impregnated with fatal venom, to repel the aggressions of injustice : for we must be equitable even to the serpent race, and confess, that they but seldom act on the offensive ; on the contrary, they are rather timid than ferocious—conscious of their own weakness, they are the emblem of prudence, and have been instanced for their wisdom by no meaner an authority than the great Founder of Christianity himself. It is only when forced by necessity, by despair of escape, or by the imperious stimulus of hunger, that they employ those empoisoned arms, so deadly to all other races, but perfectly harmless to their own.* To the more powerful species, such as the boa, similar weapons are not accorded, because they are sufficiently protected from attack by their magnitude and strength. Though some smaller species of reptiles, such as geckos, salamanders, and toads, can distil from their skin an acrid, fetid, disgusting humour, this is an harmless defence accorded to them by nature, to hinder these otherwise defenceless animals from being seized, and to prevent them from being immolated to the general hatred in which they are held by all other creatures.

In general, all the reptiles, though hideous and even terrific to the sight, cause more repugnance or horror than real evil. Nature environs them with this apparatus of terror, covers them with this revolting mask, to keep off other animals, and thus protects them by the disgust which they inspire. That their existence was necessary, is argued from this consideration, that the impure and filthy recesses which they inhabit, swarming with such a fearful multitude of worms and insects, would become more fatal and more infectious from this enormous multiplication of parasite creatures, but for the ministry of the reptile tribes, which feed upon, and purge the marshes of these vermin. In the same point of view, the reptiles, too much multiplied by a superabundance of aliment, would have invaded and overrun the earth, but nature has sent quadrupeds, such as the ichneumon and the swine, or

* This, perhaps, is a little too unqualified. The fact is not yet clearly ascertained.

legions of long-legged aquatic birds, which come to devour the reptiles in their turn. Thus the ibis performs this purgative operation on the alluvial depositions of the annual inundation of the Nile. The stork performs the same office for the marshes and stagnant .waters of Holland; as do likewise for other places the flocks of cranes, which transport themselves into the various climates of the earth. It may be superfluous, however, to refer the philosopher to these and such like conditions of existence; he knows that the system of being, in all its changes and modifications, is a system of necessity; he knows that the laws of physics are just as immutable as those of mathematics. It is nothing but our own ignorance, short-sightedness, or inattention, that ever makes us imagine them to be otherwise.

It is most particularly in warm climates that reptiles multiply to the greatest extent, that they arrive sometimes to an enormous size, and that the poison of the venomous races becomes most acrid and pernicious. There are marine tortoises in the Antilles so large, that fourteen men may stand at once upon their backs. The missionary Labat tells us, that he has been repeatedly carried in this way. A tortoise of this size would suffice for the repast of a hundred men: it is cooked in its shell, as in a great dish. Enormous crocodiles are found in Africa; and these animals, which are very small when born, continue to grow for so long a time, that they arrive at last to those exceeding dimensions. Those of the Nile, adored by the ancient Egyptians, attained even to forty feet in length; but they have been seen in Madagascar as long as sixty. The gavials, or crocodiles of the Ganges, have jaws five or six feet in length—all bristling with long, sharp, and curved teeth. The alligators, or caymans of America, are extremely numerous in the lakes: they howl ferociously night and morning. All these animals, though agile enough, cannot turn themselves with facility.

Serpents have been found in the East Indies more than fifty feet in length. When these enormous serpents creep along, one might well imagine that it was the great trunk of a pine-tree, or the mast of a man of war, advancing, and winding through the bushes, and leaving immense furrows behind it. We are told by the Roman historian, that Regulus found between Carthage and Utica a monstrous serpent, which darted on the soldiers when they went to draw water from the river. It strangled them in its folds, and stifled and poisoned them with its breath. No darts could pierce its scales, which were as brilliant as brass. Catapultas, and other machines of war, were raised against it. Its spine was thus broken by an immense fragment of rock, discharged with great violence; but it even still remained formidable to the army, whose cohorts had some difficulty in despatching it. Regulus sent its skin to Rome, which was a hundred and twenty feet in length. When Cato traversed Lybia with the wrecks of the army of Pompey, he was surrounded by a multitude of serpents, by whose deadly bites great numbers of his warriors perished, whose sufferings have been sublimely depicted by the majestic pen of Lucan. The poets, to explain the abundance of serpents and other reptiles in the Lybian sands, have invented the fable, to which we before alluded, that the blood which dropped from the head of Medusa, cut off by Perseus, was changed into these animals:—

> " Gorgonei capitis guttæ cecidêre cruentæ
> Quas humus exceptas varios animavit in angues
> Unde frequens illa est, infestaque terra colubris."
>
> Ovid, *Met.* iv.

The external coverings of reptiles present general characters not difficult of distinction. If all the viviparous quadrupeds are, more or less, covered with hair, and birds with feathers or a warm and light down, at least on some parts of their bodies, nothing similar is ever found among oviparous quadrupeds or serpents. The skin is naked in frogs and salamanders, scaly in lizards and serpents, covered with an osseous shell in tortoises. Those which have a naked skin, like all the batracians, are susceptible of absorbing a deal of water through its pores, which to these animals is a substitute for drink; for though frogs and salamanders are often plunged in water, they drink none of it; but their skin imbibes this liquid, and distributes it to the internal organs. It even appears, from recent observations, that water partly ministers to the respiration of these batracians, through the skin: for we know that the skin in all animals has the power of absorbing oxygen, and can thus receive this elemental gas contained in the water, in the same manner as do the gills of fishes. The water even collects in a sort of bladder, which it has been supposed was intended to receive the urine; but this last secretion is immediately transmitted to the cloaca, or intestinem rectum, by the ureters; and the water of the bladder in toads and frogs, which has been regarded as a venemous urine, has no more odour nor sapidity than distilled water.

Frogs and salamanders have certain glands upon the skin, from which exudes an acrid and virulent humour, which in the brown toad has the smell of garlick. A very dangerous humour, of a similar kind, exudes from the feet of geckos. The crocodile has, towards the neck, the anus, and under the axilla, certain glands which spread a musky odour; and many tortoises exhale the like. Something analogous is found about the glands of the thighs in lizards and chalcides, and near the anus of amphisbenæ, especially in the season of reproduction. Adders and serpents exhale a nauseous vapour, which produces sensations of faintness, or even a putrid and pestiferous odour, which the negroes can perceive at a very considerable distance. Their prey does not seem to digest, but by putrifying in the stomach, which produces this disagreeable effluvium—as persons who digest badly have a fetid breath, especially in the morning. This fact has, probably, given rise to the notion of the pretended fascination exercised by serpents on their prey; to which may be added the lively terror which they inspire by their fiery glances, wide-gaping mouth with trenchant teeth, upreared body, and tremendous hissings of rage and fury. What weak animal would not be terrified at that which would make the bravest man recede, were he unarmed? It is thus that it has been pretended that serpents could charm birds, and make them fall by this fascination into their mouths. It is, unquestionably, to this oppressive vapour and horrid aspect, that we must attribute the strange effects of the look of toads on dogs, and even on men themselves.

The power of another over us is very frequently founded on the weakness of our imagination. A strong individual, by his very aspect, inspires terror in the timid—as the dog is seen, by his glance alone, to arrest the game; and as we see persons blush with shame, turn pale with fear, or lose the power of articulation, at the mere look of another. In warm climates, where the imagination of the inhabitants is very ardent and susceptible—and, indeed, among the savage and semi-barbarous of all countries, even our own—we find this doctrine of the fascination of looks very generally prevalent. The envenomed glances of envy or of hatred are dreaded with regard to infants, and other delicate beings, supposed to be subjected to their influence. What has not been written and spoken, and with much truth, concerning the looks of love and their prodigious empire? Animals are not destitute of this sensibility, or rather this moral impressibility: they act upon each other by these looks and influences, as we act upon them, and as they unquestionably can upon us. A story has been related of a pretended magician, who having shut up several toads in a large box well closed, was desirous, after some time, to see the result of this confinement. On opening the box, however, whether from horror at the sight of these deformed creatures, heaped together, and fixing with fury their huge eyes upon him—or overcome by the fetid odour which they exhaled, he turned pale, and fainted ; while the reptiles, having escaped, were found croaking and jumping around him. The people who witnessed this spectacle, immediately conceived the man to be surrounded by demons, in the shape of these reptiles, punishing him for his sorceries and wickedness. Toads have been seen as large and inflated as pumpkins, covered with sanious pustules, and opening, in their obscure hole, their large projecting eyes and wide mouth. It is not wonderful that dogs should tremble to attack an hideous enemy of this kind, especially as the acrid humour exuding from the skin is highly corrosive to the mouth of the animal that is adventurous enough to bite one of these creatures.

Many species of oviparous quadrupeds are endowed with the singular property of changing colour, through certain moral affections. The case of the cameleon is known to every body, which does not, however, as has been pretended, assume the tints of the objects which surround it, but which changes the shade of its colours, through fear, anger, love, cold, heat, &c., as we shall more fully enlarge upon hereafter in the proper place. The common frog, and its kindred species, also change colour, and become more grey or brown when terrified. The green lizard, the green iguana, and many other reptiles, are liable to the same changes of colour. All the animals of this class, like most others, exhibit more lively and brilliant tints in the season of their amours than at any other period.

The skin of the majority of reptiles is armed with polished and shining scales, embellished with metallic reflexions, like those of brass or steel, relieved with gold and silver, as is seen in various lizards and serpents. They also display the azure of lapis lazuli, and the hues of the emerald, the amethyst, and the turquoise. This variegated armature is more especially splendid in spring, when these reptiles throw off the old epidermis of the preceding year. Then they may be seen, in renewed youth, gliding under the new-born vegetation, glittering in the sun-beams, and essaying their recovered vigour and agility. Among the reptiles with thick skin, this renovation takes place but once a year. In the rattle-snakes, this epidermis of the body sliding towards the tail, remains adherent there in the form of small bells—a marvellous provision of nature, who by this means warns other animals of the approach of these redoubtable serpents, whose poison is the most fatal of all. The batracians, instead of parting with a solid epidermis, detach it in shreds.

The explication of the change of colour in variable reptiles seems to belong to the following mechanism. The skin of all the multi-coloured reptiles is not at all of the nature of the feathers of some birds, which assume different tints according to the various reflections of the light, like silk of certain tissues. But in the skin of which we are speaking, the change of colour, whether partial, total, or instantaneous, is dependent on the will of the animal, or some internal emotion, without being affected by the circumstances of position, or the aspect of the light. This skin is fine, demi-transparent, and traversed by an infinity of vessels in all directions, like the dermis of all other animals. But these reptiles, respiring slowly, have a blackish or violet-coloured blood rather than red, because it is so little oxygenated. Now, according as this blackish blood precipitates itself more or less abundantly into the small capillary vessels of the skin, it will produce these shades more or less deep, varied *ecchymoses*, or extravasations, with the other natural humours which are found there. Just in the same manner, under the influence of anger, the human countenance becomes red or livid. Fear produces paleness, cold, violet-colour, diffused bile, yellowness, &c. In fact, the cameleon, and other changing reptiles, do by no means assume the tints of neighbouring objects, but according to the emotions of fear, anger, or love, the want of food, heat or cold, or the greater or less degree of respiration, they undergo these divers colourations. These even become so varied, that it is not always easy to determine what is the natural colour of the animal.

The reptiles with naked skin, such as the batracians, are closed up in this covering, as in a sack. It adheres to the flesh only toward the mouth, the eyes, the anus, and the toes. These last are always without claws in the batracians, the species of which order are also always destitute of scales, and of osseous plates or carapaces.

The reptiles possess three kinds of teeth, tortoises excepted, which have in lieu of them semi-osseous and trenchant gums. In some species these are crenulated. The teeth which are observed in the lizards are pointed, like canines, and more or less long according to the genera. The crocodiles have them very numerous, and very long, through the entire extent of the jaw. But we have already in another place entered pretty fully into the dentition of these animals.

The serpents have also sharp teeth in both jaws, and, what to them is peculiar in their class, teeth in the palatine arches, like many of the fishes. These teeth are fixed, or implanted in the bone, and not hollowed. There are many ranges of them. In addition, the venomous serpents have attached to the maxillary bones very sharp and long teeth, hollowed like a funnel, from which they can distil a yellow poison into the wound which they make. These teeth are called fangs. In the very venomous species, they stand in the upper maxillary bone only. Under the tooth is a gland which secretes the poison.

The teeth of frogs, toads, &c., are very short, and almost reduced to nothing.

Almost all reptiles live on animal substances; for scarcely any, except some tortoises, feed on algæ, fucus, &c. Accordingly, their stomach is more capacious than that of other reptiles. Frogs, and the majority of lizards, feed on insects and worms, which they catch with their gluey tongues. The larger species of lizards, such as crocodiles, swallow other animals. Serpents seek for their food animals of all species, whose size is not too disproportioned with their own. The stomach of all reptiles has but poor digestive powers, especially in the order of serpents, in which it forms a sort of membraneous funnel. All reptiles, except the tortoise, are destitute of cœca.

The skeleton of reptiles is not so hard as that of warm-blooded animals. Their bones contain less phosphate of lime, and the gelatinous matter is more abundant, especially in frogs and salamanders.

The limbs of reptiles vary in their forms, proportions, and number. The tortoises and lizards have four, and also a tail. There are no ribs in the batracians. They have four legs: but the salamanders alone have a tail. These and frogs are remarkable in their first age for breathing by gills, like fishes. Their mode of circulation is then similar to that of fishes, and their intestines are considerably extended, and destined to digest vegetable nutriment; but they are afterwards transformed into their last state, and changed not only externally, losing their gills, but also, internally, as to the constitution of their viscera. They no longer seek vegetable aliment as before—animal food becomes necessary to their subsistence. There are, however, two genera in the order of batracians, which always preserve their gills, and have internal lungs beside; — these are the siren and the proteus. Some salamanders do not lose their gills for the first year, if the weather happen to be too cold.

The lungs of reptiles extend into the abdominal cavity, without a diaphragm, except in crocodiles, which have a peritoneal membrane, which answers the same purpose. There is but a small number of gross vesicles in the lungs, and respiration is performed by a sort of deglutition of the air.

There are reptiles resembling lizards which have but two feet; and the feet of the chalcides are so small that they are scarcely perceptible. Here we have an evident gradation to the serpent family which are destitute of limbs altogether.

One species, conformed like the lizards, has on each side of its body a membrane, supported by many rays, by means of which it can run with greater quickness, or leap with greater force. This species, called the flying dragon, is not very suitably designated as a reptile, though it clearly appertains to the class.

The act of generation in tortoises, lizards, and serpents, which lay calcareous eggs, is complete, and similar to that of mammalia. In the batracians, it is performed like the fish.

The eggs have a shell more or less calcareous in tortoises, lizards, and serpents. In the batracians they have only a membraneous covering. These last are laid in the water.

The generality of reptiles being little capable of self-defence, and exposed without parental protection from their earliest infancy, and surrounded by numerous enemies, forming as they do the common food of most birds, quadrupeds, and fishes, and held in such detestation by man, their races would speedily become extinct, but for the astonishing fertility in which they are, as it were, resuscitated annually from the bosom of corruption. Nature has bestowed upon them a wonderful fecundity, and in many species has even doubled the reproductive organs, apparently with a view to increase their multiplication.

It is at the period of reproduction that the voices of reptiles are chiefly to be heard, which vary very considerably. The crocodiles, but more particularly the caymans of America, are said to howl loudly. The hissing of serpents and the croaking of frogs are well known. A traveller towards the desert shores of the Caspian and the Volga, would imagine that he heard of a sudden, in the evening, a joyous assembly of men and women laughing very heartily. He approaches; the inextinguishable laughter redoubles among the rocks, and, to his astonishment, he finds that it proceeds from an assembly of enormous black toads celebrating their nuptial orgies. Certain species of America imitate the sound of a funeral bell tolling during the night, and others the rattling noise of cymbals.

Though reptiles never sit upon their eggs, it does not appear that the sentiment of maternity is altogether non-existent among these animals. There are serpents (and those are particularly the venomous species,) which retain their eggs in their oviductus longer than other animals of their kind. These eggs disclose within, and the young ones come out alive. These animals produce in smaller number than those reptiles which lay their eggs. It is said that the female crocodile lays its eggs on a bed of rushes and sand, and that she covers them with a second and a third similar bed, with other layers of eggs, to conceal them from the watchful ichneumon. The serpents heap up theirs in some hole exposed to the sun. Small lizards have been observed carefully carrying their eggs in the mouth to warmer places, more favourable for the exclusion of the young. But the young, once disclosed, have nothing more to expect from the mother. She has no milk to offer them—she takes no care to provide them with nourishment of any kind; still, even if a great number of these young should perish, there is no fear of the extinction of the species, nature having made a sufficient provision against that in the excess of their fecundity.

The obscure recesses inhabited by the majority of the reptile tribes are, in all probability, far from being even yet thoroughly explored. How many of these still unknown beings may lie concealed in the depths of inland waters, and of vast and desert marshes, of unthreaded and impervious wilds of vegetation? How many may creep yet unheeded among the gorges of the Alpine mountains, of the Alleghanies, and of the Andes? It is in such retreats that those double-natured animals have been found, those true amphibia, such as the siren and the proteus, equally capacitated to visit the profoundest depths of the liquid element, and to sport through the humid prairie and the drowned Savannah. Of the manners, the modes of reproduction, and, perhaps, the metamorphoses of many of these, we are entirely ignorant. Let the consciousness of this ignorance, then, teach us modesty in our judgments, and diffidence in the establishment of our systematic codes. Naturalists should remember that they are not the lawgivers of nature, but the humble registrars of her operations, and that, if they do turn interpreters of her statutes, they should be especially regardful of the nature and the number of the cases on which their conclusions shall be founded. It is true that we have made great progress in the Book of Nature; but how many leaves of the mighty volume yet remain unturned!

We have now exhausted all the observations which can come with propriety in this place concerning reptiles in general. The osteology of these animals presents a very wide field of discussion, but we have avoided entering into it for two reasons; the first is, that with regard to the Chelonians and Saurians, this subject has been already pretty fully treated on, in an account of fossil reptiles; and, in the next place, all observations of that kind, which may be necessary in addition to the text of Cuvier, will come with more propriety under the particular divisions to which they are applicable.*

* M. de Blainville, considering, according to anatomical observations, that all the reptiles which are not aquatic, have an organization approaching to that of birds, while the others in their conformation more or less resemble fishes, has divided this class into two sub-classes, the first of which he names Ornithoïdes, the second, Ichthyoïdes. He has also made some changes in the order of the subdivisions; but his arrangement, though very scientific, is perhaps not so easy of general application as that of our author.

SUPPLEMENT ON THE ORDER CHELONIA.

The name of this order is of Greek origin. Χελώνη is the name given by Aristotle to the marine tortoise, which is derived from Χελυς, *testudo*, simply signifying the shell or covering. It must be confessed, that there is some objection against the employment of this term to characterize the order, as it is also applied to designate the subdivision of the turtles. *Testudinata* is a preferable appellation.

This order, strictly speaking, consists but of a single genus, that of the tortoises, and one of the most natural that can be well imagined. Linnæus divided these animals into three sections according to the difference of organization which prescribes a different mode of life to each. The first is that of the *marine turtles*, the *Chelonia* of Brongniart; the second, the *river turtles*, *Emys* of later writers; and the *land tortoises*, which are the true tortoises of modern authors.

The tortoises in general possess a very remarkable organization. Nature has compensated for the absence of those active faculties of which they are destitute, by passive means of preservation, far more complete and extensive than what she has accorded to any other genus. Invested with a solid armour, the nature of which has already been pretty fully described, they can shelter themselves within it very effectually from the attacks of all their adversaries, except man, and some few other animals of surpassing size and strength. The body of the tortoise being thus protected by its back-plate and breast-plate—called *carapace* and *plastron* by our author, and which terms we have thought it right to adopt—is unprovided with a skin, except on its extremities, and on that portion which intervenes between the two divisions of the armature. This skin is usually covered with scales, or with scaly turbercles more or less inter-approximated. Its contexture is so solid that the sharpest instruments can with difficulty make any entrance into it. It is closely fixed through its entire border, at some distance from the internal edge of the carapace and plastron, and it is susceptible of very great extension.

The head of tortoises has, in general, the form of a quadrangular truncated pyramid, that is, it presents four faces, more or less convergent, and more or less rounded, the upper one of which is formed by the cranium, the lateral by the orbit of the eyes, and the upper jaw, and the under one by the under jaw. It presents, besides, an anterior truncation, where the nostrils are situated.

The eyes of tortoises are generally small, though usually placed in very large orbits. They are conformed like those of other reptiles, but independently of the two external and horizontal lids, there is a third lower and vertical. The lower is more susceptible of elevation than the upper is of depression. It would appear that the sense of sight is not much extended in those animals, and that of hearing still less so ; but it cannot be true that they are deaf, for they have a tympanum, though concealed, which may yet be remarked from the tension and colour of the skin in that peculiar place.

The nose of tortoises is formed by two oblong holes, which are seen in the upper part of the muzzle. Some species have them placed at the end of a short and cylindrical trunk, which projects from this same part. It has also been asserted that the sense dependent on this organ is also very feeble in the chelonia ; but direct observation does not authenticate this conclusion.

The tortoises have a slender pyramidical tongue, of thrice greater length than breadth, covered with a great number of nervous papillæ, which indicates a certain degree of sensibility in the organ of taste.

The instruments of manducation in the tortoises are two jaws, or to speak with more propriety, two trenchant gums, pointed and recurved, and greatly resembling in their form and consistence the bill of parrots. Sometimes they also possess teeth, or asperities enchased in the palate, which are more or less numerous. Many of them have fleshy lips, without either teeth or horn.

The neck of tortoises is generally cylindrical, and capable of great extension. It is almost always covered with small scales, separated and very hard. It is, notwithstanding, the part of the animal the least capable of defence, and that in which it may be the most easily killed. Accordingly, we find that the tortoise very seldom projects it out of the testa, and on the slightest appearance of danger withdraws it so completely, that it cannot be seen.

The feet of tortoises are always covered with scales more or less numerous, which prove a considerable protection to them from accidents and impediments of various kinds. Many can withdraw them completely within the carapace, the apertures of which are stopped by one of their ribs, which is always more furnished with scales than the rest. The shortness of their limbs prevents the tortoises from turning themselves when they are laid upon their backs.

Though the walk of these animals is in general so slow as to have passed into a proverb from all antiquity, yet there are some species which run tolerably fast. The fresh-water and marine tortoises swim very well.

The tail of the tortoise is usually of no great length, always conical, and furnished with scales on the upper part. In danger, this tail is recurved and concealed under the hinder limbs, after they have been folded up in the testa. It is sometimes terminated by a sharp and corneous point, which has been considered as a defensive weapon. Under this tail is the anus, which also constitutes the aperture for the organs of generation, both in male and female.

Having enlarged on the osteology of the tortoise in another place, we shall content ourselves here by briefly stating that their frontal bones constitute only the vault of the orbits; that the parietals are three times as long as those; that the mammillary processes are considerable; that the very numerous bones of the face cover each other by their edges, which are basiled; that there are seven or eight vertebræ in the neck, two of which alone have spinous processes; that there are from eight to eleven in the back, and three to the sacrum, which, as has been observed in the text, are soldered to the carapace; and that the vertebræ of the tail vary in number according to the species, and their condyle is turned in a contrary direction to that of the neck.

When the plastron or breast-plate of a tortoise is removed, we find a membranous periosteum-like parchment, which is nothing in fact but the skin of the belly. When this is open we can see the different muscles which give play to the head and feet, and the peritoneum is then visible. This, again, being open, we see the intestinal canal, the liver, and the lungs, which consist of two lobes separated by the dorsal spine.

To Townson we owe the discovery of the very remarkable manner in which tortoises respire, a mode to which there is nothing analogous in other animals, because no other has, like them, an immoveable thorax.

This anatomist, on examining some muscles of the region of the flanks, placed on the sides of the hind legs, and at the extremity of the lobes of the lungs, was convinced that two of them were distinct, though strongly joined in the middle. The first originates at the carapace, near the dorsal spine, and is inserted in the peritoneum;—this one is the contractor of the lungs, or the expirator muscle. The other extends over almost the entire cavity, between the under part of the carapace, and the inside of the sternum, and is inserted on the edges of the carapace above and below. These muscles act alternately. The first, compressing the little lobe of the lungs, expels the air; the other, stopping this compression, determines the fresh air to come and supply the place of that which has issued forth. Thus, respiration takes place in this genus, as it does in those which have a genuine thorax; but it is never so complete, that is, the pulmonary cavity is not emptied at each expiration of all the air which it contained. To this cause, doubtless, as also to the direct passage of the blood from one chamber of the ventricle into another, must be attributed the faculty which tortoises possess of living without respiration at the bottom of the water, and even of not perishing for a long time, in the vacuum of a pneumatic machine. Duvernois explains this action somewhat differently. He believes that the muscles above mentioned do not act directly on the lungs, but on the viscera of the abdomen, which thereby compress the lungs.

Like other vertebrated animals provided with lungs, the tortoises have a trachea and larynx, capable of producing a voice. It is accordingly ascertained that certain species, chiefly the marine, send forth hissings, and utter cries more or less sharp. They are also said to snore in sleeping.

Anatomists, for a long time, believed that the heart of reptiles had but one ventricle and one auricle. But it is now ascertained that there are two auricles, and in the tortoises a ventricle separated by a fleshy partition, pierced with small holes; so that in reality there are two auricles and two ventricles in this genus. It is known, moreover, that the tortoises have a third ventricle in the middle. From this heart proceed three arterial trunks, the insertion of which varies according to the species. Perrault has given us the following explanation of the circulation of the blood in one of the fresh-water tortoises:—" The right and left ventricle receive the blood of the two pulmonary veins, because these veins discharging themselves into each axillary vein, mix the blood of the lungs with that of the vena cava, to carry it into the right ventricle from which issues the aorta. The anterior, or little ventricle, has no other vessel than the pulmonary artery, and this artery, as well as the aorta, has three sigmoïd valves, which hinder the blood which has issued from the heart from re-entering there, when the ventricles come to be dilated to receive the blood of the vena cava, and the pulmonary vein. The aorta, on issuing from the right ventricle, is divided into two branches, which form two crosses, which, before turning entirely downwards, produce the axillaries, and the carotids. Afterwards, the left one of these casts out three branches, the first of which is distributed into all the parts of the ventricle; the second proceeds to the liver, the pancreas, the duodenum, and the spleen; the third furnishes branches to all the intestines. The left cross then unites itself with the branch of the right, and they form together but a single trunk, which descends along the body of the vertebra, and gives out branches to all parts of the abdomen."

As to the rest, the circulation of tortoises is extremely slow. They remain lethargized during the winter, but this lethargy is merely a simple diminution of the vital force, not a suspension of certain faculties as in the hybernating mammalia. The transpiration too, of tortoises, amounts almost to nothing, and the only loss of substance they sustain is by the dejections. It was ascertained, that at the end of six months' abstinence, the only diminution of weight in a tortoise of four pounds and-a-half, was one ounce.

The bladder of tortoises is distinguished by its size, covering the entire of the intestines and all the other parts of the abdomen. The stomach is situated under the liver, and is not formed unlike that of dogs. It discharges itself into the duodenum, which, like it, has folds and interior membranes, and may be considered as a second stomach. The liver is a firm substance, and composed of two parts, which are divided into seven lobes on their edges. In the spleen, pancreas, and kidneys, there is nothing remarkable.

The marine tortoises lay their eggs about the middle of spring. Then they proceed on shore during the night, dig a hole out of the range of the high tide with their fore feet, and lay about a hundred eggs there, which they cover with the sand. This operation is repeated three times, at intervals, as is said, of fourteen days. The tortoises, though in general so timid, are so much absorbed by this occupation, as to be quite insensible to the approach of danger. At this period they can be turned over and caught with great facility. It is the females alone which come to the shore on these occasions, and therefore we cannot be astonished if tortoises are growing more rare every day in the places where they were formerly numerous. Both themselves and their eggs are sought after with avidity, and each year thus destroys the hope of future generations. This consideration suggested to Martin Moncamps, who had travelled much in the Indian seas, the idea of establishing, at the Sechelle Islands, depôts for the tortoises, where males and females might be preserved for reproduction. This idea, probably, is not one of very easy execution; but it could not have originated but from a true friend of humanity. There are, indeed, depôts of this kind in Jamaica and elsewhere, but they rather increase the destruction of these animals, by administering to the luxury of our civic banquets.

The eggs of the marine tortoises, thus abandoned in the sand to the vivifying influence of the sun's heat, do not exclude the young regularly at a fixed period. The birth of the latter depends rather on climate, season, and species. We know that at St. Vincent, one of the Cape de Verde islands, and the most northerly of those where the marine tortoises lay, the young tortoises issue from their eggs at the end of seventeen days. In the very warm climates it isprobable that they are born sooner. Still, it has been said, that twenty-four, and even forty days, were necessary for their exclusion. On this point, however, it must be observed, that there is nothing but contradiction among authors.

The eggs of tortoises are more or less round according to their species, and they have a white and a yolk. Their envelope is more or less calcareous, but never so much so as that of the eggs of birds, and it is often soft. They are cooked in the same manner as those of hens, and their flavour is not inferior, though the white does not harden so easily. They are in great request in those countries in which tortoises abound. It is even reported that dogs are trained to find them in some parts of South America.

The little tortoises, issuing out of the sand, proceed directly to throw themselves into the sea, which they will do in spite of any efforts made to turn them away from it. They walk quicker at this age than when they have grown more bulky. They at first experience considerable difficulty in plunging into the water, and, accordingly, many of them then become the prey of aquatic birds. When they do enter this element, they frequently become the victims of voracious fish, so that great numbers of the young ones thus perish. In proportion, however, as they advance in age their means of defence are strengthened; and at the end of the first year few fishes can attack them with much chance of success.

These little tortoises have a form pretty similar to that of the mother; but the carapace is at first covered with a transparent membrane, which grows brown by degrees, and forms wrinkles, or transverse folds. This skin hardens by degrees, and is afterwards divided into scaly plates.

The tortoises of the fresh-water also in general deposit their eggs at the end of spring in the sand, exposed to the sun; but they lay much fewer eggs than the others. The land tortoises, in Sardinia, lay only five or six. Observations, however, on these points are far from being exact, nor can we present our readers with any very positive results from them.

Nicholas Stenon has remarked, that the eggs inside the tortoise are very numerous, and adhere around a membrane in each of the ovaries. They are, as in hens, unequal, and proportioned to the epoch of their first development; but those which are fecundated soon acquire a similarity of bulk. They come out by the same aperture.

One would imagine, from the heavy form of the tortoises, and the want of vivacity in their motions, that their growth was slow. Nevertheless, some facts seem to prove that it is rapid. There is a story told by Valmont de Bomare, of a native of St. Domingo bringing a sea tortoise to France, and being obliged to increase the size of the vessel in which he carried it, so as to prove that the animal grew a foot in the space of a month. We confess that we cannot attach much credit to this relation.

There is every reason to believe that tortoises are very long lived. Cetti mentions a land tortoise in Sardinia that was ascertained to be sixty years of age, and which did not appear to be older than those fresh taken in the country. We are not in possession of sufficient data to establish the difference which exists in this respect between sea, fresh-water, and land tortoises.

When marine and fresh-water tortoises have remained out of the water for a certain length of time, they have considerable difficulty in replunging into it at first. This proceeds from their lungs being inflated with an unusual quantity of air, from their having lost by the dessication of the carapace (according to the observation of Lacépede) at least one-sixth of their weight, and their specific gravity being inconsiderable in proportion to the volume of water which they displace. Accordingly, they are observed, on entering the water, to discharge from their mouth and nostrils, in bubbles, the superabundant air of which they must get rid before they can sink to the bottom.

The brain of tortoises is extremely small, and appears to be scarcely necessary to their existence. The experiment of Redi is well known, who removed the brain from a land tortoise, which lived six months after the operation.

If, as every fact with which we are acquainted on the subject appears decisively to prove, the intelligence of the animal is in proportion to the capacity of the cranium, the tortoises must be placed among the most irrational and inert of the living tribes. The range of their sensations may be said to be confined to the strictest limits of necessity, or, in other words, merely to what is indispensable for the purposes of self-preservation and reproduction. All of them, whether observed in captivity or in freedom, have a stupid, hebetated physiognomy. It is said, that at the epoch of their amours, they assume somewhat more of a lively character, and that the males, according to the custom of most animals, fight furiously for the possession of the females.

"Causa teterrima belli."

But at other periods they wrap themselves up in apathy, and remain inert until some severe evil obliges them to make use of their formidable beak and claws. M. Bosc says, that he has frequently removed the breastplate from a living tortoise, and the animal exhibited no uneasiness, nor made any effort of resistance. It was not until they experienced the utmost degree of pain, in the removal of the internal organs, that they ceased to remain in a state of contraction, and made any attempt to bite or scratch. It is asserted, however, that there are some species which know how to defend themselves when an attempt is made to seize them; if so, they are assuredly by much the fewer number.

But if the tortoises do not very readily bite their adversaries, it appears that when they do bite they bite most cruelly. There is no possible means of making them let go their hold. Even killing them will not suffice. They will retain the piece unless their jaws be completely broken. The naturalist above-cited, used to avail himself of this peculiarity of the tortoise in the operations which we have mentioned. It was sufficient to present them with a piece of wood to bite, to hinder them from making any effort to revenge themselves for the pain inflicted on them.

Having mentioned that M. Bosc employed himself in stripping those animals of their testa while alive, for the purpose of stuffing and preserving it, it would be gross injustice not to state that gentleman's apology for this seemingly wanton cruelty. To the question, why did he not first kill the tortoises, whose spoils he was desirous of preserving? M. Bosc replies, because the thing was impossible. It is difficult to conceive the degree of tenacity of life in these animals. There is no means of killing them without destroying their general organization; and it was necessary for the purposes of science that their skin and carapace should remain unaltered. Privation of air, and the deleterious gases, scarcely exercise any action upon them. It was only after removing all the internal organs, that M. Bosc could put an end to their sufferings, by cutting the spinal marrow: and even this did not always succeed. A tortoise has been seen in Paris, enfeebled by a voyage of two hundred leagues, and a fast of many months, to live an entire day after the head was cut off.

Tortoises, of all the subdivisions of this order, can remain a long time without eating. Marine tortoises have been on board vessels for many months without food, before they were delivered up to the dealers and cooks. Those sent from Algiers to Paris for the use of apothecaries, arrive there after a fast of two or three months, and often remain there as long in the same condition before they are made use of. Blasius mentions one that remained at his house ten months without taking any nourishment. All those tortoises which inhabit the countries beyond the tropics, pass annually four or six months buried in the mud of marshes or in sand hills, and of course without eating. Nature has accorded to them, as she has to other hybernating animals, the faculty of accumulating, during the summer, an enormous provision of fat, on which they subsist during winter, a period in which, as we have already seen, their loss of substance amounts almost to nothing.

In India and America, the children often amuse themselves by mounting on the backs of tortoises, and being thus carried about. Some of these animals can thus carry a great many, and walk as fast as if they had no burthen. But this amusement soon becomes very fatiguing to these youngsters, because the tortoise cannot advance one of its paws without raising the corresponding side of the carapace, which occasions very violent shocks, and no small danger of being overturned to the riders, if they are not perpetually on their guard.

Pliny and Diodorus Siculus have informed us, that entire people made use of the scales of marine tortoises to shelter them from the inclemency of the weather, or to form boats, &c. They are employed at the present day in many places for similar objects. Even in the European colonies they are frequently used for domestic purposes, such as holding the drink or food of cattle, washing children, &c. They form a large dish, the shape of which is not ugly, but which cannot be kept straight, but by fixing the convex part in the ground.

In the family of *terrestrial tortoises,* the *Greek tortoise, Testudo Mydas,* is common enough in all the southern parts of Europe, and naturalists have had ample opportunities of observing and studying it. Notwithstanding this, according to Daudin, almost all of them have confounded under this specific name, three different and approximating species. Schœpff is the first observer who has properly separated these three species, and given a just description of them.

The true Greek tortoise, which was so well known by the ancient Greeks, and which Phidias placed at the feet of his statue of Venus, as the symbol of gentleness, was first ascribed by our countryman Ray, under the name of the common land tortoise. He says that it is to be recognized by the spots, or rather by black and yellow areolæ, or circles, on the back, by its osseous shell, very convex above, and flat underneath, by its small head not unlike that of a serpent, which it can elongate or withdraw at will under the carapace, and by the absence of upper eyelid, and of tympanum. Ray adds, that it passes the whole winter in a state of torpor in the ground, and that it can live a length of time without eating.

The general length of the shell in the Greek, or common tortoise, is from six to eight inches. M. Daudin has seen one of this species ten inches French measure, in entire length. It does not seem ever to exceed this last dimension.

The skin of the neck, though scaly, is very loose, and folds, occasionally, according to the inclination of the animal, in a sort of capote behind the head.

This tortoise is very common in the southern parts of Europe, as we have before mentioned, and especially in the various countries which border on the Mediterranean. It is found wild in Dalmatia, Sardinia, Portugal, on the coast of Barbary, and probably also in Egypt, if the land-tortoise named *Zolhafa*, by the Arabs, is to be regarded as synonymous with it, and of which Forskaël has given but an indication in his Fauna of Arabia. According to him, it is a foot long. The plastron, or breast-plate, of the male is flat, that of the female somewhat concave. The latter lays thirty or forty eggs, pretty nearly as large as those of a pigeon. It is rare in Cairo, but found commonly enough towards Aleppo and Mount Libanus. It is sold in the markets in Egypt, and the Greeks eat it during Lent, regarding it in the same point of view as fish: they cook its eggs for their table, and drink the blood fresh.

According to Cetti, the common tortoise seldom weighs above three pounds. It is not often brought up in the gardens of France or Germany, but very frequently in those of Italy, where Targioni Tozzetti, professor of medicine at Florence, has examined it with some attention. According to this observer, it multiplies there, grows very slowly, and may live for forty years and more. We have already mentioned one cited by Cetti, which was sixty years old and upwards. But we are told in Shaw's General Zoology, that there have been several well-attested examples of the tortoise having lived considerably beyond the period of a century. In the time of Archbishop Laud, one was introduced into the episcopal garden at Lambeth, about the year 1633, and continued to live there until 1753, when it died, as it would seem, from neglect, and not the mere effect of age. The shell is preserved in the library of the palace of Lambeth. This tortoise appears to have exceeded the usual dimensions of its species, for the shell measured ten inches in length, and six and-a-half in breadth.

The common tortoise, about the end of October, buries itself about two feet under ground, and does not emerge until April. We are told that it hybernates in this manner even in Barbary, and it is probable that it does so in all the countries which it inhabits, without being determined to this torpor merely by the severity of cold. The males are sometimes seen to trot about, and make little bounds upon the earth; but it is hard to say, whether this manifestation of vivacity is the result of anger or of love. When many males are assembled together in one place, they attack each other pertinaciously, butting with the head and biting.

This tortoise prefers for its habitual haunts, woods, and elevated lands. It feeds equally on the roots of plants, on fruits, on insects, and on worms. It breaks the shells of snails with facility, and swallows the animal. It does not touch fish, as it never goes into the water. Its character and habits are gentle, and it is easily domesticated. In gardens it is very useful, as it destroys a considerable quantity of insects and mollusca, which are pernicious, and never does any harm. If it experience hunger for many days, it may be given bran moistened with a little milk. There is no risk, however, in allowing it to fast even for months, especially when a part of the autumn is gone by, for it is then so replete with fat, that a fast and torpor of four or five months are absolutely necessary to its preservation. The tortoise of Blasius, which we mentioned before, and which lived ten months without eating, died in consequence of the rigour of the cold. On being opened, blackish, green, and yellow excrements were found in its intestines.

If the *Zolhafa* of Forskaël, belong to this species, as M. Daudin believes that it does, the German writer must be erroneous in saying that it lays thirty or forty eggs. Cetti tells us, on the contrary, that the common tortoise, in Sardinia, lays, towards the end of June, only four or five eggs, which are white, and about the size of those of a pigeon. These eggs, deposited in a hole, and covered with sand, disclose about the end of September, and the young which then come forth are not larger than a walnut shell.

That we may leave nothing unsaid respecting the manners of this tortoise, we shall quote an interesting account given by Mr. White in his History of Selbourn, of one in a domesticated state:—" A land tortoise, which has been kept thirty years in a little walled court, retires under ground about the middle of November, and comes forth again about the middle of April. When it first appears in the spring, it discovers very little inclination for food, but in the height of summer grows voracious; and then, as the summer declines, its appetite declines, so that for the last weeks in autumn it hardly eats at all. Milky plants, such as lettuces, dandelions, sowthistles, &c., are its principal food. On the first of November, 1771, I remarked that the tortoise began to dig the ground, in order to form its hybernaculum, which it had fixed beside a great tuft of hepaticas. It scrapes out the ground with its fore feet, and throws it up over its back with its hind, but the motion of its legs is ridiculously slow, little exceeding the hour hand of a clock. Nothing can be more assiduous than this creature, night and day, in scooping the earth, and forcing its great body into the cavity; but as the noons of that season proved unusually warm and sunny, it was continually interrupted and called forth by the heat in the middle of the day, and though I continued there until the thirteenth of November, yet the work remained unfinished: harsher weather and frosty mornings would have quickened its operations. No part of its behaviour ever struck me more than the extreme timidity which it always expresses with regard to rain; for though it has a shell which would secure it against the wheel of a loaded cart, yet does it discover as much solicitude about rain, as a lady dressed in her best attire, shuffling away on the first sprinklings, and running its head up in a corner. If attended to, it becomes an excellent weather-glass, for as sure as it walks elate, and as it were on tip-toe, feeding with great earnestness in a morning, so sure it will rain before night. It is totally a diurnal animal, and never pretends to stir after it becomes dark.

" The tortoise, as well as other reptiles, has an arbitrary stomach, as well as lungs, and can refrain from eating, as well as breathing, for a great part of the year. I was much taken with its sagacity, in discerning those that do it kind offices; for as soon as the old lady comes in sight, who has waited on it for more than thirty years, it hobbles towards its benefactress with awkward alacrity, but remains inattentive to strangers. Thus not only 'the ox knoweth his owner, and the ass his master's crib,' but the most abject and torpid of beings distinguishes the hand that feeds it, and is touched with the feelings of gratitude. This creature not only goes under the earth, from the middle of November to the middle of April, but sleeps great part of the summer; for it goes to bed in the longest days at four in the afternoon, and often does not stir in the morning until late. Besides, it retires to rest for every shower, and does not move at all in wet days. When one reflects on the state of this strange being, it is a matter of wonder that Providence should bestow such a waste of longevity on a reptile that appears to relish it so little, as to squander away more than two-thirds of its existence in a joyless stupor, and be lost to all sensation for months together in the profoundest of all slumbers. Though he loves warm weather, he avoids the hot sun, because his thick shell, when once heated, would, as the poet says of solid armour, 'scald with safety.' He therefore spends the more sultry hours under the umbrella of a large cabbage-leaf, or amidst the waving forests of an asparagus bed. But as he avoids heat in the summer, so in the decline of the year, he improves the faint autumnal beams, by getting within the reflexion of a fru -tree wall; and, though he has never read that planes inclining to the horizon receive a greater share of warmth, he inclines his shell, by setting it against the wall, to collect and admit every feeble ray."

On the observation of this amiable and agreeable writer relative to the longevity of the tortoise, we may remark, that it is more philosophical to consider the *necessary* conditions of life in any being, than to puzzle ourselves about the final causes of its duration. Without reference to peculiarity of constitution, we might pronounce, in a general way, that the life of any animal passed in great activity and intense enjoyment, will, of necessity, be comparatively short. The *capital* of life, if we may be permitted to use such a phrase, is sooner spent. When the functions of an animal are numerous and energetic in their action, when its sensations are lively and varied, its constitutional vivacity great, and it appears, as it were, to be perpetually under the influence of a powerful stimulus, we may naturally expect that the vital principle will be soon exhausted, unless, as is the case in some few instances, a strong counteracting principle is found in the strength, solidity, and compactness of the different systems of which its structure is composed. This principle does exist in some mammalia and birds, which are, therefore, longer lived than others; but, in general, we find that the species endowed with most activity and evident sense of enjoyment, are not at all distinguished for longevity. The result of the most extended and accurate observations on nature confirms one great truth, which may be laid down as an axiom; namely, that to procure certain advantages, a certain arrangement is necessary, and that this arrangement is quite incompatible with other and opposite advantages. Any arrangement calculated to secure the latter must exclude the former. The activity of the lively bird, incessantly on the wing; the longevity and tenacity of life in the cold-blooded reptile; and the intelligence of man, cannot co-exist in any material being, no more than the properties of a square and a circle can be united in one and the same simple figure. The only difference between the two cases is, that we see plainly the impossibility in the latter, because the entire subject is completely within our view. In the former this is not the case, because the *whole* is not within our grasp, yet, as far as we can see, the *impossibility* is equally evident in the one as in the other; and could we survey all things with the eye of a divinity, we should find that the laws of physics are as immutable as those of mathematics.

There are six or eight varieties of the common land tortoise, which have been pointed out by Schœpff and Daudin. It is not necessary to describe them here.

The *Indian Tortoise* (*Test. Indica*) has the carapace about three feet long, compressed in front, and elevating itself above the head, at the anterior edge. The general tint is deep brown; and the tail is terminated by a corneous point.

This tortoise, of which, by the way, the opportunities of observation have not been numerous, would appear to be the largest species of the genus. Perrault has described it at considerable length, and left pretty ample and useful details on its anatomy. It has been figured by Schœpff; and M. Dumeril has shown, that, in collections under the general name of *Testudo Indica*, many species have been confounded, similar in their system of coloration, but nevertheless to be distinguished by the form of the carapace.

A tortoise of this species, according to Perrault, was brought from the coast of Coromandel to Europe, and lived more than a year in Paris, towards the end of the seventeenth century. In a Dutch collection, Vosmaer observed a carapace like that of this tortoise, and which came from the Cape of Good Hope.

Perhaps the Indian tortoise has some analogy with those which were seen by Dampier on the Gallapagos islands, and which, according to him, are so plentiful there, that five or six men might subsist there for many months, without any other provision. Some of these tortoises weighed a hundred and fifty, and two hundred pounds, and their flesh was as finely flavoured as that of the most delicate chicken. It is also very possible, that certain land tortoises seen by another traveller, Leguat, at the Island of Rodriquez, in 1692 and 1693, belonged to this species. They weighed about a hundred pounds each. This last statement was verified, in 1761, by the astronomer Lacaille, who mentions that these animals assemble in troops of from two to three thousand individuals, so that their carapaces touch, and present a sort of pavement of almost a hundred paces in extent.

The *Geometrical Tortoise* (*Test. Geometrica*) is remarkable for the elegant form of its carapace, and the yellow lines agreeably disposed in rays on each of its scaly plates. It is much sought after by the curious, and specimens may be found in most collections.

M. Daudin tells us, that the largest shell of this species which he has seen, was ten inches six lines (French measure) in length, and eight inches broad. The depth, about three inches nine lines.

The geometrical tortoise closely resembles the common in its osseous covering, which is ovâl, very convex, especially towards its hinder part: also in the scaly plates of its carapace, surrounded with concentric striæ, very numerous, and similarly formed. The marginal pieces are said to be twenty-four, and sometimes twenty-six in number. The number of pieces composing the disk is said to vary, there being fourteen sometimes instead of thirteen, as is the case with a specimen in the British Museum, figured in the Naturalist's Miscellany.

This tortoise is found in Asia and Africa, at the Cape of Good Hope, and in Ascension Island. Grew tells us that it is found at Madagascar. Seba received one individual from Brazil, and another from Amboina. In the former of these countries, according to Pison, it bears the name of *Jaboti*, but this name would appear to be applied to many Brazilian tortoises. Thunberg informs us that the Hottentots make use of its shell to hold their stock of tobacco, and adds, that it is very common in the small woods near the Cape. According to Bruguière, it lays twelve or fifteen eggs.

The *Tabular Tortoise* (*Test tabulata)* is an American species, in which the shields are in general hexagonal; but some on the sides are nearly pentagonal. The central part of each shield is large and slightly granulated, the sides are deeply furrowed, and the whole has a sort of tabular or flattened appearance, as intended to be expressed by its name. The whole shell is very convex, of a deep yellowish chestnut, palest at the centre of each shield. The legs are thick, and are remarkable for having a number of blood-red spots on each.

The radiated tortoise, or *Coui*, of the French, (*T. Radiata*,) has the carapace very rounded and convex; yellow rays on one or two sides only of the scaly plates; twenty-four marginal plates; the general colour is black.

This tortoise was first observed by Daudin, in the galleries of the Museum of Natural History in Paris. This writer was ignorant from what country it came. The baron states it to be from New Holland, without giving his authority; but the native country of this species is said by Grew to be Madagascar, whose very accurate description of the animal we present to our readers.

" It was sent from Madagascar. I find the animal no where described or figured. It is above half oval, being, of all that I ever saw, the most concave; a foot long, eight inches over, and almost six inches high. The convex is curiously wrought with black and whitish pieces, alternately wedged in one against the other, and notched, as it were, with transverse incisions. Those near the margin and on the sides are composed into several pyramidal areas, or great triangles, whose bases are about two inches broad; on the back, into six angular ones, each of them convex; on the sides, and quite behind, the shell is carried somewhat inward; before and hinderly, the edges are toothed, and bended outward and upward. The inward edges are covered with shelly plates above an inch and a half broad. The concave is composed of six and forty bones. Along the middle of the back are twelve, all, except the foremost and the four last, almost square. Next to these are eight on each side, like so many contiguous ribs, together with two lesser square bones before. Next to these, eight more, as it were, under-ribs on each side. To the twelve middlemost bones, the ribs are joined by an alternate commissure, so that one of them answers to the halves of two ribs, and *vice versâ*. To these the under ribs are attached in a wonderful manner, viz. by a branched suture or indenture. For the great teeth of the under ribs being first inserted into those of the upper ribs, the indenture is afterwards repeated by lesser teeth, out of the sides of the great ones. Besides the most elegant ordering of the work, in the convex there are three things chiefly observable, which serve for the greater strength of the shell. That is to say, the convexity of the several areas on the back, the branched sutures, and the alternate commissures of the bones; answerable to the rule of nature in a human skull, and of art in laying of stones in buildings, and in covering of broader vaults, not with one arch, but several lesser ones, for the greater strength."

Dr. Shaw seems to consider that this tortoise is also a

native of Jamaica, as the *Hicatée* tortoise mentioned by Brown, seems to agree pretty well with it in characters and size, and by no means so with the geometrical tortoise. Daudin, however, considers this tortoise of Brown to belong to a variety of another species, the *marginata*.

The *Areolated tortoise* (*T. Areolata*). The individual of this species, figured by Schœpff, in his Natural History of Tortoises, was brought from the East Indies by Thunberg. Seba, on the other hand, has figured another individual, which he names the land tortoise of Brazil. The length of this animal is only three or four inches. The carapace is oblong, a little more narrowed in the front, scarcely indented at the top of the neck, and moderately convex.

The disk is covered with thirteen plates, more or less square or sexangular, moderately raised, and with their sutures depressed. The edge of these plates is surrounded by some concentric striæ. The areolæ, or central parts, are smooth on their edges, and have their centre slightly sunk in, and punctated roughly. On the centre of the areola of the two anterior vertebral plates, there is a small projection disposed length-wise, which is either effaced or non-existent on the other plates. The centre of the areolæ is yellow, the edges white, and the rest of the plates of a dirty bay.

The marginal plates are square, nearly of an equal length, and four and twenty in number. The marginal collar-plate is small, and more narrow, while that which is placed above the tail has a wider form than the others, and a little furrow on the middle, which makes it appear almost double. These plates are of the same colour as the disk, and are separated from it by a sufficiently well-marked furrow.

The plastron, or breast-plate, is yellowish and flat, truncated, and scarcely notched in front, but marked with a singular notch behind, and covered with twelve scaly plates, which are surrounded by some brownish furrows.

Seba thinks that the aquatic tortoise, found by Marcgrave in Brazil, and which is named *Jurura* by that writer, should be referred to this species. But this is a mistake. Seba is also in error in making Brazil the habitat of his tortoise. Daudin believes it to be Indian, but it is more likely that it belongs to the Cape.

Mr. Bell, in the third volume of the Zoological Journal, has described three new and beautiful species of the land-tortoise, under the names of *T. actinodes*, *T. teutoria*, and *T. pardalis.*

The actinodes has a black head, with yellow spots. On the fore-feet are numerous large black yellow spots; not so numerous, but larger, on the thighs. The shell is elevated, ovate, with the margin anteriorly notched, posteriorly denticulated; the whole of the scuta sulcated, black, with regular yellow radiations, and with large yellow areolæ. The marginal scuta are twenty-three, forming a deeply-serrated outline; the sternum is yellow. In the young state, this however appears to be the *T. stellata* of Schneider.

The teutoria has the dorsal scuta conical, acute, at the apex black, striped with yellow; the marginal scuta, twenty-four; sternum narrowed behind, with a deep acute emargination; deep uniform brown in the middle, through the whole length, and pure light yellow at the sides.

In the pardalis, the head is of a uniform dull brown, which also pervades the neck, feet, and tail, but with occasional shades of dirty yellow. The neck is very long, feet robust, furnished with strong conical and triangular scales. The shell is very deep, and formed like that of *T. Indica;* scuta, flat sulcated, of a dirty light yellow; claw with numerous large irregular black spots, as if splashed, but assuming in a slight degree a radiating direction; the areolæ of the vertebral scuta, a little elevated; those of the costal scuta, placed very near their upper margin, towards their junction

THE STARRED TORTOISE.

T. ACTINODES.

THE LEOPARD TORTOISE. *T. PARDALIS.*

with the vertebral,—which character constitutes almost the only distinction of importance between this species and *T. Indica*, excepting the colour. Marginal scuta, twenty-three.

Mr. Bell adds, that the specimen from which his description was given was then living, and had been in his possession since the commencement of the summer, during which time it had had the range of a small orchard, feeding heartily on grass, which it plucked with a movement similar to that of a goose. The neck is so extensile as to permit the head to be raised above the level of the top of the back, and thus to enable the animal to look around on all sides by merely turning the head. It is the largest known species of land-tortoise, (having the shell following the curvature two feet long,) excepting *T. Indica*.

To Mr. Bell we shall be indebted for our account of two genera (or subgenera, according to the Cuvierian plan) of land tortoises, which form a natural passage from the land to the fresh-water tortoises: these are the *pyxis* and *kinyxys*. We shall present our readers with an abstract of that gentleman's paper on the subject, in the fifteenth volume of the Linnæan Transactions, Part the Second.

This gentleman was formerly of opinion that the affinities which connect the fresh-water tortoises with those of the land, were to be found in the genus *Terrapene*, and especially in those species confounded by writers under the trivial name *Clausa*. These last, however, are true fresh-water tortoises, notwithstanding a certain approximation in their habits and structure to those of *Testudo* proper; and Mr. Bell was unable to find in the known species of the latter family the relations of which he was in search.

He was fortunate enough, however, afterwards to obtain a living specimen of a new species of tortoise (*Kinyxys Castanea*), which appeared to possess those relations. The depression, and great lateral expansion of the shell, evince an evident approximation to Emys, and the large openings for the feet indicate great facility and power of motion. Accordingly, the movements of this tortoise are far more lively than those of others of its family. The feet have the elevated form of the land tortoises, but with a clear approach to the flattened palmate form of Emys, and the claws have somewhat of the length and sharpness of those of the latter. These affinities are corroborated by the structure of the *dorsum* or carapace, which is divided into two portions, the hinder of which is moveable, and can come in contact with the hinder margin of the sternum, and thus completely secure the hind feet and tail, when they are drawn within the shell. Also, by a relaxation of the muscles which thus close the box, it may be opened, about one-half or three-fourths of an inch. This mobility is produced by the want of any bony union between the fifth and sixth ribs, which are connected only by an elastic ligament. In older specimens Mr. Bell has observed, that part of the inferior margin of the upper shell, which is opposed to the edge of the sternum, to be eroded by the force of its continual action.

The situation of the hinge, and the part which closes the shell, are indeed different in the two groups (*Kinyxys* and *Emys*) ; still the fact of a structure for this specific end, proves relationship, and added to other affinities, establishes the link required.

Two specimens of another species, in the British Museum, presented by Sir Everard Home, exhibit the same peculiarity of structure.

To this genus, Mr. Bell has given the name Kinyxys, from κινέω, *moveo ;* and ἰξύς, *lumbus.*

The other genus presents an affinity, or possibly, says Mr. Bell, only an analogical relation to the box tortoises, though it strictly belongs to the terrestrial. A perfect specimen, in which the sternum was uninjured, exhibited a perfect *land*

B. W. Hawkins Del.

BELL'S KINIXYS. *KINIXYS BELLIANA.*

tortoise, with the anterior lobe of the sternum moveable, and capable of closing the shell as completely as in any species of the fresh-water box tortoises. This unexpected peculiarity appeared sufficiently important to Mr. Bell to merit the distinct generic appellation of Pyxis.

Emys is the scientific designation of those tortoises which live habitually in the fresh-water. Their feet are mobile and palmated, their jaws horny, and beak trenchant. The tail is generally short, and the carapace scaly and solid, as is also the sternum, which is broad.

The body is naked, or covered with papillæ and scales; the head scaly, or naked, is concealed under the carapace, or on its sides. The jaws are most usually entire, and the upper covers the lower like a box. The neck is usually naked and rounded; the carapace is generally convex, and the claws are pointed.

Most part of the reptiles of this genus live in marshes, and especially inhabit hot climates, where they live on plants and mollusca.

The word Emys is derived from the Greek ἐμὺς, which signifies *tortoise.*

The species of this genus are numerous—they are properly divided into two sections. The first has the sternum or plastron immoveable and angular.

The first species is the *Speckled Tortoise* of Shaw, which might more properly be designated the *European fresh-water tortoise. Emys Europæa, Testudo orbicularis,* Linnæus. *Testudo lutaria,* Hermann, Marsigli, Brunnich. *Testudo flava,* Bonnat. Daudin, &c.

This tortoise is rather of a small size, the shell being about four or five inches in length, though sometimes it is eight long and four broad. The carapace is oval, very little convex, rather smooth, blackish, and marked with yellowish points disposed in rays. The plastron, or under plate, is as long as the carapace, rather less broad, oval, oblong, rounded in front, and truncated, and almost without indenture behind. The head is flatted at top and sides; the tail is small and scaly; the skin of the neck thick, loose, wrinkled, and smooth, not scaly, as Schœpff has nevertheless stated it to be.

In adults the carapace is smooth; while in the young, there are scales marked with furrows parallel to their edges, especially at the hinder part of the back and on the sides. Marsigli says, that these furrows are more apparent in the males than in the females.

The disk of the carapace is occupied by thirteen large plates, five of which are vertebral, and eight on two lateral ranks. The first vertebral plate is pentagonal; the three following are hectagonal; the last is irregularly pentagonal, and sometimes marked in its centre with a salient ridge. The first lateral plate is large, square, and irregular; the two following are large, pentagonal, and elongated; the last is quadrilateral, with curved sides.

The edge of the carapace is composed of twenty-five plates, all squared. The plastron, or breastplate, is composed of twelve plates.

The nostrils are small, round, and separated by a slender partition. The feet are covered with scales, and semi-palmate.

This species is by far the most extended of the genus Emys. It is to be seen in all the south and east of Europe, even as far as Prussia. It has also been observed in Poland, in Italy, in Sardinia, and in Hungary. It is rare in France. Count de Lacépède informs us that it is also found in America and Ascension Island.

It lives in muddy waters and marshes, and feeds on insects, mollusca, small fish, and plants. Its flesh is used as food, and forms an article of sale in many of the German markets.

It is reared in ponds and gardens, with bread, lettuce, leguminous plants, &c. J. C. Wulff informs us that the Prussian peasants keep these animals in troughs for two years sometimes, for the purpose of fattening them.

The eggs of this European Emys are about the size of those of a pigeon, but more oblong. The female deposits them in the sand, exposed to the sun, and, according to Marsigli, they take a year before the young are excluded.

There are several varieties of this Chelonian, one of which, it would appear, inhabits the marshes of Spain.

The *painted Tortoise (Emys picta)* is easily distinguished from the others by the colours which decorate it. It is five inches and a half in length, near four inches in breadth, and an inch and a half in thickness.

The carapace is oblong, convex, and smooth, without any appearance of furrow or points. Its plates, thirteen in number, are almost all quadrangular, except the three anterior, and two of the vertebral range, the angles of which are obtuse, and the sutures furrowed.

The edge of the carapace is trenchant, except upon the flanks, and it is composed of twenty-five plates. The breastplate (plastron) is as long as the carapace : its form is oblong. The two plates of the neck are rounded, and the two caudal ones truncated. All the plates, to the number of twelve, are marked with a stria, so that one might believe that these plates were eighteen.

The feet and tail, which is short, are covered with scales: the former are slightly palmated.

The general colour of this tortoise is brownish, deeper on the head and limbs. All the plates of the carapace are bordered with yellowish; so that the animal appears marked above with broad bands, which cross each other. The marginal plates are yellowish, with irregular and blackish concentric circles. The plastron is of a yellowish grey. On the sides of the head and jaws some yellow traits are visible, and the tail is marked with four longitudinal lines of clear yellow.

This tortoise is found in North America. It appears to delight much in deep and tranquil streams and solitary places. When the sky is serene, and the atmosphere sufficiently warm, they proceed in little troops out of the water, and go to rest on the trunks of trees and neighbouring rocks. Then, at the least noise, or on the approach of man, they replunge into the water with the greatest rapidity. The painted tortoise swims with swiftness, and walks slowly. It can live a long time under water; but it cannot exist on dry land more than a few days.

This tortoise, whose form is so elegant, and colour so agreeable, is considered as extremely voracious. It is even pretended that it will seize the young ducks which swim above it, by the feet, and drag them to the bottom of the water to devour them. Its flesh is considered by many Americans as a wholesome and delicate food.

Schœpff, in his history of tortoises, has given a complete description of this species. He has moreover applied himself to the disentanglement of the synonimy. He remarks that Seba has described it under the name of the tortoise of New Spain. This last author says that the Portuguese name it *radago d'agno*. He has given a tolerably correct figure.

The painted tortoise of Hermann and Gmelin is certainly the same species; but they have evidently described a young individual, for they state its size to be only that of a common apple. All the other characters resemble the preceding, with the exception of a double spot of blackish blue, which they have observed on each side of the carapace.

Schœpff received from Pennsylvania a young individual, whose carapace was four inches long, two inches nine lines in breadth, and one inch two lines in thickness. Its clear

LAKE ERIE TORTOISE.

EMYS GEOGRAPHICA.

colours were of a lively yellow. The breast-plate was white, the head marked with yellow points, and the feet adorned with reddish striæ. Schœpff was assured that this tortoise was aquatic; that during summer it inhabits banks of earth, and that it conceals itself from the month of October in marshy places to pass the winter.

M. Daudin gives two varieties of the painted tortoise. The *cinereous tortoise* is only a young *picta*.

Among the more rare specimens of emys in the British Museum, are the following:

The *Lake Erie tortoise; Emys geographica* of Say and Leseur. It is olive-brown, with numerous narrow dark-edged open lines on the back and margin; beneath it is yellow, with a narrow black line on the sutures of the sternal plates. The shell is ovate, smooth, and convex, with the back bluntly carinated; the sternal-plate, sterno-costal suture, and the margin, have irregular lines of different breadths; the head and limbs have numerous narrow lines, and there is a subtriangular patch on each temple. Its name sufficiently indicates its locality.

The *Emys occipitatis*, so named by Mr. Gray, appears to be hitherto unnoticed. That gentleman thus describes it:— Shell, ovate oblong, smooth, rather convex, and very obscurely and bluntly keeled, doubly indented behind; dark brown, varied with broad and narrow pale lines; those of the vertebral plates somewhat ringed, diverging on the costal ala; the margin has a series of square, pale-ringed black spots, placed on the suture, and each spot occupying half a plate. The first vertebral plate squarish; the others broad, hexangular; palish yellow, with a large black ring on each plate, with a blackish eye-like spot on each suture of the marginal plate, and a long ring-formed spot at each end of the sterno-costal suture.

In Mr. Bell's collection, and, we believe, also in that of the

Zoological Society, is a very beautiful little emys, named, by that gentleman or by Mr. Gray, *E. ornata*, which is extremely small, though evidently young. The shell is oblong and rugose; the costal and marginal shields have each a central dot, surrounded by a dark-edged pale ring; the vertebral plates have rings less regular than the others—the first is urceolate, and the second and third long and hexagonal. The marginal plates have also eye-like spots and rings; the shield is fine green, with the spots and rings yellow; the head is yellowish, lined with orange streaks: beneath it is yellow, with dark lines.

We have engraved a figure from Mr. Bell's valuable collection of another *emys*, which is named in his MSS. *E. decussata*. It may, however, be allied to the *E. serrata* of Daudin. The shell is oblong and bluntly carinated. The first vertebral plate is nearly square, but rounded at the sides, and the rest are subquadrangular. The shell is uniformly pale brown; the shield irregularly and concentrically groved beneath; yellow, convex, with a spot on each end of the sterno-costal suture, and a round spot on the suture of the marginal plates. The animal is greenish, with the cheeks and chin palely streaked. Mr. Gray has seen several specimens of this species.

The *Mud Tortoise* of Shaw, *Emys lutaria*, is remarkable for exhibiting so great difference during its non-age from the adult state. In the former, its shell has three distinct ridges or keels passing from front to rear, and is otherwise irregular in the surface; whereas in the latter it is smooth, with a single, very blunt keel. The colour is olive, varied with dark edged orange spots.

Mr. Bell, as we have seen in the text, distinguishes a group of tortoises, principally by length of neck, under the head Hydraspis. The long-necked tortoise (*Em. longicollis*) of Shaw, is among these. It is a native of New

THE FURROWED TORTOISE.

EMYS DECUSSATA.—*Bell.*

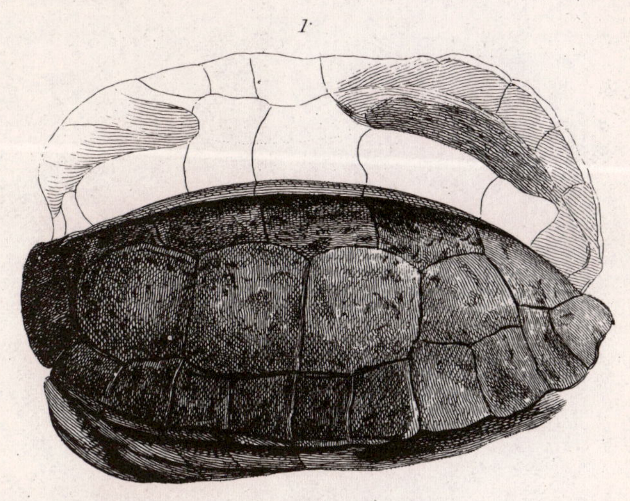

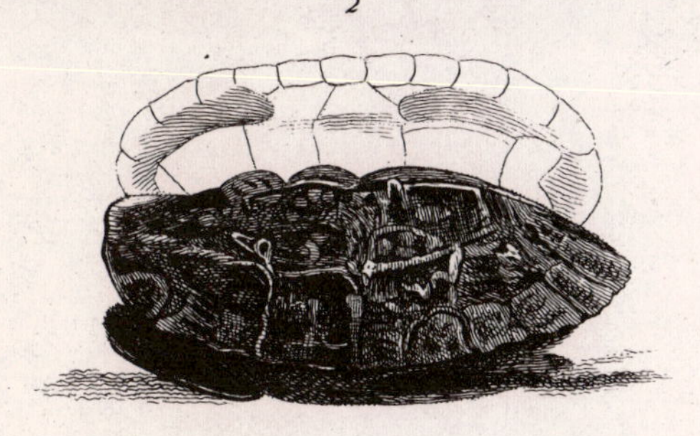

MUD TORTOISE.

EMYS LUTARIA.

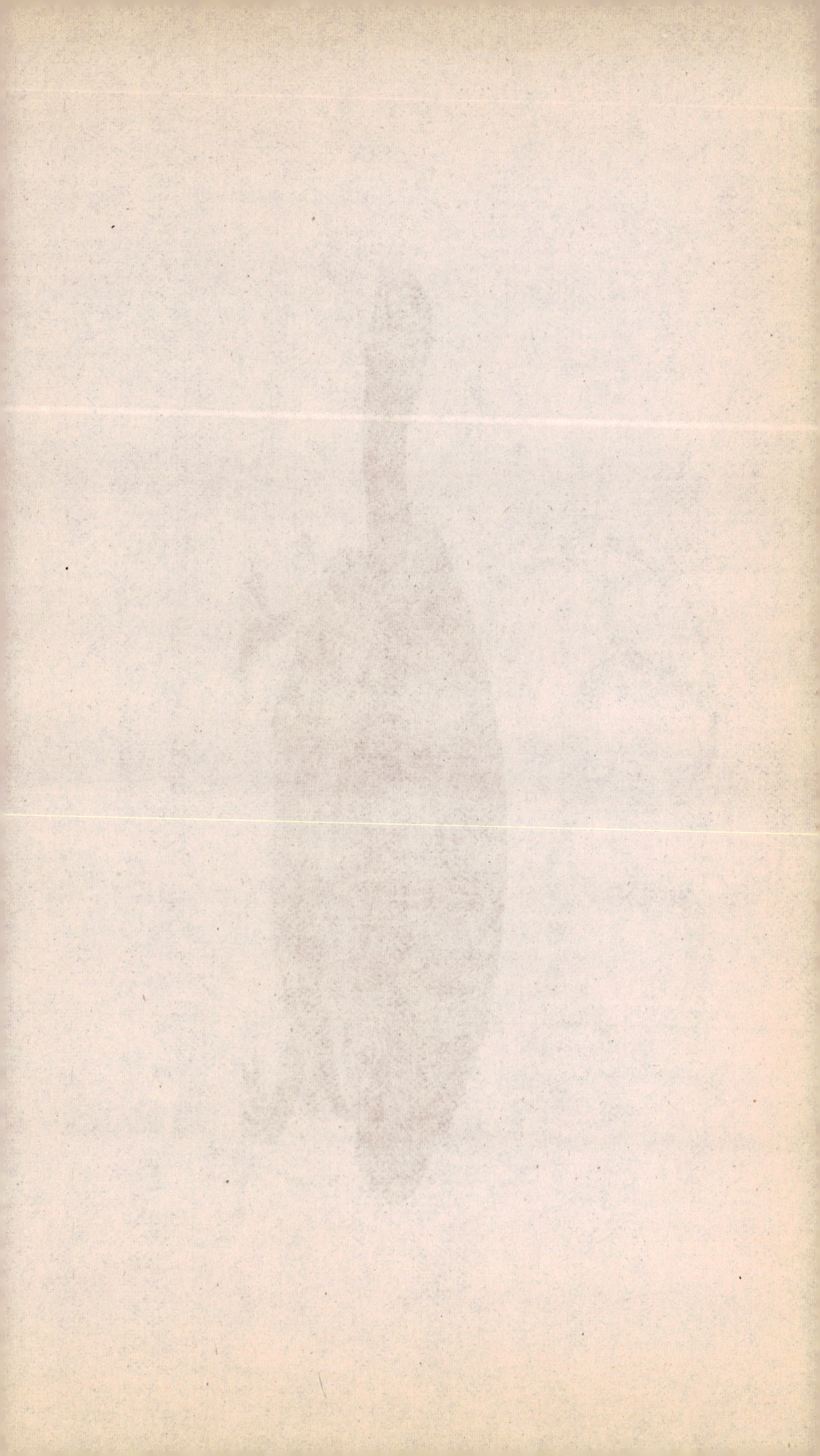

LONG-NECKED TORTOISE.

TEST LONGICOLLIS. — Sh

Holland; of a dark olive-brown colour, and of a texture resembling leather. The shell is oblong and flattish, and the first vertebral shield is very large. Nothing is known of its habits; but the head, in common with the species of this group, cannot be drawn back under the shield.

On the second division of emys, the BOX-TORTOISES, we shall again gladly avail ourselves of the excellent observations of Mr. Bell, in his monograph on this genus, in the second volume of the Zoological Journal, p. 299, &c.

This gentleman very properly observes, that when a number of species are found in any group of animals, different in some important and essentially *anatomical* character, they should be separated into a subordinate group, and receive a distinctive appellation. The box-tortoises were included in the *emydes* of Brongniart, with which they, doubtless, have much affinity. But Merrem, observing the character of their sternum, separated into two or three moveable divisions, formed them into a distinct genus, under the name of *Terrapene*. Mr. Say, the American zoologist, without appearing to have known of this separation of Merrem, has also divided them into a distinct group, which he has called *Cistuda*. Spix has applied the generic term *Kinosternon* to two Brazilian species.

Like the emydes which we have already noticed, they are fresh-water tortoises. They agree with them in the general form of the shell, the subpalmation of the toes, and in aquatic habits. In the species confounded under the name *Terrapene clausa* and its synonyms, there would seem, at first view, to be an exception to this; but though they are called in America land-tortoises, and have something similar in appearance and habits to *testudo* proper, yet their affinities with emys are sufficiently numerous and important. We are told by Schœpff that *T. clausa*, though sometimes found in dry places, and not well formed for swimming, are fond of marshes. Mr. Bell, as we have already observed, before his discovery of the two genera, Kinixys and Pyxis, imagined that the natural transition from emys to testudo, was to be found here by means of his *Terrapene Europæa.* But the true connecting link between the two families he has since, as we have seen, much more satisfactorily established.

The important character of the group of which we now speak, is the mobile structure of the sternum. This sternum, in all of them, may be considered as consisting of three portions: the posterior is that part which is covered by the two posterior pairs of plates—the middle one, by the next pair—and the anterior by the remaining anterior plates, which differ in number according as the foremost pair are either united into one single plate, or have a small supernumerary one interposed between them.

There are three different modifications which take place in this part. " In the first (the genus *Kinosternon*, Spix), the middle portion or lobe is quite fixed to the sides, the anterior and posterior lobes moving upon it. In the second form (the genus *Sternothærus*, Bell), the middle portion is fixed, as in the other, and the posterior one also connected with it by continuous bony union, the anterior lobe only being moveable. In the third (constituting the genus *Terrapene*, Merrem), the middle and posterior lobes are also immoveably connected together, but forming a single moveable valve, without any bony union with the upper shell, the anterior lobe being also moveable on the same axis. The only connexion between these two valves and the upper shell, is by means of a strong ligament, becoming cartilaginous at the axis.

" The hinge or connection between the valves is formed by a sort of articular cartilage, allowing, by its elasticity, of sufficient motion to enable the animal to open the shell, so as to move its limbs without inconvenience, or, on the other hand, to bring it into close contact with the upper shell, and thus to enclose itself, particularly in the genus *Terrapene*, within a complete box. At the angles of these valves are small processes of bone, or at least distinct muscular impressions, to which the adductor muscles are fixed; and these in the anterior valve, of *Sternothœrus Leachianus*, form long spinous processes. It is obvious, that in the genus Sternothærus, the hinder part of the shell cannot be closed, as that part of the sternum is immoveable.

"Upon the whole, then," adds Mr. Bell, "notwithstanding the affinities by which these animals are connected with the *Emydes* of Merrem, are such as forbid me to consider them as a distinct family, yet the structure which I have been describing is so striking, and appears to me of so much consequence, especially as requiring a considerable addition to, or modification of, the muscular system, that I could not look upon it as forming a less important group than a subfamily, particularly as it includes several subordinate divisions with distinct generic characters."

The following is Mr. Bell's arrangement:—

Fam. Emydidæ.

Subfam. Sternothærina.—Characters are: Toes distinct, with sharp claws. Beak horny; scales of the disk thirteen. Breastplate consisting of one or two valves, united by a ligament, moving as if on a hinge, and thus capable of partially or totally closing the shell.

Genus I. Kinosternon, Spix. Breastplate consisting of three distinct lobes, the middle one fixed, to which the anterior and posterior, which are moveable, are articulated by a ligament.

The species are, 1. *Shavianum*, called by Mr. Bell after Dr. Shaw. 2. *Longicaudatum*. 3. *Brevicaudatum*. 4 *Pennsylvanicum*. 5. *Amboinense*. 6. *Nigricans*.

Genus II. STERNOTHÆRUS, Bell. Breastplate having but one moveable valve, formed of the anterior lobe. The middle and posterior lobes immoveably connected and fixed.

The species are, 1. *Trifasciatus.* 2. *Leachianus.* 3. *Odoratus.* 4. *Boscii.*

Genus III. TERRAPENE, Merrem. CISTUDA, Say. Sternum bivalve; the two valves moving on the same anxis; the posterior valve consisting of the two posterior portions or lobes of the sternum.

The species are, 1. *Europæa.* 2. *Carolina.* 3. *Maculata.* 4. *Nebulosa.*

The Emydes, whose long tail and voluminous limbs cannot be concealed within the carapace, have given rise to the establishment of a new genus, by M. Schweigger, called *Chelydra.* They are separated by our author from the preceding, without any given denomination.

The first and best known of these is *Emys Serpentina*, mentioned in the text. The carapace is depressed, oval, and almost subquadrangular, truncated, and a little more narrow in front. The dorsal plates are thirteen, and the marginal five and twenty. All the plates of the carapace are slender, rather transparent, and of a brown colour more or less deep.

The plastron is small, and has but nine or ten plates, which are smooth and slender.

The beak is not unlike that of a buzzard, and is terminated by two wattles. The upper jaw is wider than the lower. The neck is at least as long as the body, and very retractile. The upper part of the tail is armed with a denticulated crest. It is as long as the body. This arrangement has occasioned this animal to be named, in Carolina, the *alligator tortoise.*

This Emys is found in the fresh waters of North America. It is rare, and much in estimation for the excellence of its flesh. It attains the length of about four feet, and often weighs more than twenty pounds. It is a mischievous and voracious animal, tearing young ducks and fishes, and often attacking its own species. It occasionally removes to some distance from the water. It seizes its prey, rising on its hind feet, and elongating its neck with great rapidity of motion. It is said to utter a hissing cry, and when irritated, to bite with so much violence, that there is much difficulty in forcing it to let go its hold. Schœpff reared several individuals of this species in a chamber. They always looked out for the most gloomy corners, and concealed themselves in the ashes of the chimney.

We now come to the CHELONIANS proper, or sea tortoises, vulgarly termed *turtles* in this country. M. Brongniart devoted the name Chelonia to them, and it is highly proper, as preserving an analogous signification to the original word among the ancient Greeks. It is unnecessary to add any thing here to the generic characters given in the text.

The marine tortoises are all inhabitants of the seas of warm climates, under the torrid zone, and as far as the fifteenth degree of latitude. One single species, that of Japan, lives in the fresh water.

The *Green turtle*—Shaw—*Testudo mydas*, has its scales neither imbricated nor carinated. The lower jaw is strongly denticulated. The head is comparatively smaller than that of other marine tortoises. The carapace is oval, heart-formed, and but little convex. All the scales are very transparent and agreeably shaded: they are also very slender. When the animal is in the water the colour of the carapace is a deep green, though Dr. Shaw informs us, that it is from the tinge of the fat that this reptile has been named the *green* turtle.

According to Dr. Cloquet, this tortoise exceeds all others of its genus in size and weight, being six or seven feet long, and weight seven or eight hundred pounds. But Dr. Shaw gives the palm of magnitude to another species, the *loggerhead*, or *caretta*.

In his voyage to the Canary Islands, Lemaine assures us, that near Cape Blanco, the tortoises are of such a bulk, that their carapace is not less than fifteen feet in circumference, and that the flesh of one of them would be a sufficient banquet for thirty men.

The green turtles are very common on the low arid sandy shores of both continents, principally under the torrid zone. They are never taken towards the north, or beyond the fiftieth degree of latitude, but when driven there by tempests. But they have been caught even towards the mouth of the Loire. One mentioned by Count Lacépède was taken near Dieppe, in 1762, of nine or ten pounds' weight. Dr. Shaw indeed thinks it probable that this was not a *mydas*, but a *caretta*. However, M. Cloquet has seen within these fifteen years two or three small ones which came from the same part.

They seek the neighbourhood of islands and deserted coasts, coming to land as seldom as possible, and remaining there but a very short time. After remaining thus on shore, they find great difficulty in sinking again in the water, either because they are then filled with a considerable volume of air, or, according to the opinion of M. de Lacépède, they become so dry as to lose a sixth of their weight.

At certain periods, the green turtles are observed to quit the bottom of the sea, and repair in crowds towards the mouths of great rivers. They are very timid, and never seek to defend themselves except when in the act of coition. On such occasions, they resist with fury, and brave every danger.

It is in the month of April that the females deposit their eggs in a dry place on the shore. They first of all, without being ever accompanied by the males, seek out a convenient situation, quitting the water with many precautions after the setting of the sun, but return immediately to the sea on the slightest disturbance. If this is not the case, they proceed above the line of the highest tide, excavate the sand with their fins, and, after having made a hole of about two feet deep and two wide, formed like a reversed cone, they deposit their eggs there, sometimes to the number of one hundred in a single night. During this labour, nothing can disturb them or distract their attention. At such times, they are taken with great facility.

In this manner, they lay three successive sets of eggs—an interval of fourteen days or three weeks elapsing between each set. They return to the sea, after having covered their eggs with sand.

We are told by Père Labat, that on the coast of Africa a single one of these tortoises will produce two hundred and fifty eggs, and even more.

The young are excluded generally in about three weeks, though some little variation will take place according to latitude, and the temperature of the atmosphere. The accounts of authors, however, on this subject cannot be implicitly relied on, as they abound in contradictions, though the above may be considered the average time.

The eggs are round, two or three inches in diameter, and enveloped in a soft membrane, not unlike moistened parchment. Their albuminous part does not coagulate in the fire, but the yolk hardens very well.

These eggs are excellent eating, and in great estimation.

With the very young turtles, the carapace is covered with a white and transparent skin, which grows brown by degrees, forms transverse wrinkles, then thickens, and finally is divided into scaly plates.

Dampier has remarked, that towards the season of laying, the greater number of these turtles remove for two or three months from the latitudes where they habitually reside.

They proceed to deposit their eggs at some distance from their usual domicile, and then abandon them. In this voyage the male follows the female, and does not quit her until their return. It is believed that during the whole time of their absence they eat nothing; it is certain that they are extremely lean when they do return, especially the male. The same traveller adds, that they are accompanied in their route by sharks, and an infinite number of other fishes.

The places most remarkable for the deposition of eggs by the *testudo mydas*, are the Alligator Islands, in the sea of the Antilles, and that of Ascension, in the middle of the Equinoxial Atlantic Ocean. They arrive at the former from the end of April to the month of September, and none of them can have travelled less than forty or a hundred leagues, for such is the distance from their nearest points of departure, which are the little isles southward of Cuba. Those which proceed to Ascension Island cannot have travelled less than three hundred leagues, whether they come from Africa or America.

An innumerable quantity of these turtles are found in the channels between the Gallapago islands and the Equinoxial Ocean. They proceed to the coasts of America to deposit their eggs—a distance, at the least, of one hundred and forty leagues.

In consequence of all this we may believe, that the same instinct which leads the young turtles to enter the sea the moment they are born, a fact noticed in our general observations on the order, also conducts them to the latitudes inhabited by their mothers, where they find an abundant supply of food. Another consequence of the fact which we have described, is the circumstance of these tortoises having been met by travellers in the high seas at seven or eight hundred leagues of distance from any land whatsoever.

This animal may be considered as one of the most useful

productions of equatorial climates. On distant shores it furnishes to navigators an aliment equally agreeable, abundant, and salutiferous, and an assured remedy against the ravages of scurvy.

The flesh and broth of turtle are recommended in a number of morbid affections, as in consumption of the lungs, inveterate syphilis, and a variety of cutaneous affections.

The fat is often of a deep green, but it is very finely flavoured. Leguat informs us, that, in the Island Rodriguez, the fat of the tortoises there is so highly coloured, that people at first were afraid to eat it, and that it communicates to the urine the tint of emerald.

The turtles of Batavia are not in much estimation. In Cook's Voyages we learn, that those of the river Endeavour, in New Holland, are very good. There is more or less a musky flavour about the green turtles, according to the season in which they are caught.

It would appear that, under certain circumstances, and in certain latitudes, these animals possess pernicious qualities. At the time of the voyage of Commodore Anson, in 1740, the Spaniards and Americans of the western coasts of Mexico, near Panama, regarded their flesh as poisonous. Quære, whether the species of which we are now writing was the one which they thus stigmatized?

Be this as it may, it is certain that in the European colonies, in the Antilles, and at the Isle of France, they are in the highest estimation. In Jamaica they are even preserved in parks; and their flesh is sold in the shops at a less price than that of beef and mutton.

From this last island in particular is our turtle-eating metropolis supplied with immense quantities of this luxurious food. It would be quite superfluous to descant on the enthusiastic veneration in which turtle-soup is held by our wealthy and discerning fellow citizens.

Every year many vessels proceed to the Cape Verd Islands to take in their cargo of these animals, and salt them, to transport them into America. Those consumed in the Isle of France come from the Sechelles Islands.

From the fat of the turtle oil may be derived for burning. A large one will furnish thirty pints or more.

When these tortoises are on land, they are taken by turning them on their backs, either with the hands, or, if they are too heavy and bulky, with stakes or levers. They are left in this position for a greater or less length of time. According to Père Labat, they may be kept alive in this way for fifteen or twenty days provided care be taken to wet them with sea-water four or five times every day. They grow very thin, however, under this kind of discipline.

In the midst of the ocean these tortoises are caught by the harpoon. Lord Anson reports that, in some parts of the South Sea, hardy divers will get under them when they are asleep, seize them by the lower part of the carapace, and support them long enough on the water to be enabled to hoist them into a boat. M. Laborde, a physician of Cayenne, mentions that they are caught in nets. But the most extraordinary mode of catching them is by means of a fish of the genus *Echeneis*. In 1809, when Mr. H. Salt was at Mozambique, having received a present of one of these fish, all the inhabitants assured him that they were wont to employ it, by fastening it with a cord to a boat, and that it fixed itself by the head to the breast-plate of the first tortoise it met, with so much force, that the latter could not escape. Commerson has likewise reported something of the same kind.

The shell of this turtle is too thin to be applicable to the same purposes as that of the imbricated turtle, of which we shall presently speak.

It appears that the nature of the sea tortoises was very little understood by the Europeans until after the middle of the last century. The discrepant accounts of navigators respecting the flavour and wholesomeness of the turtle's flesh as food, must be entirely attributed to their ignorance concerning the distinction of species. In fact, the introduction of the green turtle as an article of luxury into this country dates at no very distant period. In the part called the *Historical Chronicle* of the Gentleman's Magazine for 1753, is recorded :—" Friday, Aug. 31, a turtle, weighing 350 pounds, was eat at the *King's Arms* Tavern, Pall-Mall; the mouth of an oven was taken down to admit the part to be baked." Again, in the same work for the same year is noticed :— " Saturday, Sept. 29, the *Turtler*, Capt. *Crayton*, lately arrived from the Island of *Ascension*, has brought in several turtles of above 300 pounds weight, which have been sold at a very high price. It may be noted, that which is common in the West Indies is a luxury here." And, once more, in the same publication, for 1754, we read, " Saturday, July 13, the Right Hon. the Lord Anson, made a present to the gentlemen of *White's* chocolate house, of a turtle, which weighed 300 pounds weight, and which laid five eggs since in their possession. Its shell was four feet three inches long, and about three feet wide. When its head was cut off, at least five gallons of blood issued from it, and so full was it of life, that the mouth opened and shut for an hour after it was cut off."

All these paragraphs show how recent an article of luxury the turtle must have been at the period in which they were written.

We shall conclude this sketch of the history of the *testudo mydas*, with an extract concerning it from Catesby.

" Of the sea-turtles, the most in request is the *green turtle*, which is esteemed a most wholesome and delicious food. It receives its name from the fat, which is of a green colour. Sir Hans Sloane informs us, in his History of Jamaica, that forty sloops are employed by the inhabitants of Port Royal, in Jamaica, for the catching them. The markets are there supplied with turtle as ours are with butcher's meat. The Bahamians carry many of them to Carolina, where they turn to good account, not because that plentiful country wants provisions, but they are esteemed there as a rarity, and for the delicacy of their flesh. They feed on a kind of grass growing at the bottom of the sea, commonly called turtle-grass. The inhabitants of the Bahama Islands, by often practice, are very expert at catching turtles, particularly the green turtles. In April, they go in little boats to Cuba, and other neighbouring islands, where, in the evening, and especially in moonlight nights, they watch the going and returning of the turtle to and from their nests, at which time they turn them on their backs, where they leave them, and proceed on, turning all they meet: for they cannot get on their feet again when once turned. Some are so large that it requires three men to turn one of them. The way by which the turtle are most commonly taken at the Bahama Islands is by striking them with a small iron peg of two inches long, put in a socket at the end of a staff twelve feet long. Two men usually set out for this work in a little light boat or canoe, one to row and gently steer the boat, while the other stands at the head of it with his striker. The turtle are sometimes discovered by their swimming with their head and back out of the water, but they are oftenest discovered lying at the bottom, a fathom or more deep. If a turtle perceives he is discovered, he starts up to make his escape, the men in the boat pursuing him endeavour to keep sight of him, which they often lose and recover again, by the turtle putting his nose out of the water to breathe: thus they pursue him, one paddling and rowing, while the other stands ready with his striker. It is sometimes half an hour before he is tired, then he sinks at once to the bottom, which gives them an opportunity of striking him, which is by piercing him with an iron peg, which slips out of the socket, but is fastened with a string to the pole. If he is spent and tired by being long pursued, he tamely submits, when struck, to be taken into the boat or hauled ashore. There are men, who, by diving, will get on their backs, and by pressing down their hind parts, and raising the fore part of them by force, bring them to the top of the water, while another slips a noose about their necks."

It is probable that many varieties of *testudo mydas* exist, but they are far from being determined.

The *Imbricated turtle* (*T. Imbricata*) is so named from the peculiar arrangement of the plates of the disk, as is mentioned in the text. The head is smaller in proportion than in other turtles. The neck is longer; and the narrow, sharp, and curved form of the beak, has given rise to the name of hawk's-bill turtle, which is the vulgar appellation of this reptile. Its fore legs are longer than those of others of the tribe; and it is said to be able, when laid on its back, to recover its former position, which no other turtle can do.

Its volume is less considerable than that of *testudo mydas*, and its weight is said seldom to exceed two hundred pounds. Its usual length seems to be about three feet; but it has been known to measure five, and weigh five or six hundred pounds. Specimens of very great magnitude are reported to have been taken in the Indian Ocean.

This turtle is a native of the American and also of the Asiatic seas, and has been occasionally found even in the Mediterranean. It is common enough near the islands and coasts of America, under the torrid zone, in the Atlantic. It especially prefers the Alligator Islands, and those of the Bay of Honduras, the coasts of Vera-Cruz, in the Gulf of Mexico, the north of Jamaica, the coasts of Guinea, and the Indian Ocean.

The imbricated turtle feeds on the *turtle-grass*, the moss

on rocks which grows under water, &c.; and Catesby informs us, that it eats a fungus, which the Americans denominate *Jew's ear*.

The flesh is disagreeable and unwholesome. According to Dampier, it purges those violently who eat of it. Labat tells us, that, in Martinico, it produces fever, and causes boils to break out over the entire body. Notwithstanding this, its eggs are excellent eating.

But if the imbricated turtle presents us no advantages on the score of food, it is highly deserving attention for the value of its shell, which, from the remotest ages, has served for the purposes of decorating the furniture and palaces of the great. The plates, which are stronger, thicker, and more transparent than those of any other kind of tortoise, constitute, in fact, the sole value of the animal, by affording the substance which is particularly known by the name of tortoise-shell. They are half transparent, and beautifully varied with whitish, yellowish, reddish, and dark brown clouds and waves; they take a fine polish, and constitute, for all ornamental purposes, a most elegant article.

The dorsal pieces are thirteen, and the marginal twenty-five, smaller than the others. These are raised and separated from the bony part, which they cover, by putting fire under the shell. The heat causes the plates to start, which are then easily detached from the bone. Almost eight pounds of tortoise-shell may be thus gained from a large turtle, sometimes as much as fifteen or twenty; and the shell is not considered as of much value, unless the animal weigh at least one hundred and fifty pounds.

The next process is steeping the tortoise-shell in boiling water, for the purpose of softening it, and it is then put into a strong metallic mould of the figure required. When pieces are to be joined, the edges must be first scraped, and, while heated, laid over each other, and put into a strong press, whereby they are completely united. These are the modes by which gold and silver ornaments, &c. are affixed to tortoise-shell. This substance is not capable of being united, as has been erroneously imagined.

The ancients appear to have been exceedingly fond of the use of this beautiful substance, in the decoration of their houses, apartments, &c.: their doors, pillars, tables, and beds were embellished with it. This luxury was carried to an excessive degree amongst the Romans in the time of Augustus. Mr. Bruce has remarked on this subject, that—

"The Egyptians dealt very largely with the Romans in this elegant article of commerce. Pliny tells us the cutting them for fineering, or inlaying, was first practised by Carvilius Pollio from which we should presume that the Romans were ignorant of the art of separating the lamina by fire placed in the inside of the shell, when the meat is taken out; for these scales, though they appear perfectly distinct and separate, do yet adhere, and oftener break than split, where the mark of separation may be seen distinctly. Martial says that beds were inlaid with it. Juvenal and Apuleius, in his tenth book, mentions that the Indian bed was all over shining with tortoise-shell on the outside, and swelling with stuffing of down within. The immense use made of it in Rome may be guessed by what we learn from Velleius Paterculus, who says that when Alexandria was taken by Julius Cæsar, the magazines or warehouses were so full of this article, that he proposed to have made it the principal article of his triumph, as he did ivory afterwards, when triumphing for having happily finished the African war. This, too, in more modern times, was a great article in the trade to China; and I have been always exceedingly surprised, since the whole of the Arabian Gulf is comprehended in the charter of the East India Company, that they do not make an experiment of fishing both pearls and tortoises—the former of which, so long abandoned, must now be in great plenty and excellence; and a few fishers put on board each ship trading to Jeddo, might surely find very lucrative employment with a long-boat or pinnace, at the time their vessels were selling their cargo in the port, and, while busied in this gainful occupation, the coasts of the red sea might be fully explored."

The species figured and described by Bruce, does not appear, however, to be the one we have been now describing. The scales are not imbricated.

The *Loggerhead turtle (Testudo caretta)* is considered to be one of the largest, if not the very largest of its genus. A skull, supposed to belong to it, which was in the Leverian Museum, is said to have been taken from a turtle weighing more than sixteen hundred pounds, and measured more than a foot in length. In general appearance, this turtle most nearly resembles the *mydas*; but the head is considerably larger, the breadth of the shell greater in proportion, and the colours deeper and more varied. It has also fifteen dorsal segments, or scutella of the shell, and there are but thirteen in *mydas*. This number is observed to be pretty constant, and may therefore form a specific character of some certainty. These scutella, or dorsal segments in the middle range, are very protuberant, and form a row of tubercles on the middle of the back of the shield. The fore-feet are very large and long; the hinder shorter, but broad.

The habitat of this turtle is the same as that of *mydas*; but it is also found in remoter latitudes, and even in the Mediterranean, especially about the coasts of Italy and Sicily. It is of little or no value in a commercial point of view—the flesh being coarse and rank, and the shell of no utility in the arts. It affords oil, however, which may be used for lamps, &c.

The loggerhead-turtle is a powerful, fierce, and voracious reptile, and is even dangerous from its courage and ferocity. It will defend itself vigorously with its fore-legs, and is capable of breaking the strongest shells and other hard substances, with its mouth. Aldrovandus tells us of one which bit a thick walking-stick in two in a moment.

Catesby says, "The loggerhead-turtles are the boldest and most voracious of all other turtles. Their flesh is rank, and therefore little sought for, which occasions them to be more numerous than any other kind. They range the ocean over—an instance of which, among many others that I have known, happened the 20th of April, 1725, in lat. 30° north; when our boat was hoisted out, and a loggerhead-turtle struck, as it was sleeping on the surface of the water. This, by our reckoning, appeared to be midway between the Azores and the Bahama Islands; either of which places being the nearest land it could come from, or that they are known to frequent, there being none on the north continent of America farther north than Florida. It being amphibious, and yet at so great a distance from land in the breeding time, makes it the more remarkable. They feed mostly on shell-fish, the great strength of their beaks enabling them to break very large shells."

According to Schœpff, many parasite shells may be seen attached to the carapace of the testudo caretta.

The *Coriaceous turtle* (*T. coriacea*) is a very large reptile of this tribe. Individuals are reported to have been seen eight feet in length, and weighing a thousand pounds. Without reposing, however, implicit faith in such accounts, in consequence of the decided proneness of man to exaggeration, we may believe that this turtle attains the length of five feet or more. Its body is longer in proportion than those of others of its tribe, and it is particularly distinguished by its external covering, which, instead of being horny, is of a strong substance, resembling leather. This circumstance, along with the absence of a plastron or breast-plate, furnishes very sufficient reasons for a generic distinction of this animal. The head is large, and the upper mandible notched at the top. The legs are large and long, and covered with a similar substance to that on the back. The tail is short and sharp.

This remarkable species is a native of the Mediterranean Sea, and has at different periods been taken on the coasts both of France and England. In 1779, one was taken at Cette, measuring five feet five inches. In 1729, one, said to have been seven feet one inch, was caught near the mouth of the Loire. M. de Lafont, who has described it, says that it uttered dreadful howlings when it was killed. But this fact is so far from being well authenticated, that it does not seem at all probable. In July, 1756, says Borlase, in his History of Cornwall, one was taken which "measured six feet nine inches from the tip of the nose to the end of the shell; ten feet four inches from the extremity of the fore-fins extended; and was adjudged to weigh eight hundred pounds weight."

This species belongs not only to the European seas, but is also found in the South American; it occasionally appears about the African coasts, and is said to lay its eggs in the sand on those of Barbary.

The ancient Greeks were well acquainted with this animal. According to traditions preserved among them, the first lyre is said to have been fabricated of its shell. Accordingly, they consecrated this animal to Mercury, the inventor of that instrument. Notwithstanding this, Pausanius informs us, that the tortoises employed for this purpose were those of the woods of Arcadia. It may also be observed, that Rondeletius is the first modern who has attributed this employment to the carapace of the species now in question. In this supposition, he has been followed by most of his successors, *ut mos est.* The French call this animal *luth*, from this circumstance;

CORIACEOUS TURTLE.

T. CORIACEA.

or, perhaps, from some fancied resemblance in the ribs or prominences on the back of the shell to the strings of a harp.

Pennant reports the coriaceous tortoise to be extremely fat, but the flesh coarse and bad. The Carthusian friars, however, will eat of no other species. The small sea-tortoise described by Pennant seems to be the young of this animal.

The *testudo fimbriata* is a fresh-water tortoise apparently, and is said to have been once common in the rivers of the isle of Cayenne. It has grown rare, however, from having been much sought after, in consequence of the estimation in which its flesh was held. It lives on aquatic herbage, and is said to remove by night from its watery habitat in quest of pasture. M. Bruguiere had a specimen in his possession which lived for some time on herbs, bread, &c. It layed five or six eggs, one of which produced a young tortoise in the box in which it was kept.

In the Trionyx, called also *soft tortoises*, the ribs do not reach the edges of the carapace. The bones analogous to the sternal ribs are replaced by a simple cartilage. The sternal pieces of the breast-plate, too, do not fill the entire lower face of the body. They are easily distinguished from all the other tortoises.

The *Trionyx Ægyptiacus*, Groof., *Test. triunguis*, Gm., has the carapace oval, almost orbicular, carinated along the back, without scaly plates or emargination, coriaceous, striated, rough to the touch, convex, more flattened and rugous in front, widened and smoother behind. The breast-plate is flat, smooth, whitish, without scales or emarginations, and in a great measure cartilaginous. The head is smooth, depressed, and small. The feet short, and palmate.

This tortoise sometimes grows to three feet in length; and is green, spotted with white. It inhabits the Nile, and devours the little crocodiles at the moment in which they proceed from the egg. It also appears to have been described as frequenting the waters of the Euphrates.

The *Tierce tortoise, Test. ferx*, is distinguished as belonging to this tribe, by the nature of its shield, hard or osseous on the middle part only, while the edges are flexible and coriaceous. The head is small and somewhat trigonal, with the snout much lengthened. The neck, when drawn back, appears very thick, and surrounded by many folds of skin; but when stretched out, its length is equal to that of the whole shell. The legs are short and thick, and covered with a wreathed skin. The feet are strongly webbed. The tail is short, pointed, and curved inwardly. The shell beneath is marked very elegantly with branches of vessels disposed upon it.

This species inhabits the rivers of Carolina, Georgia, Florida, and Guiana. Contrary to the character of the great majority of tortoises, it possesses a very considerable portion of vigour and activity, and will defend itself when disturbed or attacked with remarkable fierceness and alacrity. It will raise itself upon its legs, dart upon its assailant, and bite with uncommon violence. It was first described by Dr. Gordon, and afterwards by M. Pennant, in the Philosophical Transactions. One specimen seen by the former gentleman weighed five and twenty pounds; but it is said sometimes to grow so large as to weigh seventy pounds. Its eggs are exactly spherical, and more than an inch in diameter. Its flesh is also reported to be extremely delicate, and equal, if not superior, to that of *testudo mydas*.

Mr. Bartram, in his travels, has described a tortoise which he calls the great soft-billed turtle, and which appears to be the same as this.

THE SECOND ORDER OF REPTILES,

OR

THE SAURIA,

Have the heart composed, like that of the Chelonia, of two auricles, and of a ventricle sometimes divided by imperfect partitions.

Their ribs are mobile, partly attached to the sternum, and have the capacity of being raised or depressed for the purposes of respiration.

The lungs extend, more or less, to the hinder part of the body. They frequently penetrate considerably into the abdomen ; and the transverse muscles of the abdomen are insinuated under the ribs, as far as towards the neck, to embrace them. Those Saurians which have this organ much expanded, possess the singular faculty of changing the colours of the skin, according as they experience the emotions of any want or passion.

Their eggs have an envelope more or less hard, and the little ones issue from them with the form which they are always to preserve.

Their mouth is always armed with teeth, and, with few exceptions, their toes with claws. Their skin is clothed with scales more or less crowded, or at least with small scaly grains. The male organ of generation is sometimes double, sometimes single, according to the genera.

All have a tail more or less long, and almost always very thick at its base. The greater number have four legs: some of them, however, have but two.

They form but two genera in the system of Linnæus—the Dragons and the Lizards; but the last should have been divided into many, which differ in the number of the feet and of the genital organs, in the forms of the tongue, tail, and scales, to such a degree, that we are obliged to divide them into many families.

The first, or that of

The Crocodilians,

comprehends but a single genus; namely,

The Crocodiles. *Crocodilus.* Br.

They are of a great size; have the tail compressed laterally; five toes before, and four behind, of which the three internal alone are armed with claws on each foot, but all are more or less united by membranes. There is a single rank of pointed teeth in each jaw. The fleshy tongue is flat, and attached to the floor of the mouth as far as within a very little of its edges, which led the ancients to believe that it was altogether wanting. The penis is single; the aperture of the anus, longitudinal; the back and tail covered with large square scales, very strong, and raised into a ridge or crest on their middle. There is a crest with strong indentations in the tail, double at its base. The scales of the belly are square, slender, and smooth ; the nostrils open at the end of the muzzle by two small crescented clefts, closed by valvules, and proceed by a long narrow canal, pierced in the palatines and the sphenoïd, into the bottom of the hinder part of the mouth.

The lower jaw being prolonged behind the cranium, gives an appearance of mobility to the upper ; and the ancients believed that this was the case ; but it moves only with the entire head.

Their external ear may be closed at will by two fleshy lips. Their eye has three lids. Under the throat are two small holes, orifices of glands, in which a musky sort of pommade is secreted.

The vertebræ of the neck rest one upon the other by little false ribs, which render lateral motion difficult. Accordingly, we find that these animals have some trouble in changing their direction, and are easily escaped from by turning : they are the only saurians which are destitute of clavicular bones ; but the coracoïd bones are attached to the sternum, as in all the rest of the order. Besides the true and false ribs, there are some which protect the abdomen without ascending as far as the spine, and which appear to be produced by the ossification of the tendinous inscriptions of the straight muscles.

Their lungs do not sink into the abdomen, like

those of other reptiles; and some fleshy fibres adherent to that part of the peritoneum which covers the liver, give them some appearance of a diaphragm. This, added to the fact of their heart being divided into three compartments, where the blood which comes from the lungs does not mix so completely with that of the body as in other reptiles, approximates the crocodiles a little more to the warm-blooded quadrupeds.

The os tympani and pterygoïd process are fixed to the cranium, as in the tortoises. Their eggs are hard, and as large as those of geese; and the crocodiles are animals whose two extremes of magnitude—those at birth and full growth—are decidedly the most different. The females watch their eggs, and when the young are excluded, they take care of them for a few months.

The crocodiles are inhabitants of the fresh water, and are extremely carnivorous; not being able to swallow their prey in the water, they drown it, and suffer it to remain in some subaqueous cavity until it putrifies, before they eat it.

The crocodiles differ so much from the lizards in general, that some recent writers have made a separate order of them. They are the Loricata of Merrem and Fitzinger, and the Emydosauria of De Blainville.

The species, which are more numerous than was formerly imagined, are referred to three distinct sub-genera.

The Gavials. Cuv.

Have the muzzle narrow, and greatly elongated. The teeth are pretty nearly equal—the fourth below passing, when the mouth is closed, into notches, and not into holes of the upper jaw. The hind feet are indented at the external edge, and palmated to the end of the toes. There are two large holes in the bone of the cranium behind the eyes, which may be perceived through the skin. They have been as yet observed only on the ancient continent.

The most known is,—

The Gavial of the Ganges. Lac. Gangetica Gm.
Faujas Hist. de la Mont St. Pierre, pl. xlvi.
Lacep. I. xv.

A species which becomes very large, and which, independently of the length of its muzzle, is remarkable for a thick cartilaginous prominence surrounding the nostrils, and thrown backwards.*

* This prominence occasioned Elian to assert, that there exist in the Ganges some crocodiles which have a horn on the end of the muzzle.—See the description and figures of it by M. Geoffroy St. Hilaire, Mem. du Mus. XII. p. 97.

Add the *Little Gavial* (*Croc. Tenuirostris.* Cuv.), Faujas loc. cit. pl. xlviii.—if, indeed, it be a distinct species.

N. B. The calcareous schists of Bavaria have produced a little fossil gavial of a peculiar species, which has been described by M. Sœmmering, in the Memoirs of the Acad. of Munich for 1814.

I have publsihed an account of the crania, and other parts of fossil crocodiles, approximating to the gavial, found at Caen, Honfleur, and other

THE CROCODILES (properly so called),*

Have the muzzle oblong and depressed; the teeth uneven, the fourth below passing into notches, and not into holes, in the upper jaw, and all the other characters of the gavials. There are species of this form in both Continents.

The Common Crocodile, or that of *the Nile*. (*Lac. Crocodilus*. L.) Geoff. Des. de l'Eg. Rept. ii. 1. Ann. Mus. X. iii. 1. Cuv. *ibid*. X. pl. 1. f. 5. and ii. f. 7; and Oss. Foss. V., part 2, same plate and fig.

So celebrated among the ancients, has six ranges of square plates, pretty nearly equal, along the back.†

places; and I have marked the points in which the osteology of their cranium differs from that of the existing gavial.—See my "Récherches sur les Oss. Foss." V. 2d part. Analogous observations have also been made in England by Mr. Conybeare. From these differences, which principally attach to the hinder part of the palate, M. Geoffroy has thought proper to make two genera of these last animals, which he names THELEOSAURUS and STENEOSAURUS; and nevertheless he seems to think that the present gavials may descend from them, and that the differences may result from the change of atmospheric circumstances.—Mem. du Mus. XII.

* Κροκοδειλος, *that which fears the shore*—a name given by the Greeks to a lizard common among themselves. They afterwards applied it, on the score of resemblance, to the crocodile of Egypt, when they travelled into this latter country.—Herodot. lib. ii. M. Merrem has changed the name of this sub-genus into CHAMPSES, which, according to Herodotus, was the Egyptian name of this animal.

† There are found, from Senegal as far as the Ganges and beyond it, some crocodiles very similar to the common, some of which have the

The double-crested Crocodile. (*Cr. Biporcatus.* Cuv.) Ann. Mus. X. i. 4. and ii. 8. et Ossemen's Foss. V., 2d part, same pl. and fig.

With eight ranges of oval plates along the back, and two projecting crests on the upper part of the muz-

muzzle a little longer, and more narrow, and others some varieties in the plates or scales, which furnish the upper part of their neck; but it is very difficult to distribute them into distinct species, in consequence of the intermediate shades. The small isolated scales which form a transverse range immediately behind the cranium, vary in number, from two to four or six. The approximating plates which form the buckler of the nape are generally six in number; but there is sometimes a smaller one, at a little distance from each anterior angle of this buckler, and sometimes contiguous to the buckler. M. Geoffroy gives the name of *Cr. Suchus* to those which have the muzzle more narrow and elongated; *C. Marginatus* to those in which are reckoned six scales in the range behind the cranium, some of which have six plates in the buckler, and some eight; *Cr. Lacunosus* is his name of an individual which has but two scales behind the cranium, and six plates in the buckler; and, finally, by *Cr. Complanatus*, whose characters attach to certain proportions of the head.

These different crocodiles also exhibit some variations of detail in the muzzle, and in the lateral scales of the back. But as to this last point, and still more as to the muzzle, the varieties must be much more numerous; and M. Geoffroy acknowledges that *nothing is more fugitive than the forms of crocodiles.* So much is this the case, that I cannot venture to elevate to a specific rank some crocodiles sent from Bengal by M. Duvauçel, although their head is more convex than that of the rest of this genus.

There is another point of discussion between the philosophic naturalist just cited and myself. He supposes that the species or variety with the more narrow muzzle continues smaller, that it is gentle and inoffensive, and that its littleness causes it to be cast ashore during the period of the inundations, of which it is thus a precursor; and, according to these notions, he thinks that it was particularly to it that the Egyptians rendered religious homage, and that the name of *suchus* or *suchis* appertained to it as a species. I believe, on the contrary, that I have proved from Cicero zle; it is found in many islands of the Indian seas, and probably also in the two peninsulas. It has been principally received from the Sechelles Islands.

The slender-muzzled Crocodile. (*Croc. Acutus.* Cuv.) Geoffr. Ann. Mus. II. xxxvii.

With a longer muzzle, gibbous at the base, and the back-plates arranged in four lines: the external ones are disposed irregularly, and have more projecting ridges. This species is of St. Domingo and the other great Antilles. The female places her eggs in the

and Aristotle that the crocodiles venerated in Egypt were not less ferocious than others. It is also certain that the crocodile with narrow muzzle was not tended exclusively by the priests; for, according to the very accurate researches of M. Geoffroy himself, it is ascertained that the three embalmed crocodiles now in Paris are, none of them, the *suchus*, but the *marginatus*, *lacunosus*, and *complanatus*. Finally, every thing leads me to suppose, that *Souc*, or *Souchis*, which, according to M. Champollion, was the Egyptian name of Saturn, was also the proper name of the crocodile which was kept at Arsinoe—as Apis was the name of the sacred ox at Memphis, and Mnevis that of the ox of Hermopolis.

On this point of antiquity, the different writings of M. Geoffroy may be consulted, and his recapitulation of them all in his great work on Egypt, as well as my "Researches on the Fossil Bones," tom. v., 2d part, p. 45. This last article having been written previously to M. Geoffroy's recapitulation of his own papers in the work on Egypt, I could not then employ the argument derived from the embalmed crocodiles with which M. Geoffroy himself has furnished me, and which I deem a singular corroboration of my views of the subject.—*Cuvier.*

*** We do not deem ourselves at liberty to omit any of our author's valuable notes on the text; but much in them relating to fossils, and to the comparison with living species, will be found in our own compilation on the same snbject.—*Ed.*

earth, and uncovers them at the precise moment of exclusion.*

THE CAYMANS. ALLIGATOR. CUV.

Have the muzzle broad and obtuse; the teeth uneven, and the fourth below enter into holes, and not notches, in the upper-jaw. Their feet are only semipalmate, and without indentation. They have as yet been known only in America.

The name of *Cayman*, or *Caiman*, is given to the crocodile. The French colonists employ it to designate the most common species of crocodile around their habitations. The English and Dutch use in the same sense the word *Alligator*, corrupted from the Portuguese *Lagarto*, itself a corruption of *Lacerta*.

The Spectacled Cayman. Croc. Sclerops. Schn. Seb. I. civ. 10. Cuv. Ann. Mus. X. i. 7 and 16, and ii. 3.

Thus named from a transverse ridge which unites in front the projecting edges of its orbits. It is the

* The slender muzzled crocodile has been particularly observed by M. Descourtils.—Add, *The lozenged crocodile, Cr. Rhombifer.* Cuv. Ann. Mus. XII. pl. i. 1.;—*The helmeted crocodile, Cr. Galeatus,* Perrault, Mem. pour Servir à l'Hist. des An. pl. lxiv.—if indeed this species, known by this figure only, be well authenticated;—*The double-shielded crocodile, Croc. biscutatus,* Cuv. Ann. Mus. X. ii. 6., and Oss. Foss. t. V. part ii. pl. vi. —of which but one or two individuals have been seen,—*The crocodile with armed nape, Cr. Cataphractus,* Cuv. Oss. Foss. t. V. 2d part, pl. v. f. 1 and 2.

most common species in Guiana and Brazil. The nape is armed with four transverse bands, with very strong scales. The female lays in the sand, covers the eggs with straw or leaves, and defends them with courage.*

The Pike-muzzled Cayman. Croc. Lucius. Cuv. Ann. Mus. X. i. 8, et 15, and ii. 4.

Thus named from the form of the muzzle; is also distinguished by four principal plates which it bears on the nape. It inhabits the southern parts of North America. It sinks into the mud, and falls

* There are also Caymans of several kinds, which have this transverse crest in front of the orbits, and which perhaps, like the crocodiles that approximate to the common, form different species, but difficult to characterize well.

Some have the muzzle shorter and more rounded, and the transverse ridge concave in front, and prolonged on each side over the cheek. In them I reckon thirteen teeth on each side above. The cranium is not widened behind. The body is green, punctated, and spotted with black, with black bands on the tail.

Others have the head and teeth of the same character; but their body is black, with narrow and yellowish bands, as in the *Jacaré Noir* of Spix, pl. iv.

Others, again, have the muzzle less wide, and the concave ridge less prolonged. I find in them fifteen teeth, and their neck is better armed. I would readily take them for the *Cr. Fissipes* of Spix, pl. iii.

Finally, there are some with the muzzle still less wide, the cranium widened a little behind, the transverse ridge convex in front, and not prolonged upon the cheek. The scales of the back have their ridges less projecting, and the bands of the tail are less marked. Such are the *Cr. Punctulatus* of Spix. Unfortunately, M. Spix has not insisted on the characters taken from the transverse crest.

Necks and Foot of Crocodiles.

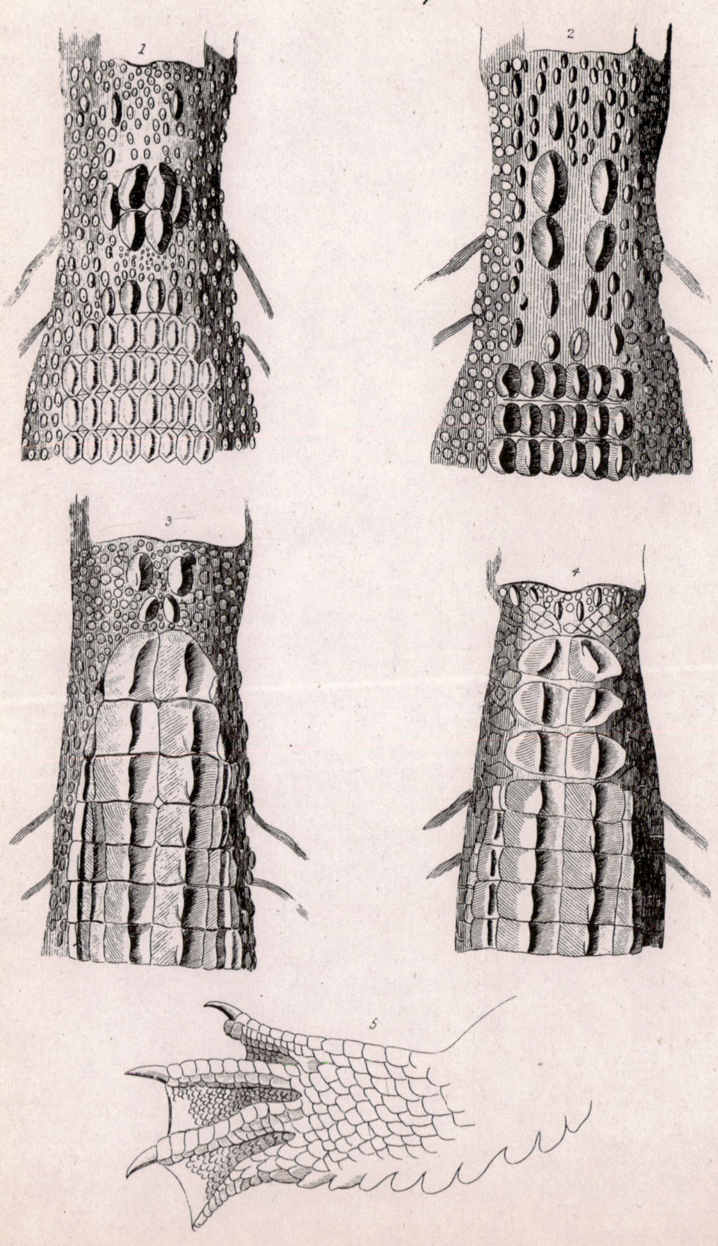

1 . C .biporcatus. 2 . C .lucius.

3 . C .tenuirostris. 4 . C .Gangeticus.

Necks and Foot of Crocodiles.

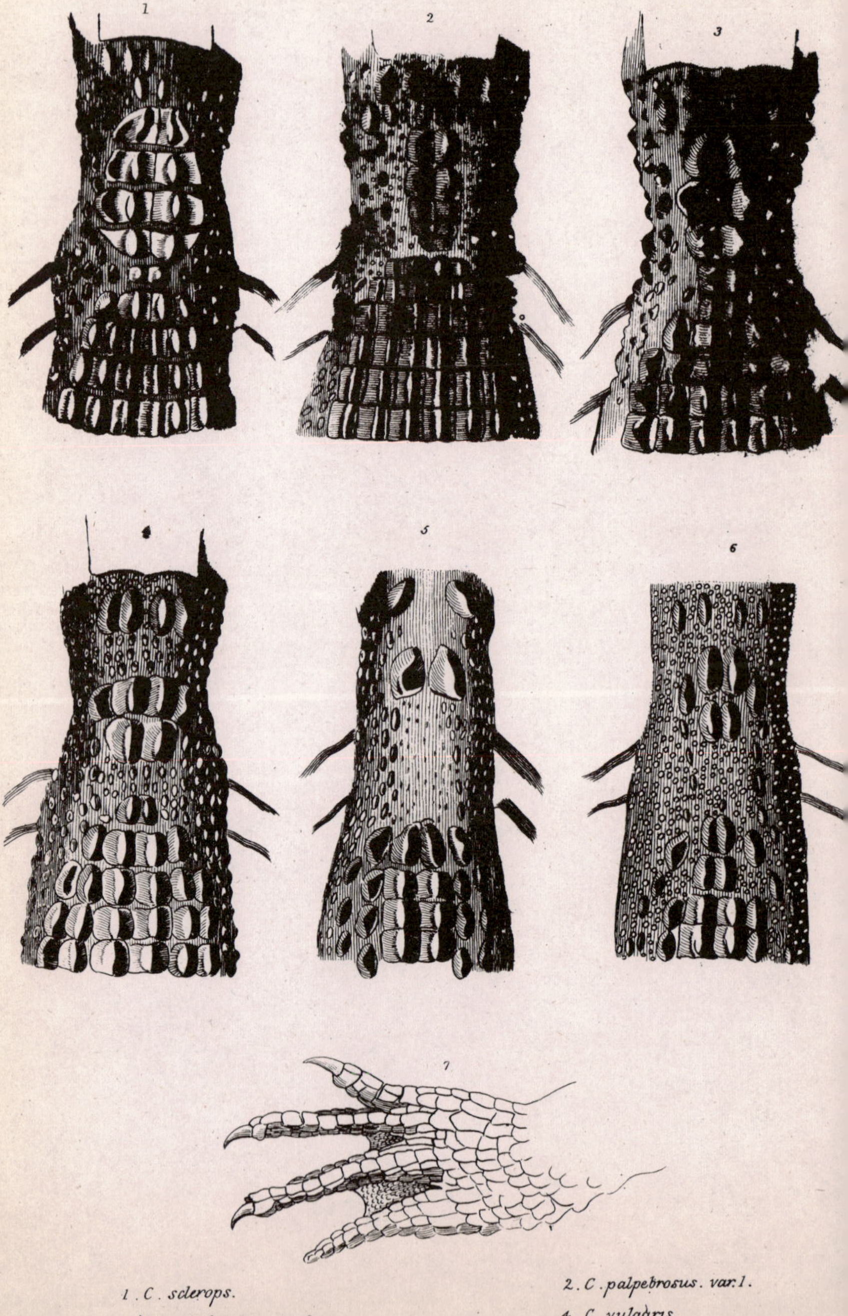

1 2 3

4 5 6

7

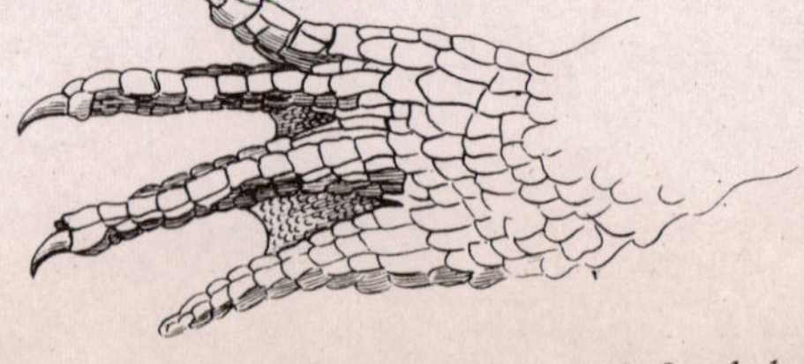

1 . *C . sclerops.*

2 . *C . palpebrosus . var.1.*

3 . *C . palpebrosus. var.2.*

4 . *C . vulgaris.*

5 . *C . biscutatus.*

6 . *C . acutus.*

into a state of lethargy during severe cold. The female deposits her eggs in layers, with beds of earth.*

The second family, or that of

THE LACERTIANS,

Is distinguished by the slender extensible tongue, terminated by two threads like that of the adders and vipers. Their body is elongated. Their walk rapid. All their feet have five toes, separated and unequal, especially those of the hinder feet. The scales are disposed under the belly and around the tail, by transverse and parallel bands. Their tympanum is on a level with the head, and but slightly sunk in, and membraneous. A production of the skin, cleft longitudinally, which closes by a spincter, protects their eye. Under the anterior angle is a vestige of a third lid. Their false ribs do not make an entire circle. The males have a double penis. The anus is a transverse cleft.

The species being very numerous and varied, we subdivide them into two great genera,—

* See on this species the Mem. of Dr. Harlan, Ac. Nat. Sc. Philad. iv. 242.—Add, *The Cayman with osseous eyelids*, *Cr. Palpebrosus*, Cuv. Ann. Mus. X. pl. i. 6 & 7, et ii. 2; and the *Cr. Trigonatus*, Schn. Seb. I. cv. 3, or the *Jacaratinga Moschifer*, Spix, pl. i. This species has the lid occupied entirely in its thickness by three osseous laminæ, of which the other crocodiles have scarcely a vestige.

The MONITORS, recently called, by a singular mistake, TUPINAMBIS,*

Are that genus in which there are species of the largest size. They have teeth in both jaws, but want them in the palate. The greater number of them may be recognized by their tail, laterally compressed, which renders them aquatic. As in the neighbourhood of waters they are sometimes near the crocodiles and caymans, it is reported that they advertise, by a hissing noise, the approach of those dangerous reptiles. This assertion, in all probability, has given rise to the names of *safe-guard* and *monitor* in some of the species. But this assertion is any thing but certain.

They are divided into two very distinct groups. The first, or that of

THE MONITORS, properly so called,

May be known by small and numerous scales upon the head and limbs, under the belly, and around the tail, which last has a keel underneath, formed by a double range of projecting scales. Their thighs are destitute of the range of pores which we find in many other Saurians. They all belong to the an-

* Marcgrave, speaking of the safeguard of America, says that it is named *Teyu-guacu*, and, among the Topinambos, *Temapara* (*Temapara Tupinambis*). Seba has taken this last word for the name of the animal, and all other naturalists have copied him.

cient continent, though Seba, and Daudin, who follows him, gives some true monitors as Americans; but this is a mistake.

There are two species, inhabitants of Egypt, which may be considered as the types of this subdivision.

The Monitor of the Nile. Ouaran of the Arabs. *Lacerta Nilotica.* L. Mus. Worm. 313. Geoff. St. Hil. Gr. Ouv. sur l'Egypte. Reptiles, pl. 1. f. 1.

With strong and conical teeth, the hinder of which become round with age; brown, with paler and deeper points, forming divers compartments, among which are remarked transverse ranges of large ocellated spots, which on the tail become rings. The tail, round at the base, is surmounted by a keel on almost its entire length. This reptile attains the length of five or six feet. In Egypt, the vulgar pretend that it is a young crocodile, excluded in a dry soil. The ancient Egyptians had it engraven on their ornaments, perhaps, because it devours the crocodile's eggs.*

* To this species are approximated by the form of the teeth, and even the arrangement of spots—which, by the way, are very similar in all the monitors—*M. Ornatus,* Daud. An. Mus. II. xlviii.;—*Lac Capensis,* Sparmann;—*M. Albogularis,* Daud. Rept. III. pl. xxxii.

It is of this subdivision that M. Fitzinger makes his genus VARANUS. Under this name Merrem comprehended all my monitors properly so called.

The Land Monitor of Egypt, Ouaran el hard of the Arabs. *Lacerta Scincus.* Merr. Geoff. Egypt. Rept. iii. 2.

With compressed, trenchant, and pointed teeth; tail almost without keel, and remaining round for a greater length. Its habits are more terrestrial; and it is common in the deserts which border upon Egypt. The jugglers of Cairo employ them in the performance of tricks, after they have plucked out their teeth. This is the *terrestrial crocodile* of Herodotus, and, according to Prosper Alpin, the true *scincus* of the ancients.*

Africa and the East Indies produce a great number of monitors with trenchant teeth like the foregoing, but the tail of which is still more compressed than that of the Nile.

The most common in the Indian Archipelago is

The Double-banded Monitor. Lacerta Bivittata, Kuhl.

White underneath, black above; with five transverse ranges of white spots, or white rings. A white band along the neck; and an angle formed by the white of the breast, which rises obliquely over the shoulder. There are some three feet long.†

* M. Fitzinger makes of this species his genus Psammosaurus.

† To this species attach, from the distribution of the colours, the *T. Bigarré,* Daud. *Lac. Varia,* Shaw. Nat. Misc. 83, J. White, 253, of New

The other group of monitors is distinguished by angular plates upon the head, and large rectangular scales under the belly and around the tail. The skin of their neck clothed with small scales, makes two transverse folds. They have a range of pores under the thighs.*

We may establish in them some subdivisions.

The first, or that of

THE DRAGONS,

Has, for its distinctive character, scales raised with ridges as in the crocodiles, forming crests on the tail, which is compressed.†

Holland;—an approximating species of manilla, *M. Marmoratus;*—the *T. Elegant,* and *T. Etoile,* Daud. III. xxxi., and Seb. I. xciv. 1, 2, 3; xcviii., xcix. 2; II. xxx. 2; xc., cv. 1, &c., which form but a single species, originally of Africa. We must add, the *T. Cepedien,* Daud. III. xxiv., or *Lac. Exanthematica,* Bosc. Act. Soc. Nat. Par. pl. v. f. 3, oscillated throughout;—*M. Bengalensis,* Daud.; *M. Indicus,* id.;—a species uniformly blackish, from Java, *M. Nigricans,* Cuv. &c.

After every comparison, I have reason to believe now that the figure of Seba, I. pl. ci. f. 1, of which Linnæus has made his *Lacerta Dracœna,* but which is very different from the *Dragonne* of Lacepède, is the *M. Bengalensis.* The original of Seba's is in our Museum.

M. Fitzinger reserves to these species with compressed tail, the generic name of TUPINAMBIS.

* Merrem has made of this second group his genus TEIUS.

† M. Spix has made of this subdivision his genus CROCODILURUS, which name Mr. Gray has changed into ADA.

THE GREAT DRAGON. *Mon. crocodilinus*, Merr. Lacep. Quadr. Ovip. pl. ix.

Has also scales elevated with ridges scattered over the back. The teeth in the bottom of the mouth grow round with age. It attains from four to six feet in length, and inhabits Guiana, in burrows, near marshes. Its flesh is eaten.

The Lizardet, Daud. *Lac. Bicarirnata*, L. *Crocodilurus Amozonicus*, Spix, pl. xxi.

Is smaller, and has no raised scales on the back. It is found in many parts of South America.

The second, or

THE SAFE-GUARDS,

Has all the scales of the back and tail without keels. The teeth are notched; but with age, those of the hinder mouth grow round also.*

Some, more especially, called SAFE-GUARDS, have the tail more or less compressed. The scales of the belly are more long than broad. They live on the banks of waters.

Such, most particularly, is

* To these M. Fitzinger reserves the name MONITOR.

The great American Safe-guard. Teyu-guazu, Temapara, &c. (*Lacerta Teguixin.* Lin. and Shaw. Seb. I. xcvi. 1, 2, 3; xcvii. 5; xcix. 1.

With yellow points and spots, disposed in transverse bands, on a ground which is black above, yellowish underneath. There are yellow and black bands on the tail. In Brazil and Guiana, it attains to six feet in length. It goes at a rapid rate on land, and takes refuge in the water when pursued. It dives there, but does not swim. It feeds on insects of all kinds, smaller reptiles, and eggs, &c. It inhabits holes, which it digs in the sand. Its flesh and eggs are eaten.*

Others, called AMEIVA, do not differ from the preceding, but in having a round tail, and nowise compressed, furnished, as well as the belly, with transverse ranges of square scales. Those of the belly are more broad than long. There are American lizards, similar to ours in exterior, and which represent them in that country. But, independently of the want of molar teeth, most of them have no collar, and all the scales of the throat are small. The head

* The individuals which are dried or preserved in spirits, take a bluish or greenish tint in their clear parts; and it is thus that they are represented by Seba. But when living, as we have seen them, the clear parts are more or less yellow. Prince Maximilian de Nieud has well represented this animal in his eleventh book.

Add, the *Tupin à taches vertes* of Daudin, if it be not a simple variety of the Safe-guard. Spix names it *Tup-monitor*, pl. xix. It is his *T. nigropunctatus* which is the true Safe-guard.

is also more pyramidal than in our lizards, and they have not, like them, an osseous plate over the orbit.*

Under the name of *lacerta Ameiva*, many species have been confounded, some of which are still rather difficult to distinguish. The most extended (*Teyus Ameiva*, Spix, xxiii. Pr. Max. de Wied. v. lib.) is a foot and more in length, and has the back more or less picked out and spotted with black and vertical ranges of white ocellæ, bordered with black on the flanks.

There is another (*Teyus cyaneus*, Merr. Lacep. I. xxxi. Seb. II. c. v. 2), pretty nearly of the same size, bluish, with round white spots, scattered over the flanks, and sometimes on the body.

The young individuals of these ameiva and of some others, have blackish stripes on the sides of the back: we must be careful not to multiply the species.†

* The name *Ameiva*, according to Marcgrave, designates a lizard with forked tail, which can only be an accidental circumstance. Edwards having had an individual of the above division, in which this accident was observable, applied the name of it to the entire species. Marcgrave compares this individual to his *Taraguira*, which, according to his description, would rather appear to be a *Polychrus*.

† Such appears to me to be the *Teyus Ocellifer*, Spix, xxv.

Add. *Am. Litterata*, Daud. Lib. I. lxxxiii.—*Am. Cæruleocephala*, id. Seb. xci. 3:—*Am. Lateristriga*, Cuv. Seb. L. xc. 7.;—*Am. Lemniscata*, (*Lac. Lemnisc.* Gm.) Seb. I. xcii 4.—*Teius Tritæniatus*, Spix, xxi. 2.—*T. Cyanomelas*, Pr. Max. 5th book.

I know not by what confusion of synonimy, Daudin has placed the *Am. Litterata* in Germany. It belongs to America, like all the rest. The *Am. Graphique* of Daud. Seb. L. lxxxv. 2, 4, is the *M. Bengalennis:* his *Am. Cargus*, Lib. L. lxxxv. 3, is the *Cepedian* Monitor: his *Goitreux*, Seb.

We may separate from the ameiva, certain species which have all the scales of the belly, legs, and tail, raised by a keel;* and others, in which the scales of the back are themselves carinated, so that the flanks alone have small grains.† These species approximate still to the lizards by having a collar under the neck.‡

THE LIZARDS, proper,

Form the second genus of the lacertians. They have the bottom of the palate armed with two ranges of teeth, and are otherwise distinguished from the ameiva and safeguards, because they have a collar

II. ciii. 3, 4, does not differ from the *Litterata;* finally, his *Tête Rouge*, Seb. I. xci. 1, 2, is a common green lizard. He has probably been led into error by the illuminated plates of Seba. The *lac. 5-lineata* appears to me to be a *L. Cœruleocephala*, a part of the tail of which being broken, was reproduced with small scales, a case of constant occurrence after such accidents. The oscis of this new portion of tail is always a cartilaginous stem without vertebra. On such accidents species cannot be found, as Merrem has done in his *Teyus Monitor* and *Cyaneus*.

* One of them has, in one sex, two small spines on each side of the anus, which has given rise to the genus CENTROPYX of Spix, xxii. 2.

† The *striped lizard* of Surinam, Daud. III. p. 347, Fitzinger makes of it his genus PSEUDO-AMEIVA.

‡ It even appears to me that the Centropyx has teeth in the palate, otherwise both these kinds of lizard have the head of the Ameiva; no bone on the orbit, &c.

N.B. Fitzinger makes a genus (TEYUS) of the *lacerta teyou* of Daudin, which would seem to have but four toes on the hind feet; but this rests only on an incomplete description of d'Azara, and does not appear to me sufficiently authentic.

under the neck, formed by a transverse range of broad scales, separated from those of the belly by a space, in which there are only some small ones, as under the throat, and because a part of the bones of the cranium advances over the temples and orbits, so that the entire upper part of the head is furnished with an osseous buckler.

They are very numerous, and our country produces several species confounded by Linnæus under the name of *Lacerta Agilis*. The handsomest is the *great green lizard*, (*Lac. Ocellata*, Daud.) Lacep. L. xx. Daudin III., xxxiii. of the South of France, Spain, and Italy, more than a foot long, of a fine green, with lines of black points forming rings or eyes, and a kind of embroidery, the young of which, according to Mr. Milne Edwards, is the *Lizard Gentil* of Daudin III. xxxi. ;—the *Lac. Viridis* of Daudin III., xxxiv. whose *B. Bileniata* is only a variety, according to the same observer.—The *Lac. Lepuim*, id. ib. 2, whose *Lac. Arenicola* is but a variety ;—the *Lac. Agilis*, id. xxxviii. 1, is found in all our neighbourhood. The South of France produces the *Velox* Pallas, to which Daudin's *Bosquien* and some new species must be referred.*

* I add, with hesitation, the *Lac. Sericea*, Laur. II. 5; *Argus*, id. 5; *Terrestris*, id. III. 5.

The *Tiliguerta* of Daudin is a mixture of an American ameiva with the green lizard of Sardinia, badly described by Cetti. The *Cœruleocephala*, *Lemniscata*, and *Quinque-Lineata*, are ameiva. The *Sex Lineata*, Catesby, is a *seps*.

Algyra, Cuv.

Have the tongue, teeth, and pores in the thighs of the lizards; but their dorsal and caudal scales are carinated, and those of the belly smooth and imbricated; and they have no collar.*

Tachydromus. Daud.

Have square and carinated scales on the back, under the belly, and on the tail. They want the neck-collar, and also the pores in the thigh; but on each side of the anus is a small vesicle open with a pore. Their tongue is the same as in the lizards; their body and tail are very much elongated.

The Iguanians

Are a third great family of the Saurians, which have the general form, the long tail, and the free and inequal toes of the lacertians. The eye, ear, pores, and anus are similar; but their tongue is fleshy, thick, not extensible, and emarginated only at the end.

They may be divided into two sections. The first, that of the ACŒMIANS, have no teeth in the palate.†

* *Lacerta Alegyra.* Linn.

† *Iguana* is a name originally of St. Domingo, according to Hernandez, Scaliger, &c. The inhabitants must have pronounced it *hinana*, or *igoana*.

We place, in this first section, the following genera :—

STELLIO. CUV.

Which have, with the general characters of the family of the iguanas, the tail surrounded by rings composed of large and often spinous scales.

Their sub-genera are as follow :—

CORDYLUS. GRONOV.*

Have not only the tail, but also the belly and back furnished with large scales, on transverse ranges. Their head, like that of the common lizards, is provided with a continuous osseous buckler, and covered with plates. In many species, the points of the

According to Bontius, it is originally of Java, where the natives pronounce it *leguan*. If this be the case, the Spanish and Portuguese must have transported it into America, and transformed it into iguana. They apply it to the *safe-guard*, as well as to the true iguana. This name ha also been sometimes given, as well as that of *guano*, to some monitors of the Old Continent. Attention should be paid to this in reading the accounts of travellers. I am even of opinion, that the *leguan* of Bontius is nothing else but a monitor.

* According to Aristotle, "the *Cordylus* is the only animal which has both feet and gills. It swims with its feet and tail, which resemble that of the Silurus, as far as we may compare small things with great. This tail is soft and broad. It has no fins: it is an animal of the marshes, like the frog. It is a quadruped, and comes out of the water. Sometimes it gets dry, and dies."

It is evident that these characters can only apply to the larva of the aquatic salamander, as M. Schneider has well observed. Belon has described this salamander under the name of Cordylus, but his printer added, by mistake, the figure of the safe-guard of the Nile. Rondelet has applied this name to the great *Stellio* of Egypt, or *Caudiverbera* of Belon, because, caudal scales form spinous circles. There are also some small spines to those of the sides of the back, of the shoulders, and of the external part of the thighs : the latter have a single line of very large pores.

The Cape of Good Hope produces many of these, confounded for a long time under the name of *Lacerta Cordylus*, L. These well-armed Saurians, a little larger than our common green lizard, feed on insects.*

The Common Stelliones†

Have the spines of the back moderate, the head swelled behind by the muscles of the jaws, the back and thighs bristling here and there, with scales larger than the rest, and sometimes spinous. Small groups of spines surround the ear; the thighs are des-

in the figure, he mistook the ear for the cleft or opening of the gill. Between Rondelet and Linnæus, the Cordylus has thus passed as synonymous with the *Caudiverbera*. Its special application to the genus above mentioned is wholly arbitrary. Merrem has changed it into Zonurus.

* Daudin has referred to the Cordylus many synonimes of the Stellio, as he has referred to the Stellio many synonimes of the Gecko. We have four species—*Cord. Griseus*, Cuv. Seb. L. lxxxiv. 4;—*C. Niger*, which has the ridges of the scales softer, Seb. II. lxii. 5;—*C. Dorsalis*;—*C. Microlepidotus*.

† The Stellio of the Latins was a spotted lizard, living in the holes of walls. It was considered as venomous, hostile to man, and cunning. It was probably the *Tarentole*, or *Tuberculous Gecko*, of the south of Europe, *Geckotte* of Lacep., as various authors have conjectured, and lastly, M. Schneider. Nothing justifies the application made to the actual species. Belon, I believe, was the first who was culpable of this misnomer.

titute of pores; the tail is long, and finishes in a point.

We know but of one species,—

The Stellio of the Levant. (*Lac. Stellio*, L.) Seb. I. cvi. f. 1; and better, Tournefort, Voy. au Lev. I. 120, and Geoffr. Disc. de l'Egypte, Rept. ii. 3. —*Koscordylos* of the modern Greeks; *Hardem* of the Arabs.

A foot long, of rather an olive colour, shaded with black. Very common in the Levant, and more especially so in Egypt. According to Belon, it is its excrements which are gathered for pharmaceutical purposes, under the names of *Cordylea*, *Crocodilea*, or *Stercus Lacerti*, and which were formerly in request as a cosmetic. But it appears that the ancients rather applied this name and this virtue to those of the monitor. The Mahometans destroy this animal, because they say that it mocks them by bowing the head as they do when at prayers.

DORYPHORUS. Cuv.

Are destitute of pores, like the stellines, but have not the trunk bristling with small groups of thorns.*

* *Stellio Brevicaudatus*, Seb. II. lxii. 6. Daudin IV. pl. 47. *St. Azureus*, Daud. id. 46.

Uromastix. Cuv. *Stellions Batards*. Daud.

Are only stelliones which have not the head swelled out, and all the scales of whose body are small, smooth, and uniform; and those of the tail still larger and more spinous than in the common stellio: but there are none underneath. There is a series of pores under the thigh.*

The Uromastix of Egypt. Stellio Spinipes. Daud. Geoffr. Rept. d'Egypt. pl. ii. f. 2.

Two or three feet long; body swelled out; altogether of a fine meadow green. There are small spines on the thighs. The tail is spinous only on the upper part. It is found in the deserts which surround Egypt. It was described of old by Belon, who has asserted, but without proof, that it was the *land crocodile* of the ancients.†

* The name of *Caudiverbera*, and that of ὀϱγμάστιξ, are not ancient. They were forged by Ambrosius for the great Egyptian species of which Belon had said, "*Caudâ atrocissimè diverberare creditur.*" Linné has applied the first to a *Gecko*, and other writers to Saurians, altogether different. Add, *Urom. Griseus*, from New Holland;—*Ur. Reticulatus* of Bengal;—*Ur. Acantinurus*, Bell., Zool. Journ. I. 457. if, indeed, it be a distinct species.

N.B. The Stellio, with flat tail, of New Holland, is a phyllurus.

† It is an uromastix which has been described by M. de Lacépède, Rept. II., 497, under the name of *Quetzpalco*; but which is that of a different Saurian, of which we shall speak lower down. Add, *Ur. Ornatus*, Ruppel.

AGAMA. Daud.

Have a great resemblance to the common stelliones, especially in their inflated head; but they are distinguished by the imbricated, and not verticillated scales of the tail. Their maxillary teeth are nearly the same as those of the others, and they have no palatine teeth.*

THE COMMON AGAMA.

Some scales raised into a point or tubercles, bristle in various parts of the body, and especially in the neighbourhood of the ears; with spines, sometimes grouped, sometimes isolated. A range of them is occasionally seen upon the nape; but they do not form the spangled crest which characterizes the *calotes*. The skin of the throat is flabby, folded crosswise, and susceptible of inflation.

There are some species in which the thighs have a series of pores.

The Ocellated Agama of New Holland. Ag. Barbata,

Is very remarkable from its magnitude and extraordinary figure. A series of large spiny scales pre-

† *Agama*, from ἄγαμος (unmarried). It is not easy to conjecture why Linnæus should have given this name to one of those lizards. Daudin has extended it to the entire subgenus into which this species should enter, and believes that Agama is the national name.

dominates in transverse bands along its back and tail, and approximates it to the stelliones. Its throat, susceptible of being much inflated, is furnished with scales elongated into points, which form a kind of beard. Similar scales bristle on the flanks, and form two oblique crests behind the ear.

Under the belly are yellowish spots, bordered blackish.

From this must be distinguished the

Muricated Agama, of the same country. *Lac. Muricata.* Sh. Gen. Zool. vol. iii. part i. pl. lxv. f. 11. White, p. 244.

In which the raised scales are disposed in longitudinal bands, and between them are two series of spots, paler than the ground, which is a blackish brown. The size of this reptile is also considerable.

Other species have no pores in the thighs.

Ag. Colonorum. Daud.
Seb. I. cvii. 3.

Brownish, with long tail, having a small range of small spines on the nape. It comes from Africa, and not from Guiana, as has been asserted.*

* Nothing can equal the confusion of synonimes cited by authors under the different species of lizards, but especially those of the *Agamæ, Calotes,* and *Stelliones.* For instance, in the case of the Agama, Daudin, after Gmelin, quotes Seb. I. cvii. 1 and 2, which are *Stelliones,* Sloane, Jam. II.

There is at the Cape a smaller agama, with moderate tail, varied with brown and yellowish, and having all the upper part bristling with raised and pointed scales. (*Ag. Aculeatea*, Merrem. Seb. I. viii. 6, lxxxiii. 1 & 2, cix. 6). Its belly sometimes has an inflated form.*

AGAMÆ ORBICULARIÆ. Daud. in part.

Which are only agamæ; those with the inflated belly have the tail short and slender. Such is

The *Tayapaxin* of Mexico. Hern. 327. *Lac. orbicularis*. L.

With spiny back and belly sown with blackish points.†

cclxxiii. 2, which is an *anolis*, Edw. ccxlv. 2, which is also an anolis, and this same figure is again cited by him and Gmelin under the *Polychrus*. Shaw even copies it to represent the *Polychrus*, with which it has nothing in common. Seb. I. cvii. 3, which is the true *Ag. Colonorum* of Daud., is quoted by Merrem under *Ag. Superciliosa;* and Seb. I. cix. 6, which is his *Aculeata*, is cited under *Orbicularis*, &c.

* The *Agame à Pierreries*, Daud. IV. 410. Seb. I. viii. 6, is but the young of this spinous agama, more varied in colours than the adult. Add. *Ag. Atra*, Daud. III. 349, rough, blackish, with a yellowish line along the back;—*Ag. Umbra*, Daud., which is not the *Lac. Umbra* of Linnæus, but is distinguished by five lines of very small spines predominating on the back.

† I do not think that this subgenus can be preserved. The species just named does not appear to me to differ from the *Agama Cornuta* of Harlan. Annat. sc. Phil. IV. pl. xlv. except, perhaps, in sex. Daudin has reprinted it in its place, III. pl. xlv. f. 1. The adult is our *Trapelus* of Egypt.

Trapelus. Cuv.

Have the form and teeth of the agamæ, but their scales are small and without spines. They have no pores in the thighs.

The *Trapelus* of Egypt. *Trap. Ægyptius.* Geoff. Rept. d. Eg. pl. v. f. 3, 4. The adult, Daud. III. xlv. 1, under the name *orbiculaire.*

Is a small animal which sometimes has the body swelled, and is remarkable for changes of colour, more rapid than those of the cameleon. The young is entirely smooth. The adult has some scales a little larger, scattered over the body among the others.*

Leiolepis.

Have the teeth of the agamæ, the head less swelled, and are entirely covered with very small, smooth and crowded scales. They have pores in the thighs.†

Tropido-Lepis. Cuv.

Are similar to the agamæ in teeth and form, but uniformly covered with imbricated and carinated

* This subgenus is also rather difficult to separate with precision from certain agamæ of a clumsy form, and but slightly furnished with spines.

† We have one species of Cochin-China with a long tail, blue, with white stripes and spots. (*Leiol. Guttatus.* Cuv.)

scales. Their series of pores is very much marked.*

LEPOSOMA. Spix. TROPIDOSAURUS. Boié.

Do not differ from tropidolepis, but by the want of pores.†

CALOTES. Cuv.

Differ from the agamæ, only because they are regularly covered with scales disposed like tiles, often carinated, and terminated in a point, as well over the body, as on the limbs and tail, which is very long. Those on the back are more or less raised and compressed into spines, and form a crest of variable extent. They have no pores visible on the thighs, which, in addition to their teeth, distinguishes them from the iguanas.‡

* *Ag Undulata,* Daud. A species of all America, remarkable for the white cross which it has under the throat on a ground of black blue. The *Ag. Nigricollaris,* Spix, xvi. 2, and *Cyclurus,* xvii. f. 1, are at least greatly approximating to it.

† Spix has expressed himself with little exactness in saying that the scales of his Uposama are verticillated, which has deceived M. Fitzinger. The genus *Tropidosaurus* has been made by Boié, after a small species of Cochin-China, in the Royal Cabinet.

‡ Pliny tells us that the stellio of the Latins was named by the Greeks *Galeotes Colotes,* and *Askalabotes.* It was, as we have observed, the *Gecko of the walls* (Tarentola, or *Lac. Facetanus.* Aldrov.) The application made of it by Linnæus to his Lac. Calotes is arbitrary ; it was suggested to him by Seba. Spix comprehends our calotes in his genus Lophyrus, which is not the same as that of Dumeril.

The species most common (*Lac Calotes*, L. Seb. I. lxxxix. 2, xciii. 2, xcv. 3, and 4; Daud. III. xliii.; *Agama Ophiomachus*, Merr.) is of a beautiful clear blue, with white transverse marks on the sides, and two ranges of spines behind the ears. It comes to us from the East Indies. It is called cameleon at the Moluccas, though it changes its colours but little. Its eggs are spindle shaped.*

Lophyrus. Dumeril.

Have the scales of the body like the agamæ, and a crest of spangled scales still higher than that of the calotes. Their tail is compressed, and they have no pores in the thighs.

* Add, *Ag. Gutturosa*, Merr., or *Cristatilla*, Kuhl, blue without bands, and small scales on the back, Seb. I. lxxxix.;—*Ag. Cristata*, Merr. Seb. I. xciii. 4, and II. lxxvi. 5, reddish brown, with scattered, blackish brown spots, of which the *Agame Arlequimé*, Daud. III. xliv., is the young;—*Ag. Vultuoso*, Harl. nat. sc. Philad. IV. xix. All these species come from the East Indies. The *Lophyrus Ochrocollaris*, et *Margaritaceus*, Spix, xii. are calotes of America. The first is the same as the *Agama Picta* of Pr. Max. The *Loph. Panthera*, Spix, pl. xxiii. f. 1, is the young. Add to these American calotes *Loph. Rhombifer*, Spix, xi. whose *Lophyrus Albomaxillaris*, id. xxiii. f. 2, is the young;—*Loph. Auronitens*, sp. pl. xiii.

We may separate from the other calotes a species of Cochin-China, with smooth back without apparent scales, with belly, limbs, and tail covered with carinated scales (*Cal. Lepidogaster*, Nob); the *Ag. Catenata*, Pr. Max, fifth book, may belong to this group.

N.B. We must remark that the artist of Seba has given to most of his iguanas, agama, and calotes, extensible and forked tongues, altogether from his own imagination.

A remarkable species is,

The Lophyrus, with Forked Casque. Agama Gigantea. Kuhl. Seb. I. c. 2.

Which has its dorsal crest very high on the nape, and formed of several ranks of vertical scales. Two osseous ridges proceed from the muzzle, and finish each in a point over the eye on its side joining the temple. This singular Saurian appears to come from the Indies.*

THE GONOCEPHALA. Kaup.

Approximate pretty closely to these lophyri. Their cranium also forms a kind of disk, by means of a ridge, which terminates above each eye by an indentation. They have a crest upon the nape. Their tympanum is visible.†

THE LYRIOCEPHALA. Merrem.

Unite to the characters of the lophyri that of a tympanum, concealed under the skin and under the muscles, as in the cameleons. They have also a dorsal crest, and a carinated tail.

In the species which is known (*Lyriocephalus Margaritaceus*, Merr.; *Lacerta Sentata*, L.), Seb. cix. 3, the osseus crest of the eyebrows is still more

* It is not easy to say why Kuhl has given to this Saurian the epithet of gigantic. Its size does not exceed that of the Agama and Calotes, which are the nearest to it.

† Isis, 1825, I. p. 590, pl. iii.

marked than in the lophyrus, with forked casque, and terminates on each side behind with a sharp point. Larger scales are scattered among the smaller ones over the body and limbs. On the tail are imbricated and carinated scales. A soft, though scaly swelling, is on the end of the muzzle. This truly singular species is found in Bengal, and other parts of the East Indies. It lives on grains.*

BRACHYLOPHUS. CUV.

Have small scales, a tail somewhat compressed, a crest on the nape and back a little projecting, a small fanon, a series of pores on each thigh, and, in a word, much of the appearance of the iguanas; but they are destitute of palatal teeth. Those of the jaws are denticulated.

Such is,

The Banded Iguana. Brogn. Essai et Mem. des Sav. etr. I. pl. x. f. 5.

Of the Indies, deep blue, and with clear blue bands.

* M. Fitzinger forms of this *Lyriocephalus* of the PNEUSTES of Merrem, and of the PHRYNOCEPHALUS of Kaup, a family which he names PNEUSTOIDEA, and approximates to that of the cameleons. The PNEUSTES rests only on an incomplete and vague description of Azara, II. 401, on which, also, Daudin has established his *Agame à quene prenante*, III. 440. D'Azara says that the ear is not visible, probably on account of its smallness. The PHRYNOCIPHALUS includes the *lacerta guttata*, and the *lacerta uralensis* of

Physignathus

Have, with the same teeth, the same scales and pores, a head very much swelled behind, without cuticular appendage; and a crest of large pointed scales on the back and tail, which is much compressed.

We are acquainted with one large species of Cochin-China (*Physignathus cocininus*, Nob.); blue, with strong scales, and some spines on the swelling of the sides of the head. It lives on fruits, &c.

Istiurus, Cuv. Lophura, Gray.*

Have, as a distinctive character, an elevated and trenchant crest, which extends over a part of the tail, and which is sustained by high spinous apophyses of the vertebræ. This crest is scaly, like the rest of the body. The ventral and caudal scales are small, and approach somewhat to a square form. The teeth are strong, compressed, and without denticulations: they have none in the palate. Their thighs have a range of pores. The skin of the neck is loose, but without forming a cuticular appendage or dewlap.

Lepechin, Voy. I. p. 317, pl. xxii. f. 1 and 2, which make but one species. M. Kaup assures us, that it has no exterior tympanum. (Isis of 1825, I. 391.) Not having seen these animals, I hesitate to class them.

* I have changed this name of *Lophura*, which approaches too much to that of *Lophyrus*.

The Crest-Bearer. Lacep. *Lac. Amboinensis,* Gm. Schlosser Monog. Copie Bonnet Erpet. pl. v. f. 2.

Has no crest but on the origin of the tail, and has spines on the fore part of the back; lives in the water, or under the shrubs on its banks. Eats grains and worms: we have found in its stomach leaves and insects. Its size sometimes approaches four feet. Its flesh is eaten.

THE IGUANAS, proper. IGUANA, Cuv.

Have the body and tail covered with small imbricated scales. All along the back is a range of spines, or rather of raised scales, compressed and pointed; and under the neck, a compressed and pendant cuticular appendage, the edge of which is sustained by a cartilaginous production of the hyoïd bone. Their thighs have the same range of porous tubercles as those of the lizards proper, and their head is covered with plates. Each jaw is surrounded with a range of compressed triangular teeth, with indented edges. There are also two smaller ranges on the posterior edge of the palate.

The Common American Iguana. (*Lac. Iguana,* L. *Iguana Tuberculata,* Laur.) Seb. I. xcv. 1; xcvii. 3; xcviii. 1.

Yellowish green above, marbled with pure green; the tail ringed with brown. In spirits it appears blue, changing into green and violet, and pointed with black; more pale underneath. It has a crest of large dorsal scales in the form of spines, and a large round plate under the tympanum, at the angle of the jaws. The sides of the neck are furnished with pyramidal scales, scattered among the others. The anterior edge of its cuticular appendage is denticulated like the back. This reptile is from four to five feet in length, and common throughout all the warm parts of America, where its flesh is considered delicious, though unwholesome, especially for those afflicted with the venereal disease, the pains of which it renews. It lives principally upon trees, sometimes visits the water, and feeds on fruits, grains, and leaves. The female lays eggs in the sand as large as those of a pigeon, agreeable to the taste, and almost without white. The Mexicans name it *Aquaquetzpallia* (Hern.); and the Brazilians, *Senembi* (Margr.)

The Slate-coloured Iguana. Daud. Seb. I. xcvi. 4.

Of an uniformly rather violet blue, paler underneath. The dorsal spines are smaller. As to the rest, it resembles the preceding. In both, there is an oblique, whitish spot on the shoulder. This one comes from the same country, and is probably only a variety of age or sex.*

* I have even every reason to believe that this conclusion should be extended to the iguanas of Spix, pl. v. vi. vii. viii. & ix.; they appear to me to be only varieties of age in the common species.

The Iguana with naked Neck. (*Ig. Nudicollis*, Cuv.) Merr. Besler, tab. xiii. f. 3. *Ig. Delicatissima*, Laur.

Resembles the common, especially in the dorsal crest, but has no large plate under the tympanum, nor tubercles scattered over the sides of the neck. The upper part of the cranium is furnished with convex plates, and the occiput is tuberculous. The cuticular appendage is moderate, and has but few denticulations, and only in front. *Laurenti* says that its habitat is the East Indies, but this is a mistake. We have received it from Brazil and Guadaloupe.*

The Horned Iguana of St. Domingo, Lacep. *Ig. cornuta*, Cuv. Bonaterre, Encyc. Method. Erpetolog. Lezards. pl. iv. f. 4.

Tolerably like the *common iguana*, and still more like the last; but it is distinguished by a conical osseous point between the eyes, and two raised scales on the nostrils. There is no large plate under the ear, nor tubercles in the neck; but the scales on the branches of the jaw are embossed.

The Iguana, with armed tail, of Carolina. (*Ig. cychlura*, Cuv.)

Is destitute, like the preceding two, of a large plate under the ear, and small spines on the neck; but

* I suspect the *Amblyrhyncus cristatus*, Bell. Zool. Journ. I. Sup. pl. xiii., to be an individual, badly preserved, of my *Ig. nudicollis*.

some scales larger than the rest, and a little carinated, form, from space to space, cinctures upon the tail.*

Ophryessa. Boié.

Have small imbricated scales, a dorsal crest, somewhat projecting and prolonged over the tail, which is compressed; maxillary teeth denticulated, and teeth in the palate, all which circumstances approximate them to the iguanas; but they have no cuticular appendage, nor pores in the thighs.

Lac. Superciliosa. L. Seb. I. cix. 4. *Lophyrus Xiphurus.* Spix, X.

Thus named in consequence of a membranous keel which forms its brow, is an American species, fawn-coloured, with a festooned brown band along each flank.

The Basilisks. Basiliscus. Daud.

Are destitute of pores, and have palatal teeth, like the ophryessæ. Their body is covered with small scales. There is on their back and tail a continuous and elevated crest, supported by the spinous processes of the vertebræ, like that of the tail of istiurus.

* It also appears to me that this iguana is the same that Dr. Harlan (An. of Se. Nat. de Philad. iv. pl. 4) calls *cychlura carinata;* but then, as in the *amblyrhyncus,* there must be some error relative to the palatine teeth. I am satisfied that these teeth exist in all my iguanas.

The species known (*Lac. Basiliscus.* Linn.) Seb. I. c. i. Daud. iii. xlii., is recognized by a membranous prominence of its occiput, in the form of a hood, supported by a cartilage. It is an animal of Guiana, grows to a considerable size, is blueish, with two white bands, one behind the eye, the other behind the jaws. They are both lost towards the shoulder. It feeds on grains.*

POLYCHRUS.

Have, like the iguanas, teeth in the palate, and pores in the thighs, though not much marked; but their body, covered with small scales, has no crest. The head is covered with plates. The tail is long and slender. The extensible throat can form a dewlap or cuticular appendage at the will of the animal. Like the cameleons, they possess the faculty of changing colour. Their lungs, accordingly, are voluminous, filling almost the entire body, and being divided into many branches, and their false ribs like those of the cameleon, surround the abdomen, and unite to form entire circles.

The Polychrus of Guiana. (*Lac. Marmorata.* L.) Lacep. I. xxvi. Seb. II. lxxvi. 4. Spix, XIV.

Reddish grey, marbled with transverse irregular bands of a brown red, and sometimes mingled

* It has been hitherto erroneously believed, on the testimony of Seba, the basilisk of the East Indies.

with blue. The tail is very long. Common in Guiana.*

ECPHIMOTES, Fitzinger,

Have the teeth and pores of polychrus, but small scales on the body only; the tail, which is thick, has large ones pointed and carinated. The head is covered with plates. They have the somewhat short and flatted form of certain agamæ, rather than the lank form of the polychri.

The most common species (*Agama Tuberculata*, Spix, XVI. or *Tropidurus Torquatus*, Pr. Max,) is ash-coloured, sown with whitish drops, and has on each side of the neck a black semi-collar. It lives in Brazil.†

The QUETZPALCO, OPLURUS, Cuv.

Have also, with the teeth of the polychrus, the forms of the agamæ, but they want the pores on the thighs, and the scales of their tail pointed and carinated, give it some relation with that of the stelliones. Their dorsal scales are also pointed and carinated, but very small.‡

But one species is known, and belongs to Brazil.

* Add, *Polychrus Acutirostris.* Spix, XIV.

† The tropidurus of Prince Max de Wied, is not, as he thought, the quetzpalco of Seba, though that also has some black semi-collars.

‡ This name, quetzpalco, given by Seba to this species, appears corrupted from the Mexican *aqua quetz pallia,* which appears to be a name of the iguana. The Quetzpalco of Lacep. Rept. in 4° II. 497, is an Uromastix: but it is the figure of Seba's animal that he cites.

OPLURUS TORQUATUS, CUV.

With a black semi-collar on each side of the neck.

THE ANOLIS. ANOLIUS. CUV.

Have with all the forms of the iguanas, and especially of the polychrus, a distinctive character, very peculiar. The skin of their toes is widened under the antepenultimate phalanx into an oval disk, striated cross-wise underneath, which enables them to attach themselves to different surfaces, where they can also fasten very well, by means of very crooked claws. They have, moreover, the body and tail uniformly shagreened by small scales, and most of them have a dewlap or goitre under the throat, which swells and changes colour when they are under the influence of anger or sexual desire. Several of them are equal to the cameleon in the faculty of varying the colours of their skin. Their ribs unite in entire circles, as in the polychri and cameleons. Their teeth are trenchant and denticulated, like those of the iguanas and polychri, and they even have some in the palate. The skin of the tail has some slight folds or sinkings, each of which comprehends some circular ranges of scales. This genus appears to be peculiar to America.*

* *Anoli*, or *Anoalli*, the name of these saurians in the Antilles. Gronovius has very gratuitously transferred it to the ameiva. Rochefort, from whom he has taken it, gives as a figure of it, a copy of the *Teyuaguaca* of Marcgrave, or the great safeguard of Guiana. Nicholson appears to announce that this name is applicable to many species, and that which he

There are some which have a crest upon the tail supported by the spinous processes of the vertebræ, as in the istiuri and basilisks.*

The Great Crested Anolis. (*An. Velifer*, Cuv.)

A foot long; a crest on one half of the tail supported by a dozen or fifteen rays. The gular appendage extends under the belly. The colour is a blackish ashen blue.

Its habitat is Jamaica and the other Antilles. We have found berries in its stomach.

The small Crested Anolis. (*Lac Bimaculata*, Sparrm.)

One half smaller than the preceding; some kind of crest, of a greenish colour, picked out with brown towards the muzzle and on the flanks—of North America and the different Antilles.

An. Equestris. Merr.

Fawn-colour, shaded with ashen lilac; a white band over the shoulder. The tail is too fleshy to permit the apophyses of the crest to be distinguished. This reptile is a foot long.

Others have the tail round or only a little com-

describes appears to be the *Anolis Roquet* of Lacep. (*Lac. Bullaris*, Gm.) which in truth was sent from Martinique to our Museum under the name of *Anolis*. M. Moreau de Jonnes has even authenticated that at the present day this is the only name under which it is known.

* They have been confounded one with another, and with some of the subsequent species under the name of *Lac. Principalis and Limaculata*. L.

pressed. Their species are numerous, and have been in part confounded, under the name of *Roquet Goitreux*, *Rouge-gorge*, and *Anolis*. (*Lac. Strenuosa et Bullaris*. Lin.) They inhabit the warm parts of America and the Antilles, and change colour with a prodigious facility, especially in warm weather. Their cuticular appendage swells when the animal is angry, and grows as red as a cherry. These animals are not so large as our grey lizard, and live particularly on insects, which they pursue with agility. The individuals of this species cannot meet, it is said, without fighting with great fury.

The species of the Antilles, or the *Roquet* of Lacep. I. pl. xxvii. (it is now particularly the *Lacerta Bullaris*. Gm.) has the muzzle short, dotted with brown, and the eyelids projecting. Its ordinary colour is greenish. Except its round tail, it much resembles the little crested anolis.

The Striped Anolis, Daud. IV. xlviii. 1, differs only by having series of black marks on the flanks. It appears the same as the *Lacerta Strenuosa* Lin. Seb. II. xx. 4. Its length is a little more considerable than the preceding.

The Anolis of Carolina. (*Iguane goitreux*, Brong.) Catesby, II. xxvi.

Is of a fine golden green, with a black band on the temple. Its muzzle is elongated and flatted, which gives it a peculiar physiognomy, and constitutes it a very distinct species.*

It is to this family of the IGUANIANS with palatal teeth, that an enormous fossil animal belongs, known under the name of the animal of Maestricht, for which the name of MOSASAURUS has been recently fabricated.†

The fourth family of the saurians, or

THE GECKOTIANS,

Is composed of nocturnal lizards, so much resembling each other that they might be left in one genus.

THE GECKOS, Daud. STELLIO, Schn. ASCALABOTES, Cuv.‡

Are saurians, which have not the lank form of those of which we have hitherto treated, but are, on the contrary, flatted, especially about the head,

* Add the *Anolis with white points*, Daud. IV. xlviii. 2;—*An. Viridis*, Pr. Max, 6th book;—*An. gracilis*, id. and many other species, of which, unfortunately, I have no figures to cite.

† See upon this animal, the Fifth Volume, Second Part, of my Researches on the Fossil Bones.

Among the fossils many reptiles have also been discovered, of large size, which appear to approximate to this family, but the characters of which are not known completely enough to enable us to class them with safety.

Such are the GEOSAURUS discovered by M. Soemmering, the MEGALOSAURUS of M. Buckland; the IGUANODON of M. Mantell, &c. I treat of them at length in the Fifth Vol., Second Part, of my Researches, &c.—Cuv.

See our Treatise on Fossil-Remains, "Fossil Reptiles," 23d Part of the Animal Kingdom.

‡ *Gecko*, a name given to a species from the East Indies, and imitated, from its cry, as another species has been named *tockaie*, at Siam, and a third, *geitje*, at the Cape, ἀσκαλαβώτης, the Greek name of the Wall-Gecho.

and have the feet middling, and the toes nearly equal. Their walk is heavy and creeping. Their eyes are very large, and the pupil contracts from the influence of light like that of the cats. This constitutes them nocturnal animals, and during daylight they remain in obscure places. Their eye-lids, remarkably short, are withdrawn between the eye and the orbit, which gives to their physiognomy a different aspect from that of the other saurians. Their tongue is fleshy and not extensible, and their tympanum a little sunk in. Their jaws are furnished all round with a range of very small crowded teeth. Their palate is destitute of teeth. Their skin shagreened above with very small grained scales, among which are often observed some thicker tubercles; has, underneath, some scales less small, flat, and imbricated. Some species have pores upon the thighs. The tail has circular folds, like that of the *anolis*; but when it has been broken, it shoots again without folds, and even without tubercles, when it has them naturally, which has caused the species sometimes to be multiplied.

This genus is numerous, and spread through the warm regions of both continents. The gloomy and heavy aspect of the geckos, and a certain resemblance to the salamanders and toads, has caused them to be held in abhorrence, and accused of being poisonous, but without any real proof of the fact.

The majority of them have the toes widened, over all or a part of their length, and furnished underneath with very regular folds of skin, which enables them to adhere so well to different bodies, that they are observed to walk on ceilings. Their claws are retractile in various manners, and preserve their edge and their point. This character, united to that of their eyes, might justify a comparison between the rank of the geckos among the saurians, and that of the felinæ in the carnassial mammifera. But their claws vary in number according to the species, and are entirely wanting in some.

The first and most numerous division of the geckos, which I shall call

Platydactylus,

Have the toes widened over their entire length, and furnished underneath with transverse scales.

Among these platydactylous geckos some have no claws whatsoever, and their thumbs are very small. There are handsome species, all covered with tubercles, and painted with lively colours. Those which are known come from the Isle of France.

Some are destitute of pores in the thighs.*

There is one of them violet above, white underneath, with a black line on the flanks. (*G. Inunguis*, Cuv.)

Another is grey, all covered with eye-like spots, brown and white in the middle. (*G. Veillatus*, *d'Oppel.*)

* To this division Mr. Gray has reserved the name of *Platydactylus*

Some others, on the contrary, have these pores very marked.*

Such is

The Cepedian Gecko, Peron.

Of the Isle of France; rose-colour, marbled with blue, and a white line along each flank.

I know not, however, if the pores in this first subgenus be not a mark of sex.

Other platydactyla are destitute of claws on the thumbs, and on the second and fifth toes of all the feet. They have no pores in the thighs.†

Such is

The Wall-Gecko, (*Lacertus Facetanus*,) Ald. 654. *Tarente* of the Provincals; *Tarentola*, or rather *Terrentola* of the Italians; *Stellio* of the ancient Latins; *Geckotte*, Lacep.; *Gecko Fascicularis*, Daud.

Deep grey; the head rough; all the upper part of the body sown with tubercles, each of them formed of three or four smaller and approximating tubercles. The scales of the under part of the tail like those of the belly. An hideous animal, which conceals itself in the holes of walls and heaps of stones, and covers its body with dust and ordure. It appears that the same species inhabits all around

* Mr. Gray has made of this division his genus *Phelsuma;* the *Lacerta Gietje* of Sparrman should belong to it. It is thought very venomous at the Cape.

† Of this division Mr. Gray has made his genus *Tarentola.*

the Mediterranean, and as far as Provence and Languedoc.

There is in Egypt and Barbary an approximating species with simple and round tubercles, more projecting on the flanks. (*G. Egyptiacus*, Cuv.) Egypt. Rept. pl. v. f. 7.*

The greater number of platydactylous Geckos want claws upon the four thumbs only. They have a range of pores in front of the anus.†

Such are

Gecko, Lacep. I. xxix. *Stellio Gecko*, Schneid. Seb. I. cviii.

Rounded tubercles, not very projecting, are spread over the upper part of the body, the red colour of which is sown with round and white spots. The under part of the tail is furnished with square and imbricated scales. Seba says that it belongs to Ceylon, and pretends that it is to it in particular that the name of *Gecko* is given on account of its cry, but Bontius attributed it long before to a species from Java. Probably the name and cry are common to many. We are assured that this is found in the whole Indian Archipelago.

The banded Gecko. Lizard of the Pandang, in Amboyna. (*Lacerta vittata*, Gm.) Daud. Ill. l.

Brown, a white band on the back, which is bifurcated

* This figure, entitled *Var du Gecko Annulaire*, has too many claws.

† This division is particularly named *Gecko* by Mr. Gray.

over the head, and on the root of the tail, around this last are white rings. This species belongs to the East Indies. In Amboyna it rests on the branches of the arbustum, called *Pandang*.*

There are some of these platydactyles with four claws, whose body is bordered with an horizontal membrane, and feet palmated.

One of the most remarkable is

Lacerta homalocephala. Crevett. Soc. des Nat. de Berlin, 1809, pl. viii.

Which has the sides of the head and body augmented with a white membrane, which is cut into festoons on the sides of the tail, its feet are palmated. It is found in Java and Bengal.†

There is another species in the Indies, with the head and body bordered, and the feet palmated, but without festoons to the tail, and pores to the anus. (Pteropleura. *Horsfieldii*, Gray, Zool. Journ. No. X. p. 222.)

Finally some plactydactyles have claws on all the toes.

We have one species of them quite smooth, with palmate feet. (*A. Leachianus*, nobis.)

A second division of the Geckos which I shall call

* N.B. Daudin erroneously gives claws to the thumbs of these two Geckos.

† M. Fitzinger makes of this his genus Ptychozoon. Mr. Gray separates from it his Pteropleura, in consequence of the absence of the pores.

HEMIDACTYLUS,

Have the base of their toes furnished with an oval disk, formed underneath by a double rank of chevron shaped scales. From the middle of this disk rises the second phalanx, which is slender and supports the third, or the claw at its extremity. The known species have all five claws, and a range of pores on each side of the anus. The scales of the under part of their tail are in the form of broad bands, like those of the belly of serpents.

There is a species in the south of Europe (*G. Verruculatus*, Nob.) of a reddish grey, the back all sown with small conic tubercles, a little rounded. The tail has some circles of similar tubercles. It belongs to Italy, Sicily and Provence, as does the *G. Fascicularis*.

A very similar species (*G. Mabuia*, Nob.) with tubercles still smaller, and those of the tail more pointed, of a greyish colour, clouded with brown, and brown rings on the tail, is extended through all the warm parts of America, and enters houses. It is known in our (the French) islands, under the name of *Mabouia des murailles*.*

There are some in Pondicherry and Bengal so very similar, that one might be tempted to

* As far as we can judge by the figure, the *Thecadactylus Pallicaris*, and the *Gecko Aculeatus*, Spix, XVIII. 2, and 3, may be only the same as this *Mabina* at different ages. M. Moreau de Jonnes has given a monograph of it, but confounds it with different species.

believe that they had been transported thither in ships.*

There is also found in the Indies a hemidactylus, with edged body (*G. Marginatus*, Nob.) Its feet are not palmated, its tail is flatted horizontally, and the edges are trenchant and a little fringed. It was sent from Bengal, by M. Duvaucel.

The third division of the Geckos, which I shall call

THECADACTYLUS,

Has the toes widened throughout their entire length, and furnished underneath with transverse scales, but there is also a deep longitudinal furrow, in which the claw may be entirely concealed.

Those with which I am acquainted want claws on the thumbs only, they have no pores in the thighs, and their tail is furnished underneath and above with small scales.

The Smooth Gecko (*G. Lævis D. Stellio Perfoliatus*, Schn. *Lac. Rapicauda*, Gm.) Daud. IV. li.

Known in our (the French) islands, by the name of *Mabouia des Bananiers*. Grey, marbled with brown, very small, grained, without tubercles above; small scales underneath; its tail naturally long, and surrounded with the usual folds, breaks very easily, and

* To this division also belong the *G. à Tubercules Trièdres*, and the *G. à Queue Epineuse* of Daud. The first is the same as *Stell. Mauritanicus* of Schn. The *Stell. Platyurus* of Schn. also closely approximates to it.

is renewed sometimes in a very swelled state, and in the form of a small radish. Such accidental monstrosities have given rise to the appellation of *G. rapicauda.**

The fourth division of *Geckos*, which I shall call

PTYO-DACTYLUS,†

Has the ends of the toes only dilated into plates, the under part of which is striated like a fan. The middle of the plate is cleft, and the claw placed in the fissure. All the toes have claws very much crooked.

Some have the toes free, and the tail round.

The House Gecko. (*Lac. Gecko*, Hasselquist.) *Gecko Lobatus*, Geoffr. Rept. Egypt, III. 5. *Stellio Hasselquistii*, Schneid.

Smooth, reddish-grey, picked out with brown; the scales and tubercles are very small. This species is common in the houses of the different countries which border on the Mediterranean, to the south and east. At Cairo it is named *Abou Burs*, (*Father of the Leprosy*,) because they pretend that it communicates this malady by poisoning with its feet the provisions, especially those which are salted, of which it is very fond. When it walks on the skin, it pro-

* The *Gecko Squatydus* Herm. should belong to this division, if it be not the same with the *Lævis*. The *Gecko of Surinam*, Daud., is only a younger and better coloured individual of *Lævis*.

† From πλύον, a fenn.

duces redness there, but probably only in consequence of the fineness of its claws. Its voice somewhat resembles that of frogs.

Others have the tail edged on each side with a membrane, and the feet semi-palmate. They are probably aquatic. These are the *Uroplates* of Dumeril.

The Fringed Gecko. (*Stellio Fimbriatus*, Schn. *Tête-plate*, Lac., or *Famo-Cantrata* of Madagascar.) Brug. Lacep. I. xxx. Daud. IV. lii.

Has not only a border to the sides of the tail, but extending along the flanks, where it is fringed and slashed. It is found at Madagascar, according to report, on trees, where it jumps from branch to branch. The people of that country are in much dread of it, but without any foundation.*

The Peruvian Gecko. (*Lac. Caudiverbera*,) Lin. Feuillée, I. 319.

Has no fringe on the sides of the body, but only on those of the tail, on which it has also a membranous vertical crest. Feuillée discovered it in a spring on the Cordilleras. It is blackish, and more than a foot in length.

We may make a fifth division,

* According to the description of Bruguiere, the Sarroubé of Madagascar has all the characters of the famocantrata, except the fringe, and the thumb, which is wanting on the fore feet. M. Fitzinger makes of it his genus SARRUBA.

The Spheriodactylus,

Of certain small geckos which have the ends of the toes terminated by a little pellet, without folds; but the claws are always retractile.

When the pellet is double, or emarginated in front, they seem closely related to the ptyodactyli without edges. Those which we know come from the Cape, or the East Indies.

Such is

The G. Porphyré. Daud.

Reddish-grey, marbled and picked out with brown. Daudin was mistaken in supposing this gecko to belong to America, and to be synonymous with the mabina.

The pellet is more frequently simple and round. The species are of America.

Such is

The Spitting-banded Gecko. Lacep. Rept. I. pl. xxviii. f. 1.

A small species, prettily marked with transverse brown bands, contrasting with a red ground. It frequents houses in St. Domingo, where it also receives the name of *Mabuia.* In the same island is an approximating species, but of an uniform ash-colour. *Id.* ib. f. 2.

Finally, there are saurians which with all the characters of the geckos, have not the toes enlarged.

Their claws, however, five in number, are retractile.

Some have the tail round, the toes striated underneath, and indented at the edges.

These are

The Stenodactyli.

There is one in Egypt, the *Sten. Guttatuis*, Egypt. Rept. pl. v. f. 2.* Smooth, grey, and sown with white spots.

Others have the toes slender and naked. Those which have the tail round are

The Gymnodactyli of Spix.

There are some in America, with regular series of little tubercles. *Gymnodactylus Geckoides*, Spix, X. xiii. 1, appears to be one of them.

Others have the tail flatted horizontally in the form of a leaf. I name them

Phyllurus.

But one species is known, which belongs to New Holland, (*Stellio Phyllurus*, Schn., *Lacerta Platura*, White, New South Wales, p. 246, f. 2, where it is referred without any sufficient reason to the *Stelliones* of Daudin,) grey, marbled with brown above, all bristling with small and pointed tubercles.

We are obliged to establish a fifth family of

* Under the improper name of *Agame Ponetué*. It is produced again Suppl. pl. i. f. 2, and an approximating species, f. 4.

CAMELEONIANS.

For the single genus of

CAMELEONS. CHAMÆLEO.*

Which is very distinct from all the other saurians, and cannot even be easily allocated with just propriety in their series.

They have the entire skin shagreened, with small scaly grains; the body compressed, and the back as it were trenchant; the tail round and slender, five toes on all the feet, but divided into two parcels, one of two, the other of three. Each parcel is united by the skin as far as the claws; the tongue is fleshy, cylindrical, and capable of considerable elongation; the teeth are trilobed; the eyes large, but almost covered by the skin, except a small hole opposite to the pupil, and each of them is moveable, independently of the other. There is no visible external ear, and the occiput is raised in a pyramidical form; their first ribs are united with the sternum, the following ones are continued, each of them to its correspondent one, to enclose the abdomen in an entire circle; their lungs are so vast that when this organ is inflated their body appears transparent, which occasioned the ancients to assert that they lived on air. They feed, however, on insects, which they catch with the viscous extremity of their tongue: this is the only

* Χαμαιλέων (little lion) the name of this animal among the Greeks, and especially in Aristotle, who has described it perfectly well. Hist. An. Lib. ii. cap. xi.

part of their body which they move with any swiftness, all their other movements are tardy in the extreme. The bulk of their respiratory organ is in all probability the cause of this peculiar faculty of changing colour, which is not influenced, as has been supposed, by the bodies on which they happen to be placed, but by their wants and passions. The lungs, in fact, render them more or less transparent, constrain the blood more or less to flow towards the skin, and even colour this fluid with a greater or less degree of liveliness, according as they are full or empty of air. They remain continually on trees.

The Common Cameleon. Lacerta Africana. Gm. Lacep. I. xxii. Seb. I. lxxxii. 1, lxxxiii. 4.*

Of Egypt and Barbary, found also in the south of Spain, and again as far as the Indies; has its hood or capote pointed, and raised in a crest in front. The grains of the skin are equal and crowded, the upper crest denticulated as far as one half of the back, the lower as far as the anus.

The capote of the female projects less, and the denticulations of the crests are smaller.

Another species tolerably similar, and from the Sechelles Islands (*Cham Tigris*, Cuv.), has the casque like the female of the common species; the grains of the body are fine and equal, and it is distinguished by a compressed and denticulated sort of wattle under

* The *Cam Trapes*, Eg. Rept. iv. 3; *Cham. Carinatus* Merr. *Cham. Subcroceus*, id.?

the end of its lower jaw. Its body is sown with black points.

Another related species from the isle of Bourbon (*Cham Verrucosus*, Cuv.) has some thicker grains scattered among the others, and a series of warts parallel to the back, at two-thirds of its elevation; the capote is the same as that of the common female, the denticulations of the back are stronger, those of the belly less marked.

The Dwarf Cameleon. (*Lacerta Pumila*, Gm. *Chamæleon Pinnilus*, Daud. IV. 53. *Cham. Margarituceus*, Mer. Seb. 82, 4, 5.)

Has the capote thrown backwards, warts scattered over the flanks, limbs and tail; under the neck are numerous shreds of skin, compressed and finely denticulated, which vary according to the individuals. It is found at the Cape, the Isle of France, and the Sechelles Islands.*

The Cameleon of Senegal. (*Lacerta Chamæleon* Gm.) *Ch. Planiceps*, Merr. Seba. I. lxxxiii. 2.

Has the capote flatted, and almost without a crest, of a form horizontally parabolic. It is also found in Barbary, and even in Georgia.

A species of the Isle of France (*Cham. Pardalis*, Cuv.) has the casque flat, like that of Senegal, but its muzzle has a small prominent edge in front of the

* I believe that the *Cham. Seichellensis* of Kuhl, is only a female of *pumilus*.

mouth, thicker grains are scattered among the others, and its body is sown irregularly with round black spots, edged with white.

Another species (*Cham. Parsonii*, Cuv.), Phil. Trans. LVIII., with a flat casque a little truncated behind, has the crest of the brow prolonged and raised on each side, over the end of the muzzle, in a lobe almost vertical. The grains are equal, and there is no denticulation either above or below.*

Finally,

The Cameleon of the Moluccas, with forked nose. (*Cham. Bifurcus Brong.*) Daud. IV.

Has the casque flat and semi-circular, two large compressed and projecting prominences in front of the muzzle, which vary in length, probably according to the sexes; the grains are equal, the body is sown with crowded blue spots, and there is at the bottom of each flank a double series of white ones.

The sixth and last family of the Saurians is that of the

Scincoidians.

To be recognised by its short feet, tongue not extensible, and equal scales, which cover the body and the tail like tiles.

The Skinks (Scincus, Daud.)

Have four feet, rather short; a body almost of a

* I am not acquainted with the *Cham. Dilepis*, Leach, or *Bilobus* Kuhl.

length with the tail, without any swelling at the occiput, and without crest or gular appendage, covered with uniform shining scales, arranged like tiles, or like those of the carps. Some of them have a spindle form, others almost cylindrical, and more or less elongated, resemble serpents, and especially some of the angues with which they have also many internal relations, and which they connect with the family of the iguanas by an uninterrupted series of shades; as to the rest, their tongue is fleshy, triflingly extensible and emarginated, and their jaws are furnished all round with small and crowded teeth; in the anus, penis, eye and ear, they resemble, more or less, the iguanas and lizards. Their feet have all the toes free and unguiculated.

Certain species have some teeth in the palate, and a denticulation on the anterior edge of the tympanum.

We should distinguish in the number, in consequence of its trenchant and somewhat raised muzzle,*

Lac. Scincus, Lin.; *Scincus Officinalis*, Shn.; *El. Adda* of the Arabs, Lacep. I. xxiii.; Bruce. Abyss. pl. 39; Egypt. Rept. Supp. pl. 2, f. 8.

Six or eight inches long; the tail shorter than the body. The latter is of a silvery yellowish colour, with some transverse blackish bands. It lives in

* It is of this species only that M. Fitzinger composes his genus SCINCUS. The others form his genus MABOINA.

Nubia, Abyssinia, and Arabia, from whence it is brought to Alexandria, and thence into all Europe. It exhibits an extraordinary promptitude in sinking into the sand when pursued.*

Among those which have the muzzle blunt, we may remark a species extended throughout all India, (*Sc. Rufescens*). Greenish; a yellowish line along each flank, and each of the scales raised with three small crests.

One of the south very much spread in the neighbourhood of the Cape (*Sc. Trivittatus*). Brown; three paler lines along the back and tail; black spots between the lines.†

And especially a large species of the Levant, named (*Sc. Cyprius*, Cuv.) *Lac. Cyprius Scincoides*, Aldrov. Quad. Dig. 666; Geoff. Desc. de l'Egypt. Rept. Pl. III. f. 3, under the name of *Anolis Gigantesque*. Greenish, with smooth scales, and tail longer than the body; a pale line along each flank.

Other skinks, the TILIQUA, Gray, have no teeth in the palate.

* The Greeks and Latins named the *Land Crocodile*, which is a monitor, *Scincus*, and to which they attributed many virtues. But since the middle ages, the species above-mentioned has been generally sold under this name, and for the same uses. The Orientals, in particular, regard it as a powerful aphrodisiac.

† Add: *Scincus Erythrocephalus*, Gilliams, Sc. Nat. Phil. I. xviii.; *Sc. Bicolor*, Harlan, ib. IV. xviii. 1; *Sc. Multi Seriatus*, Nob. Geoff. Eg. Reptil. IV. f. 4, under the name of *Anolis Pavé*. We are also inclined to refer to this subdivision, though we have not had the means of procuring it, the large skink called *Galley Wasp*, in Jamaica, Sloane II. pl. 273, f. 9. (*Lacerta Occidua*, Sh.)

There is one very much spread in the south of Europe, Sardinia, Sicily, and Egypt, *Sc. Variegatus*, *Sc. Ocellatus*, Cuv. Daud. IV. lvi.; Geoff. Egypt. Rept. pl. V. f. 1, under the name of *Anolis Marbré*, and better, Savig. *Ib.* Suppl. pl. II. f. 7, which has on the back, flanks, and tail, some small round black spots, each of them marked with a trait of white. A pale line for the most part predominates on each side of the back.

There are many species in our Antilles, one of which is improperly named *Anolis de Terre* and *Mabouia*, Lacep. pl. xxiv. Smooth, greenish-brown; some blackish points scattered over the back. A brown band, badly terminated, proceeding from the temple to the shoulder and beyond.*

The Moluccas and New Holland have some species of this division, remarkable for their size.†

SEPS. (Daud.)‡

Differ from the skinks only by having their body

* The figure of Lacepède is exact, the tail excepted, which is too short, the individual having had it broken, as frequently happens to all the lizards. Add, the *Sc. à Fleures Noirs*, Quoy and Gaimard, Voy. de Freyc. pl. 42; *Sc. Bistriatus*, Spix, xxvi. 1.

† *Lac Scincoides*, White, 242; *Scincus Nigroluteus*, Quoy and Gaimard, Frey. 41; *Scinc. Crotaphomelas*, Per. and Lacep., &c.

N.B. I have been able to name but very few species of skinks, because they are so badly characterized in authors, that I find it almost impossible to indicate the synonimy with any certainty. It is the genus which stands most in need of a monograph.

‡ *Seps* and *Chalcis* were among the ancients the names of an animal which some represented as a lizard, others as a serpent. It is very probable that they designated the seps with three toes of Italy and Greece. *Seps* comes from σηπειν, to corrupt.

still more elongated, altogether similar to that of an anguis, and their feet still smaller, the two paws of which are more remote from each other. Their lungs begin to exhibit some inequality.

There is one species with five toes, the hinder of which are unequal. (*S. Scincoides*, Cuv.)

One with five toes, pretty nearly equal and short. (*Anguis Quadrupes*, Lin.; *Lacerta Serpens*, Gm.) Block. Soc. des Nat. de Berlin, tom. II. pl. 2, of the East Indies.*

One with four toes, the hinder of which are unequal. (*The Tetradactylus Decresiensis*, Ber.)†

And one with three, otherwise very like the preceding. (*Tridactylus Decresiensis.*) Both were from the island of Decrès, and are viviparous.

One with three very short toes, and very short feet, named in Italy *Cecella*, or *Cicigna*, (*Lacerta Chalcides*, L.) Grey, with four longitudinal brown stripes, two on each side of the back. It is also viviparous, and moves with rapidity without the assistance of its feet; lives in meadows, and feeds on spiders, small snails, &c.‡

Our southern provinces have one very similar, but

* Mr. Gray has made of this his genus Lygosoma. Mr. Fitzinger leaves it in his Mabuia, or skinks without palatal teeth.

† It is to this species that Fitzinger reserves the generic name of Seps. He calls it *Seps Peronii*.

‡ Merrem, on the contrary, had made his genus Seps of this species only. Fitzinger now calls it, after Oken, Zygnis, and joins with it the *Trydactylus* of Peron, of the Isle Decrés, which much more approaches the *Tetradactylus* of the same island.

with eight or nine brown rays equally distributed. (*Zygnis Striata*, Fitz.)

We might separate from the others a species whose scales, all carinated and pointed, are pretty nearly disposed in verticillæ. (*Lac. Anguina*, Lin.; *Lac. Monodactyla*, Lacep. Ann. Mus. II. lix. 2, and Vosmaer; Monogr. 1774, f. 1, under the name of *Serpent Lizard.*) Its feet are small stilets, not divided. It inhabits the environs of the Cape of Good Hope.*

The BIPEDES. (BIPES, Lacep.)

Are a small genus which do not differ from the seps, but because they entirely want the fore-feet, having only the omoplates and clavicles concealed under the skin, and the hind feet alone being visible. There is but a step from them to the angues.

Some have a range of pores in front of the anus. They form the genus PYGOPUS of Merrem.

I have dissected one brought from New Holland by the late Peron, *Bipede Lepidopode*, Lacep. An. du Mus. tom. IV. pl. lv., which has the scales of the back carinated, and the tail twice as long as the body. Its feet at the exterior exhibit only two small oblong and scaly plates; but by dissection may be found a femur, a tibia, a peroneum, and four bones of the metatarsus, forming toes, but without

* It is the genus MONODACTYLUS, Merr. or CHAMÆSAURA, Fitz.

phalanges. One of the lungs is one half less than the other. It lives in mud.*

Others have not this range of pores. There is a small species of the Cape described long since (*Anguis Bipes*, *Lin. Lacerta Bipes*, Gm. Seb. I. lxxxvi. 3), whose feet terminate each in two unequal toes.†

Brazil produces another. (*Pygophus Cariococca.*) Spix, xxviii. 2, larger, with undivided feet, like those of the lepidopode, but more pointed, with scales all smooth. It is greenish, with four longitudinal blackish lines.‡

CHALCIDES, Daud.

Are like the seps, very elongated lizards, and similar to serpents, but their scales, instead of being disposed like tiles, are rectangular, and form like those of the tail in the common lizards, transverse bands, which do not encroach one upon the other.

Some have a furrow on each side of the trunk, and the tympanum very apparent. They are related to cordylus, as the seps are to the skinks, and con-

* The figure of Lacépède is taken from an individual whose tail had been broken and reproduced. In general, in all this class, an observer is very likely to be deceived, as to the proportional length of the tails.

† This is the genus BIPES, Merr., or SCELOTES, of Fitzinger. The *Seps Gronovien*, or *Monodactyle* of Daudin, of which Merrem has made his subgenus PYGODACTYLUS, is but a badly preserved individual of the other; another genus ought to be suppressed, as Merrem himself suspected. The *Seps Sexlineata*, Harlan. Sc. Nat. Phil. IV. pl. xviii. f. 2, is only a variety.

‡ The *Pyg. Striatus*, Spix, xxviii. 1, appears to me to be only the young.

duct in many points of view to the ophisauri and pseudopodes.

One species, with five toes, is known from the East Indies. (*Lac. Seps. Lin.*)

One with four. (*Lac. Tetradactyla*, Lacep. Ann. du Mus. II. lix. 2.)*

Others have the tympanum concealed, and lead directly to the bimanes, and through them to the amphisbenæ.

There is one species with five toes.†

One of Brazil with four before and five behind. (*Heterodactylus Imbricatus*. Spix, xxvii. 1.)

One with four on all the feet.‡

One whose toes, to the number of five before and three behind, are reduced to small tubercles, so little visible that the species has been regarded, sometimes as having three toes, and sometimes but one. It belongs to Guiana.§

CHIROTES, Cuv.

Resemble the chalcides in their verticillated scales, and the amphisbenæ still more in the obtuse form of

* It is the genus TETRADACTYLUS of Merrem, or SAUROPHIS of Fitzinger.

† This forms the genus CHALCIDES of Fitzinger.

‡ This is the genus BRACHYPUS of Fitzinger.

§ On the first supposition, namely, of the three toes, it is the *Chalcide* of Lacep. pl. xxxii.; the *Chamæsaura Cophias* of Schn.; the genus CHALCIS of Merrem, and the genus COPHIAS of Fitzinger. On the other hypothesis, it is the *Chalcide Monodactyle* of Daudin, or the genus COLOBUS of Merrem, but all these genera are reducible to a single species.

the head; but they are distinguished from the first because they want hind feet, and from the second, because they have fore feet.

But one is known, from Mexico.

Bipede Cannelé, Lacep.; *Chamæsaura Propus*, Schn.; *Lacerta Lumbricoides*, Shaw, Lacep. I. xli.

With two short feet with four toes each, and a vestige of a fifth, pretty completely organized internally, and attached by omoplates, clavicles, and a small sternum; but the head, vertebræ, and in a word, all the rest of the skeleton, resembles that of the amphisbenæ.

It is about eight or ten inches long, and is as thick as one's little finger; flesh-coloured, covered with about two hundred and twenty semi-rings on the back, and as many under the belly, which meet in turning on the side. It is found in Mexico, where it lives on insects. Its tongue but little extensible, is terminated by two small corneous points. The eye is very small, the tympanum covered by the skin, and invisible without. In front of the anus are two lines of pores. I have found but one large lung, and a vestige of a small one, as in the majority of serpents.*

* The genera which terminate this order of Saurians, interpose, in various manners, between the common Saurians and the genera placed at the head of the order of Ophidians; so much so, that many naturalists do not think that these two orders should be separated, or that one should be formed, including, on the one hand, the Saurians, with the exception of the crocodiles, and on the other, the Ophidans of the family of anguis.

But among the fossils of ancient calcareous formations, there exist two genera, far more extraordinary, and which, with the head and trunk of a Saurian, have feet supported on short members, and formed of a multitude of small articulations, assembled in a kind of oar or fin, like the paddles or fore-feet of the cetacea.

One of these genera, the Ichthyosaurus, had a large head, on a neck rather short, enormous eyes, a moderate tail, an elongated muzzle armed with conical teeth, adhering in a groove. Different species have been discovered in England, France, and Germany, some of which are very large.

The other, the Plesiosaurus, had a small head, supported on a long, serpent-like neck, composed of more conical vertebræ than in any known animal. Its tail was short. Debris of it have been also found on the continent.

These two genera, the discovery of which is mainly owing to the researches of Sir Everard Home, Mr. Conybeare, Dr. Buckland, &c., inhabited the sea. They should form a very distinct family; but from what is known of their osteology, they approach more to the common Saurians than to the crocodiles, to which M. Fitzinger associates them in his family of Loricata, which allocation is so much the more gratuitous, as we are neither acquainted with their scales, nor tongue, the two characteristic parts of the Loricata.

PLATE 5. OF THE "REGNE ANIMAL."

1. *Anolis velifer.* 3. *Gecko inunguis.*

2. *Anolis equestris.* 5. *Gecko cepédien.* of Peron. 4 *Gecko ocellatus.*

SUPPLEMENT ON THE SAURIA.

Before we proceed to consider in succession the different genera of Saurians, we shall enlarge a little on the text, in our general remarks on the characteristics of this very numerous and important order of reptiles. It received its present name, first from M. Alexandre Brongniart, and comprehends the animals designated by Linnæus under the collective appellation of *amphibia reptilia*. It is easy to distinguish these animals, in the vertebrated class, by unequivocal characters common to them all, and derived from their conformation and habits. For the sake of making a distinct impression, we thus briefly recapitulate those characters;—

A Saurian is an animal with an elongated body, scaly, or shagreened, without carapace, sometimes almost apodal, but more frequently with four, and rarely with two feet, the toes of which are furnished with crooked claws. The eyelids are moveable; there is a distinct tympanum, the branches of the jaws are soldered, and armed with enchased teeth. The orifice of the cloaca is a transverse cleft: the heart has two auricles, and there are ribs and a sternum.

The reptiles of the order of Saurians, have been grouped by some zoologists into three distinct families, founded upon the conformation of the tail. The first have the tail flatted above, or on the sides. These are called Uronectes. The second have the tail rounded, conical, and distinct, and are denominated Eumerodes. The third, which receive the name of ETROBENAE, have the tail likewise rounded and conical, but this part is not distinct from the rest of the body.

A superficial examination is sufficient for the purpose of distinguishing a Saurian from every other reptile. Nevertheless, there are some of them, which exhibit relations to genera, of orders, and even classes from which they are otherwise greatly removed,—Thus for example, though the Saurians differ from the Ophidia by the presence of limbs, and the existence of moveable eyelids; from the Batracians, by not being subject to metamorphoses; from the Chelonia, by the want of carapace, and the existence of teeth; and from the fishes by not having gills; yet there are many points in which they approximate to all those animals. The anguis seems a link to connect them with the first, and the *emys serpentina* with the third—while the crocodiles and dragons, attach them to the second, and the fourth. The study therefore, of the interior organization of these animals is indispensably requisite for their proper classification.

There is perhaps no other class of animals, which exhibit more marked and numerous varieties in the way of locomotion than the Saurians. This might naturally be expected, when we consider that there is no other in which the species differ so much in their modes of life and the nature of their habitations. The sojourn of some of them seems fixed to the waters, and these are organized for swimming.—Such for instance, are the crocodiles, the caymans and the gavials. Others, like the green lizard, seek out dry and elevated grounds, while some, as the basilisk, prefer the neighbourhood of aquatic places. Some inhabit the hollows of rocks, and others live in the centre of woods—some, such as the iguanas, climb with swiftness to the extremity of the branches of trees; and others, (the dragons) spring from one to the other, with a rapidity of motion resembling flight;

certain of them may be seen to run with the quickness of an arrow, launched by the hand of a vigorous archer; others walk with difficulty, and appear as it were to trail themselves along the ground.

Whatever may be the nature of the movements which these animals are adapted to exercise, it is in those climates over which the sun exerts the most genial influence, that their various modes of locomotion are most completely and energetically developed. The greatest degree of atmospheric heat seems necessary to *bring out*, as it were, all their motile forces. In such countries therefore, as Egypt, so nearly approaching to the tropics, on the burning coast of Africa, on the torrid banks of the Senegal, the Nile, and the Gambia, in the intertropical solitudes of the New World, and in the Archipelago of the Moluccas, and the Antilles, where the sun maintains an everlasting reign, we find the saurian races enjoying all the plenitude of their existence, and distinguished for the power, suppleness, and agility of their motions.

We might also assert, that a humid atmosphere, an aquatic soil, a certain superabundance of water, are indispensable to the numerous production of these animals. Egypt, which we have just cited, and where the saurians appear to rise as it were from the earth in all parts, is not merely a hot climate. An immense river in its periodical inundations covers the whole face of the country with a humid slime. The marshes and savannahs of South America, the mighty tracts inundated by the Oronoco and the Amazons, the island shores of the Equatorial Atlantic, are all subject to similar conditions of temperature and humidity. They are all bathed by tepid waters, and the innumerable legions of lizards and other saurians which inhabit them, enjoy a degree of activity far superior to that which distinguishes the torpid beings of the same class in our more northern countries.

It is not necessary here to enter into any details respecting the forms, the connections, and relations of the organs of locomotion in the saurians. This seems to have been done sufficiently already for a work of the nature of ours. But some further remarks on the variety of modes in which these organs are applied to their destined purposes, may not be devoid of interest.

We may first observe, that the progressive movements of reptiles in general present more difficulties in the explication than those of the two higher classes of vertebrated animals. These difficulties do not arise so much from the want of proper attention having been given to the subject, as from the enormous difference existing between the locomotive system in man and in these animals. The greatness of this difference renders it almost impossible to establish such an approximation or comparison as can tend to a proper elucidation of the subject.

In the reptile class, the majority of the saurians are true quadrupeds, but oviparous quadrupeds, and in the mode of station very different from the viviparous quadrupeds of the mammiferous class. Their thighs are directed outwards, and the inflexions of the feet are made in a direction perpendicular to the rachis, so that the weight of the body acts as a very long lever, and thus hinders the straightening of the knee, whose articulation remains constantly bent, which causes the belly to drag on the ground between the legs; and this is equally the case with the thoracic and abdominal members.

In the mammiferous quadrupeds, the case is quite different. The legs bend both in front, and behind in planes pretty nearly parallel with the spine, and but little remote from the central plane of the body, where the weight is particularly exerted.

When the saurians which are provided with only two limbs, or the mere rudiments of limbs, such as the chalcides, bipeds, &c., repose upon the ground, they form circles with their body above or around each other, and the head is raised above those circumvolutions.

Certain saurians climb with a wonderful degree of facility. In this point of view the cameleon, among the reptiles, seems to be as well provided as the quadrumana among the mammalia, in consequence of its hands, claws, and prehensile tail.

The lizards, anolis, the geckos, and tupinambis in general, walk and run with great agility; others swim by the assistance of their limbs and the central part of the tail, while others push forward or backward by an impulsion of the body, or the alternate application of one or more of their lower parts against the ground, in a manner very analogous to that of the serpent tribes.

The dragons, which belong to this order, are the only reptiles possessing the capacity of flight. For this purpose they are provided, on each side, with a membrane between the feet, which unfolds like a fan, and bends at the will of the animal, by the assistance of osseous radii articulated on the dorsal vertebræ, and which are substituted for the first six false ribs.

All physiologists are agreed to consider, in animals in general, the actions of walking and running so intimately connected, that it would be difficult to establish any certain distinction between them. The saurian quadrupeds do not deviate in this respect from the common rule, and there exists no difference between their walking after a certain mode, and their running; the latter, as in the mammifera, being most commonly performed by the complicated mechanism of the walk and leap.

When, by the assistance of an alternate motion of the feet, these animals desire to transport their body from one solid place to another, different phenomena take place, according as their feet are of similar length, or of unequal dimensions. But it never happens, in this kind of progressive motion, that the body is entirely suspended above the ground on which the limbs are placed.

In all such cases the entire body of the animal may be compared to a spring with two branches, one of which leans against a resisting object. If these branches, after having been drawn together by any force, are restored to their primitive liberty, the power of their elasticity will tend to separate them widely; but the branch resting against the obstacle, not being able to overcome it, transmits the motion, which takes place entirely in the opposite direction, so that the centre of gravity of the spring is removed from the obstacle with a greater or less degree of velocity.

In these animals, as in the mammifera, the flexor muscles constitutes the force which compresses the spring; the extensors represent the elasticity which separates its branches, and the resistance of the ground or water, is the obstacle. All the saurians whose feet are nearly equal in length, walk with great vivacity. The lizards, one species of which has received the epithet of *velox*, and another that of *agilis*, the ameivas, the monitors, the agamæ, the anolis, and many others are in this predicament. In them, in the most natural mode of walking, the body is in *equilibrio* on one of the fore-feet and on the opposite hind-foot, so that the centre of gravity does not move, following a right line, but advances between two parallels, in the intervals of which it describes oblique lines, which go from one to the other, forming zigzags. The impulses communicated to the trunk counterbalance each other reciprocally, and the trunk itself moves in the diagonal of a parallelogram, of which it may be considered to form the sides.

The rectitude of direction in the walk is continually changed in those lizards, in which the greater or less breadth of the feet must be taken into consideration, and the variable degree of separation in these parts, which permits them, includes a basis of sustentation of greater or less extent, or accommodates itself more or less to the inequalities of the soil.

If these animals, although their limbs be of equal dimensions, support with effort, on feet too small or too weak, a heavy, thick, or overlong body, they then walk with slowness, constraint, and embarrassment. Such is the case with the crocodiles and chalcides.

Some saurians leap with great agility, as, for instance, the iguanas and tupinambis.

Thus we find the movements of all the genera of saurians contribute to enliven the scene of the animated world, either amid the verdure of the earth, or on the bosom of the rapid river, or the tranquil lake. The lizard, which seems almost to swallow the ground in the rapidity of its course; the flying dragon hovering from branch to branch, and shooting into the air; the dracæna, bathing in the limpid stream; the anguis (for after all it may be said to belong to this order) gliding beneath the dry leaves and bushes; the gavial, cleaving the waves with the rapidity of an arrow, all present in their varied and lively evolutions, some of the most interesting spectacles in this sublunary world, where nothing remains unmoved, and where the power and the marvels of nature should excite our continual and unlimited admiration.

We have already had occasion to hint that the organs of sensation vary considerably in the four grand orders of the reptile class. As for the saurians, they have as great a number of senses as the best conformed animals among the vertebrata. Like the mammifera and birds, they have five senses; while in fish, that of smell seems more than doubtful; and as we descend among the inferior animals, we find sight, or hearing, or taste, reduced to a nullity, until at last we find the scope of sensibility limited to a simple faculty of touch.

However, though the saurians have all their senses, yet, with the exception of that of sight, they are all feeble in comparison to those of mammalia and birds. Their perception of impressions must consequently be much more limited, their internal emotions less strong and less frequent, their want of communication with the external world less urgently renewed, and less completely satisfied; their feelings colder, their apathy more remarkable, their instinct ill determined, and their volitions ill decided; and such, in fact, are the general characteristic of these numerous animals:

It is to this union of causes, under the immediate dependance of a principal one, that we must connect a fact, which we have already noticed in our observations on the class; namely, that among the animals of which we are writing, the muscular irritability is energetic, out of all proportion with the small developement of sensibility, with the little delicacy of most of the senses, and with the relative smallness of volume in the brain. The weakness of the senses which characterizes these animals suffices, in all probability, to produce such changes in their internal organization, as may cause the decrease of swiftness, of sensibility, and of internal heat observable in the descent from one genus to another.

If to such considerations we unite that of the small abundance of blood in reptiles, the long time which it takes in circulation, without passing through the lungs, which, moreover, according to some anatomists, receive no other blood than what is necessary to their support, and may be opened, cut, and lacerated, without producing immediate death, we may easily conceive how, every year during a determined term, these animals fall into a state of torpor rendered inevitable by circumstances of temperature to which birds and mammalia seem insensible.

Internal and external causes, therefore, unite to diminish the interior activity of the saurians, and to deprive them, at certain seasons, of the brutal instinct, the physical propensions, by which they are habitually influenced, to remove all sexual desire, to silence the calls of appetite, and, in a word, to place them in that lethargic kind of sleep which is the most prolonged and perfect image of death, which beings in which the vital spark is not extinct, can exhibit.

Being less sensible at all times, less animated by lively passions, less agitated within, and consequently less active without, pretty well sheltered against violent dangers, and having few accidents, comparatively, to dread, the saurians without loss of life, may be deprived of feet, tail, and other important parts, and even, after a time, reproduce them. So extraordinary a phenomenon suffices to prove how little the different parts of these beings are dependant on each other. Their nervous system constitutes a whole, the various pieces of which are far from being so closely connected as in birds and mammalia. We may also observe that, notwithstanding the burning climates which they usually inhabit, the saurians, being less influenced by internal heat, have much less occasion to drink than the animals of the two classes above them.

The slowness of the circulation and low temperature of blood in these reptiles, while it explains the fact of their not dying immediately when deprived of their head, also agrees with their astonishing facility of supporting want of food. This is so great that crocodiles have been known to pass a whole year without taking any kind of nourishment.

But though they have the power of resisting blows and

accidents which only affect certain parts of their body, they speedily succumb under the influence of external causes which attack the entire of their economy with energy and perseverance; this is in consequence of the want of reaction in their internal faculties. Thus an atmosphere more cold than temperate, immediately renders them feeble and ill, and often kills them. Thus we find that the gigantic crocodiles, the iguanas, the basilisks, and all the large sized races of the family of the saurians, frequent, in both worlds, only the rivers, the Savannahs, the hot and humid forests, or the burning sands of the torrid regions. To such haunts these larger species appear to be confined; and if any among them are found to frequent countries more or less remote from the equator, their dimensions grow progressively smaller, and their individuals become less and less numerous.

The deficiency of the sensitive system among the saurians, prevents the individuals of the same species from forming any thing like a true society, though they may often be found united in troops more or less numerous. No kind of labour, no operations of building, of hunting, or of war, says Count Lacépède, ever result from their congregatings. They never construct an habitation; and when they make choice of one on the banks of rivers, in rocks, or hollow trees, it is no commodious dwelling for a number of them united, but a purely individual retreat, in which they make no change, and use it solely for the purposes of concealment.

If they are ever found to hunt or fish together, it is only that they are simultaneously drawn by the same want, or attracted by the same bait. If they ever defend themselves in common, it is only when they are attacked in common.

Notwithstanding the almost untameable ferocity of some of these reptiles, and the discouraging stupidity of others, yet are many of them susceptible of being tamed, and rendered familiar. The priests of Memphis, according to the ancient historians, reared crocodies in a kind of domestication, which they were wont to parade in public in certain religious ceremonies. It also appears, according to the report of Bruce, that on the western shores of Africa, the negroes bring up crocodiles, which become so gentle as to let the children play with them, and ride upon their backs.

The encephalon, in the surians, occupies but a small portion of the cavity of the cranium; so that the figure and extent of the latter are no exact indications of the form and volume of the former. As for the rest, it is terminated, above, and in front, as in the other reptiles and all vertebrated animals, by olfactory lobes, and cerebral hemispheres; but like the encephalon of fishes, it has no decussations or general commissures, although it is placed much higher, by the predominance of the organ which represents the cerebrum. This predominance is so great, that the other parts of the encephalic mass appear to have no relations with each other but through its medium, and are fixed more closely to it than the various lobes are one with another. In general, its parts are all smooth, and without convolutions.

It is not easy to establish in a general way its comparative weight in proportion to that of the body, because the weight of the first always remains pretty nearly the same, while that of the latter is subject to variations of very great extent, according to the condition of the animal. It may be affirmed, however, that absolutely speaking, its volume is wonderfully smaller than that of the encephalon of the warm-blooded animals, that is, of those belonging to the classes mammalia and aves.

The dura mater, which is destitute of all falciform and other kinds of folds, is constantly adherent to the internal face of the cranium, and is separated from the encephalon by a mucous pulp more or less solid. · The pia-mater, as in the other vertebrated animals, is formed by a vascular net-work, very complicated and delicate.

The cerebral hemispheres are placed in front of the optic beds, and do not cover them. Their existence is evident, and they cannot be confounded with any other portion of the encephalic mass, which is by no means the case with all the fishes. Their most usual form is nearly that of a triangle with the base turned behind. As they are more perfect than those of fishes, so are they less variable in their forms through the different species. In the cameleon they form an oval, and are fixed by their base to the cerebral pedicles. Their interior is, as usual, hollowed by a ventricle in which is an hemispheric tubercle, one portion of which represents the *corpus striatum*.

As to the organs of sensation which are special in the saurians, we may remark that they all have two eyes placed to the right and left of the head, and tolerably large and projecting in proportion to the volume of the body. These eyes are mobile and lodged in orbits. They are always furnished with lids which vary in number, in figure, and in the degree of mobility. In the crocodiles, for instance, we find three eyelids, two horizontal, and one vertical. The first two close exactly, and have a swelling at their edge, but no eye-lash. The third, semi-transparent, moves from front to back, and is capable of covering the entire eye. It has but a single muscle, which represents the pyramidal muscle in birds, and which fixed in the same manner to the posterior part of the orbit towards the bottom, turns round the optic nerve, repasses under the eye, and sends its tendon into this lid. The common lizards have instead of eylids, a kind of circular veil, extended in front of the orbit, and pierced with an horizontal cleft closed by a sphincter, and dilated by a levator and a depressor muscle. The lower part of this veil has a cartilaginous disk, smooth, and round, like that of birds. There is found, moreover, in these reptiles, the rudiment of a third eyelid, but without any peculiar muscle. The third eyelid is totally wanting in the cameleon, and the aperture in the veil is so small, that the pupil can scarcely be seen through it. The gecko has no mobile lid, and the eye is protracted only by a slight fold of the skin. The same appears to be the case with the skink.

Nothing positive is known concerning the lachrymal apparatus of the saurians. In the lizards and the cameleon, the sclerotica contains osseous laminæ, analogous to those in birds. But these lamina do not form an anterior disk, and only surround the lateral portion of the sclerotic.

In the crocodile, the ciliary processes are very fine, and well marked, each of them ending nearly in a right angle. These processes are not found in the lizards.

The iris of the saurians has some analogy with that of fishes in the metallic tints with which it shines. That of the crocodile exhibits a most beautiful vascular net-work. In this reptile, the pupil resembles that of the cat, while it is rounded in the cameleon and the lizards, and rhomboidal in the gecko.

In general, in all the animals of this order, the pupil is susceptible of dilatation and contraction, so as to receive the necessary quantity of light. Accordingly they can distinguish objects equally well in the obscurity of the darkest and in the most brilliant light of sunshine. There is nothing remarkable in the vitrea of these animals, nor yet in the crystalline aqueous humours.

In the crocodile the globe of the eye is preserved in its orbit by means of six ordinary muscles arranged as in the fish, and moreover, by four smaller muscles which nearly embrace the optic nerve, and spread over the sclerotic.

The organ of vision is remarkably active in the saurian tribes. Inhabiting for the most part the sea-shores, and banks of rivers in the torrid zone, where the sun is scarcely ever veiled by clouds, and where the luminous rays are incessantly and intensely reflected by the waves and the sands, a strength of this organ was peculiarly indispensable to these animals, to prevent their eyes from being injured and speedily destroyed by the excessive action of light to which they are perpetually subjected.

Like other reptiles the saurians have an auditory organ composed of a vestibulary sac, a vestige of cochlea, and three semi-circular. But the crocodile alone has any appearance of an external meatus auditorius, because the skin forms a kind of thick cover above the tympanum. This serves to explain a passage in Herodotus, who tells us that the Egyptians were in the habit of suspending jewels in the ears of the crocodile. The osseous labyrinth presses closely on the membranaceous, investing it throughout with a thin and hard lamina. The auditory apparatus of these animals is altogether very imperfect, and we may conclude that their powers of hearing are equally so. This conjecture is strengthened by the fact that they are either dumb, or only emit hoarse, confused, and disagreeable sounds. Their olfactory organs are still more imperfect, and the sense of taste is in all probability less developed than even that of smell. The tongue, in fact, of the greater number of saurians, though singularly extensible and mobile, is terminated by two long points which are semi-cartilaginous and corneous, and though it is soft and humid, its surface is smooth. In the crocodiles it is so fixed by the edges and point that it seems to be wanting, and was always thought to be so by the ancients.

Nor can the sense of touch be supposed to give rise to any great number of distinct impressions in these animals, covered as they are with hard scales, or a corneous epidermis. This skin falls at least once every year in the springtime, either in an entire piece, like a sheath, or in dry, horny and colourless scales. Many of them, besides, have their toes joined so that they cannot be applied but with difficulty to the surface of bodies; and though some others have them very long, and separated from each other, yet the under part is furnished with such thick scales, that a very small degree of sensibility must be left in it.

The corpus mucosum, under the epidermis of the saurians, is very livelily and variously coloured. There is great tenacity in the dermis, which is closely adherent to the muscles, and variable in thickness.

In many species may be observed, under each thigh, a very regular range of small pores, from which a viscous humour exudes. This peculiarity belongs to all the genuine lizards.

The saurian reptiles live generally on small quadrupeds, birds, mollusca, worms, and insects. They never drink. Their digestion is remarkably slow, and they eat but seldom, especially during the cold season. If some among them, such as the crocodiles, make a considerable consumption of food, it must be attributed to the immense bulk which they have to support. One repast, however, will suffice them for many days.

From a necessary consequence of defective mastication in these animals, the salivary glands should constitute a less important apparatus in their organization than in that of the mammifera. In some of them the tongue is composed in a great measure of a thick glandulous mass, formed by a number of small tubes united by their base, and separated towards the surface of the organ. These constitute so many papillæ, which render the surface somewhat bristly when they are fine. The sides of the mass are pierced with a multitude of small holes, which give passage to a humour secreted by the gland itself. This disposition is particularly observable in the flat-headed geckos, in some skinks, and in the iguana. In the tupinambis this gland appears to be replaced by two others, which are elongated, granular, and situated under the skin along the external face of the branches of the lower jaw, and the humour of which is poured to the external side of the teeth in the same jaw. On that side they are immediately covered by the palatine membrane.

In the saurians, in general, the tongue is capable of considerable elongation, and the mechanism which produces this movement intimately allied to the act of deglutition, is itself inseparable from that of the different pieces which compose the hyoïd apparatus.

In the majority of the saurians the epiglottis is wanting, as is likewise the veil or covering of the palate. Their pharynx is but little wider than the œsophagus, and is furnished with no muscle for the purpose of moving it, or causing it to change form. The mucous membrane which covers it exhibits numerous longitudinal folds. In the œsophagus there is nothing very remarkable, but that its diameter is very great, in comparison to the stomach, and that it appears very dilatable. The stomach is generally of an oval and very elongated form, and its parietes are usually slender and transparent, like those of the intestinal canal. Its muscular membrane seems to have but little sensibility, at least in a portion of its extent.

Proportionally to the body, the intestinal canal of the saurians is very short. This is evidently connected with the carnivorous regimen of these reptiles. Its form and dimensions, however, vary considerably according to the species.

The growth of the saurians is rather slow, because these animals live a long time, and the lethargy to which they are subject during the winter seems to suspend their existence. Certain species, such as the iguanas, and above all, the crocodiles, attain to a remarkable size. The advanced age to which they arrive is not surprising in cold-blooded animals, which transpire with difficulty, easily dispense with food, and possess the faculty of repairing such external losses as they happen to sustain.

In man and the more complicated animals of the vertebrated division, circulation is a function of the first importance. The blood, proceeding from the left ventricle of the heart, spreads itself by the arteries through the whole body, takes the capillary system in its way, passes into the veins, returns to the heart, enters into the right auricle of that organ, then into the corresponding ventricle, which, in its turn, sends it into the pulmonary artery to be distributed into the lungs, from which it issues by the pulmonary veins, to repair to the left auricle and ventricle, and proceed from thence in its round anew. In this course the blood evidently describes a double circle, one in the lungs, the other in the entire body.

This is not at all the case with the reptiles in general, and with the saurians in particular. The heart, as we have already observed, in all these animals is so disposed, that at each contraction it sends into the lungs but a portion of the blood which it has received from all the other parts of the body. Therefore the pulmonary circulation of the saurians is but a fraction of the great circulation. This fraction is more or less considerable, according to the genera, and productive of effects more or less marked. But as we have already enlarged on this topic sufficiently, and shown the consequences of this sluggish circulation on the habits, passions, intellectual and physical constitution, we shall content ourselves by remarking here, that the quantity of blood in the saurians is very inconsiderable in comparison with that of mammifera and birds. Hasselquist, who dissected a crocodile at Grand Caïro in 1751, reports, that the quantity of this fluid which flowed from the grand artery was very small indeed.

We shall conclude these general observations on the saurians, which have already, perhaps, been too far extended, by noticing a singular fact not before mentioned, and which serves to explain some peculiar phenomena in these reptiles, which had hitherto remained incompletely developed by our modern physiologists. The phenomena in question we have mentioned before, namely, the capacity of these animals to continue so long without sustenance, and the yearly lethargy into which they fall, and which is infinitely more profound than the hybernal sleep of some mammifera. Even yet our curiosity is not completely satisfied on these two points ; but until the researches of M. L. Jacobson, of the academy of Copenhagen, our acquaintance with their nature was very slender indeed. Nothing was demonstrated concerning the nature of that arrangement of organization on which they might be supposed to depend.

After some particular researches the learned Danish anatomist, whom we have just cited, recognized in the reptiles a special arrangement of certain vessels, which constitutes a peculiar *venous system*.

Nature has established this system in all reptiles in a manner more or less marked. The rudiments of it are to be found in the tortoises and crocodiles—but it is only in the other saurians, the ophidians, and batracians, that it is completely developed.

It is composed of the veins of the abdominal members, the pelvian or caudal veins, the hinder veins of the kidneys, the veins of the oviductus, a great portion of the veins of the skin, of those of the muscles of the abdomen, and of those of certain organs peculiar to the reptiles. These veins combine, and form one or many trunks, which proceed either into the vena porta, or the liver, or into both. What especially distinguishes this system is, that in it a part of the veins of the organs of locomotion, and of the skin, proceed to distribute themselves into the liver. There is no other example of this among the vertebrated animals.

Certain special organs appear connected with this venous system in a peculiar manner, and are regarded by M. Jacobson as proper for secreting and preserving a nutritive juice destined to be reabsorbed in the rigorous months of the severe season, during the hybernal slumber of these animals. These organs are formed of two membranaceous and vascular sacs, which are situated at the lower part of the abdomen, between the muscles and the peritoneum. In the ophidians, in which these had been remarked, though but imperfectly described before the time of M. Jacobson, they constitute two fatty bodies, which occupy the anterior paries of the abdomen, and receive the arteries even of the aorta, while the veins which originate from them constitute a part of the system in question.

In the saurians these organs are smaller, and situated lower down. They also appear to become developed only at a certain period of the life of the animal. Be this, however, as it may, the venous system of which we speak varies considerably.

To compose it, all the veins of the muscles, and of the skin of the pelvian extremities, enter by different apertures into the cavity of the pelvis, and reunite there in two trunks, which join on each side with the hinder renal vein. This last is peculiar to the reptiles, commences in the kidney, by roots which have no communication with those of the other renal veins, and accompanying the sciatic nerve, proceeds along the external edge of the kidney, receiving in its progress the veins of the oviductus, and the subcutaneous dorsal veins, as far as the cavity of the pelvis, where it unites with the trunk formed by the crural veins, to proceed to the lower face of the abdomen, and receive the blood of the veins of the bladder.

This principal trunk thus spreads as far as the anterior part of the abdomen, receives the veins of the muscles of the parietes of this cavity, and proceeds between the greater lobes of the liver, to unite to the vena-porta.

Some variations only are observable, proceeding from the situation of the kidneys, the size of the caudal veins, and the extent of the lower paries of the abdomen. The veins of the upper part of the muscles of this region form a separate trunk, which goes directly to the liver. In the crocodiles we find two of those venous trunks, which proceed to the liver, and the same is the case with the caymans.

We shall now dismiss the subject of the saurians in general, and proceed to make a few remarks on the CROCODILES, the first genus of this order.

The very name of crocodile excites the idea of an animal alike formidable for its size and ferocity; of an animal unequalled in its own orders, and which is the tyrant of the fresh waters of the equinoxial zone, both in the old and the new world. "In the length of its body," says the eloquent Lacépède, "it surpasses both the eagle and the lion, those haughty monarchs of the air and the forest; and if we except the elephant, the hippopotamus, some cetacea, and some few enormous serpents, it has no equal in nature."

If the crocodiles exceed the majority of other animals in size, and in extent of offensive power, they are also better protected by nature with defensive arms. Their skin is almost entirely covered with small bucklers, which are proof against the sword and the musket bullet. They possess, moreover, a terrific aspect, which principally proceeds from the fierce glances of their eyes. Their mouth, furnished with long and numerous teeth, appears like a mighty gulph ready to swallow all that approaches. Their grave and almost ominous pace concurs to the general effect which they produce on the imagination. Yet they are ferocious only when pressed by necessity, and a crocodile satiated with food is an enemy whom the weakest have no cause to fear. This was well known to the ancients, and has been long ago remarked by Aristotle. The ancients believed, and the moderns have repeated the assertion, that the upper jaw of the crocodile alone was moveable. But it has been for some time clearly ascertained that this is an error, and that in the crocodile, as in all other animals, it is the lower jaw only which is in this predicament. The origin of this opinion is, that the crocodiles having the centre of movement in the jaws, situated beyond the cranium, they cannot open them without raising the head. From the same cause they are unable to masticate their food: they can only break or crush, and then swallow it.

With the exception of the summit of the head, the entire body of the crocodile is covered with scales. The crocodiles can therefore only be wounded in those parts where the scales are separated, that is, at the junction of the thighs with the body, in the eyes, or mouth.

The brain of the crocodile is remarkably small, the œsophagus very ample, and capable of great dilatation. They have no bladder, and their organs of generation are internal, near the anus; above the jaws, two glands are observable, which contain an oily matter of a strong musky odour.

Of the osteology of the crocodiles we have already treated pretty much at length, in our account of the fossil remains, and shall therefore say nothing about it in this place.

Spring is the period in which the sexes seek each other. Concerning all that has relation to the reproduction of these animals, little is known with any degree of certainty. The crocodiles lay two or three times in the year, but at short intervals, about twenty eggs at least, and sometimes many more, according to the species, and bury them in the sand, near the lakes and rivers which they inhabit. They are left to be hatched by the sun. Those of the crocodile of the Nile are about twice as large as the egg of a goose; but the eggs of the cayman are hardly equal in size to those of the turkey. They are both whitish, and the shell is of a nature perfectly similar to that of birds' eggs. Their eggs are good eating, though they have a strong smell of musk, and are in great estimation in all the countries inhabited by these animals.

As soon as the young are born, they hasten to cast themselves into the water, but the greater number of them become the prey of tortoises, of voracious fish, of amphibious animals, and even, as is said, of the old crocodiles. Those which survive, feed, for the first year, only on the larvæ of insects, and on very small fish. This fact was verified by M. Bosc, on a nest of these young reptiles, consisting of fifteen individuals, which he caught in a marsh near his residence in Carolina. He observed that they ate only living insects, and that it was necessary that those insects should put themselves in motion before the young crocodiles would dart upon them, which they then did with great voracity, frequently disputing with each other for the same object. But when this naturalist was daring enough to take them into his hands, they never attempted to do him any injury.

Even at the conclusion of the first year, the crocodiles are still but small and feeble animals. It is only in the course of the second, that they acquire teeth which are formidable, and that the cranium becomes sufficiently thick to sustain blows with impunity.

The duration of the life of the crocodile is not exactly known; but there are facts which tend to prove that it is equal, if not superior, to that of the life of man. They do not moult or cast their skin, and accordingly they thus escape a crisis which is fatal to the majority of reptiles. The number of enemies also, capable of destroying them, after they have arrived to the maturity of their strength, is very small. Like other reptiles, they can remain a long time without eating, and such of them as do not inhabit climates the most approximating to the equator, bury themselves during the entire winter, so that the causes of death are much less frequent with them than with the majority of other beings.

The shores of great rivers, and the midst of marshy lakes, are the favourite haunts of the crocodiles. They are sometimes to be found there in very numerous troops. In such places they find, on the one hand, shelter and security, and on the other, they live on frogs, fish, water-birds, and, in fine, on all the animals which they can secure. Even dogs, swine, and oxen, cannot always escape their voracity. It is reported that they will seize them by the muzzle, or the legs, when they go to drink, and drag them into the water. M. Bosc used to amuse himself by making them issue from their retreats, and run towards him, by causing his hunting dog to bark and yelp on the banks of rivers. He usually fired at them, but sometimes would suffer them to approach near enough to strike them with a stick. Of this they did not appear to be much afraid, but they never attempted to attack him. On the contrary, they would withdraw gravely as soon as they perceived that there was no prey for them in his immediate neighbourhood. Though heavy, they swim with very great facility; but their movements are very slow on land. As soon as the negroes of Carolina perceive any of them, considerably removed from their retreat, they intercept their return, pursue them, kill them with hatchets, and banquet upon the tail. M. Bosc has often found dead ones thus mutilated, which shed so infectious an odour of ammonia, that it was impossible to remain near them. Even the vultures, to which the most corrupted flesh is a regale, abandon that of these reptiles, when it has arrived at a certain degree of decomposition.

In Carolina, the crocodiles make holes, or very deep burrows, where they pass the entire winter, and sometimes the entire day during summer. These holes are usually placed in the marshes which accompany almost all the rivers, but sometimes also on the edge of small pools, in the midst of woods. M. Bosc has repeatedly tried to catch these animals with snares of various kinds set at the entrance of these holes, but to no effect. Every morning he found his snares broken, and the crocodiles had come out, and re-entered safe and sound. They can, however, be taken in those countries with small living quadrupeds, or birds, attached to a thick hook, and fastened to a tree by means of an iron chain.

In Florida, where the population is less numerous and the heat more considerable, the crocodiles are found in much greater abundance. Bartram, in the relation of his voyage on the river St. Jean, states that he has seen the waters covered with them for considerable spaces. They so impeded his navigation, as to force him several times to interrupt it. He has been witness of great destruction of fish, and mutual combats between those reptiles.

The crocodiles during the whole summer, but more especially directly after they have quitted their burrows, utter loud bellowings, as powerful as those of an ox, and which cannot be compared with any other cry. They may be heard, sometimes in great numbers, to reply to each other in this manner, of an evening, causing the marshy forests to re-echo with a frightful din. This is so great as to prevent a person from sleeping in the neighbourhood, unless he has been accustomed to it, as M. Bosc declares himself to have experienced many times. They also make a loud noise by striking their jaws one against the other, and which may be heard at a very considerable distance. This rattling of the jaws often takes place even in the day-time, and the motives of the animals for making it, are by no means very apparent.

In Egypt and Senegal, the crocodiles are less numerous, but more strong and dangerous than in America. Notwithstanding this, the inhabitants attack them boldly, body to body. As soon as they perceive a crocodile out of the water, they go up boldly to him, and either kill him with spears, or put into his mouth, which he opens to devour, a piece of iron pointed at both ends, which hinders him from shutting his mouth again. They then drown him, in performing which operation, the arm is enveloped with a strong piece of leather. Some are even bold enough to attack the crocodile while in the water, where he enjoys every advantage of using his tremendous powers. They dive beneath him, and plunge a poignard into his belly.

The Dutch used to keep a great number of crocodiles in the ditches of Batavia, to prevent the desertion of their soldiers, most of whom were enlisted by force, and to oppose the nocturnal attacks of the people of the country, who supported their yoke with impatience.

In Africa, the tigers, and the cougars in America, make war upon the crocodiles: but they seldom attack any but the young ones. A great many animals, and principally the mangoustes in Africa, and the otters in America, seek out and devour the eggs of the crocodile. It is said that the female will keep guard upon the place where they are concealed, but this assertion does not appear to be very well founded.

In the body of the crocodiles are found the bezoar stones which formerly were held in such high estimation for their supposed medical virtues. At present, however, like all the other bezoars, they are banished from the *materia medica*, by all enlightened practitioners.

The crocodiles, as is well known, were objects of worship to the ancient Egyptians, who consecrated to them the town of Arsinoë, and interred them in the tombs of their kings.

The *Egyptian crocodiles* inhabit the Nile, the Senegal, and in all probability most of the other African rivers. At the present day it is found in the Nile, only towards the region of Upper Egypt, where it is extremely hot, and where this animal never falls into a lethargic state. Formerly, when it was wont to descend the branches of the river which water the Delta, it used to pass the four winter months in caverns, and without food. Of this fact we are informed by Pliny and other ancient naturalists.

According to the relations of Hasselquist, and many other travellers, the crocodile of the Nile must be the largest animal of its kind. In Upper Egypt, these animals have been found thirty feet and more in length. The ancients pretend to have seen some of them six and twenty cubits in length, but this is probably nothing but exaggeration.

These crocodiles have a strong odour of musk, which they even communicate to the waters which they frequent. The negroes, notwithstanding this, use their flesh as food, as did likewise the ancient inhabitants of Elephantina, according to the testimony of Herodotus. Moore informs us that among several African nations the eggs are esteemed the highest possible delicacy. They are used as food both in Egypt and the East Indies, as those of the cayman are in Florida and the other parts of America.

It would seem that other species inhabit the Nile as well as this of which we are now speaking, and that at least one variety of the common crocodile is to be found in that river. M. Geoffroy St. Hilaire found a head of this embalmed in the caverns of Thebes. It is a little flatter and more elongated than that of the common crocodile. Two entire individuals and two heads of the same form are in the museum of Paris. One of the first was presented by Adamson, who calls it the green crocodile of the Niger.

We have already noticed the superstitious veneration of the old Egyptians for this ferocious reptile. At Memphis the sacred individual was reared with the greatest care, and nourished with abundant food. Sacrifices and offerings were presented to him ; he was adorned with trinkets, and lodged in a lake or basin in the midst of the temple : thus treated, the crocodile lost its ferocity, and became so tame as to be led about in religious processions and ceremonies.

In that country, so vaunted for wisdom, persons have been known sufficiently foolish and infatuated, as to rejoice when any of their children were devoured by the crocodile. But in some districts of Egypt, these animals were held in abhorrence, and hunted and killed, also through a sentiment of religion; because they believed that Typhon, the murderer of Osiris and the genius of evil, had transformed himself into a crocodile.

Herodotus informs us that there was a law obliging the people of Apollonopolis to eat these animals, because the daughter of King Psammeticus had been devoured by one of them. In the city of Heraclea divine honours were paid to the ichneumon, because that animal was regarded as the sworn enemy of the crocodile.

Presages were also drawn from the sacred crocodiles. If the animal received favourably the elements presented to him, it was considered an happy omen—but if he refused to eat, it was regarded as an inauspicious augury.

Diderot remarks, that it is only necessary to set in motion the imaginations of men, to make them yield credit to the grossest extravagance and absurdity. This remark, though not very complimentary to our species, is unfortunately but too well founded in truth. Thus respecting the crocodile a number of absurd fancies were most implicitly believed. It was supposed, for instance, to have had no tongue—to have had as many teeth as there are days in the year. It was imagined that there were certain times and places in which it ceased to be dangerous. If any one had had the hardihood to maintain that a crocodile had attacked an Egyptian, though even on the Nile and in a bark of papyrus, he would have run a narrow risk of being stoned as a blasphemer.

It is supposed that one reason why the ancient Egyptians venerated this animal, was, that the fear of the crocodiles arrested the course of the Lybian and Arabian robbers, who, but for these reptiles, would have been continually passing and repassing the river and its canals.

The Indians of Timor, of Java, of Ceram, of Sumatra, and of the majority of the Sunda Islands, believe that when their women lie-in, they bring into the world a little crocodile, a twin of their human infant, and that the midwife receives this animal with the utmost care, and carries it immediately to the river. The family take care to provide aliment for their amphibious relation, and the twin goes, particularly at certain epochs of his life, to perform this fraternal duty, and that under pain of being struck by disease or death in case of omission. At Celebes and Bantam many of the inhabitants rear and feed crocodiles in their houses.

In the year 58 before Christ, the edile Scaurus exhibited at Rome five crocodiles of the Nile, and subsequently the Emperor Augustus had the Flavinian circus filled with water, and exhibited there to the people thirty-six crocodiles, which were killed by an equal number of men, who were habituated to fight with these animals.

In 1681, a living crocodile was brought to the menagerie

of Versailles, and within some years many young individuals have been seen in Paris.

The crocodiles of the Senegal, the Niger, and the Gambia, seem to augment in length, in proportion as we advance farther inland. In the Senegal, near Ghiam, Brue saw one twenty-five feet in length. Barbot, in the same river, and in the Gambia, observed some that were thirty feet long. Jobson tells us that in the Gambia, where they are called *bumbos*, they send forth cries which may be heard at a considerable distance, and which sound as if they came from the bottom of a well.

Adamson found hundreds of these animals in the Senegal. They appeared all at the same time, on the top of the water, like floating trunks of trees. But when the boat approached them, they exhibited fear, and dived. When they perceive any animal drinking on the bank of the river, they proceed immediately towards it, seize it by one leg, and drag it in to devour it.

Père Labat informs us that the crocodiles are often taken with hooks placed in the abdomen of a dog, and fixed to a chain of iron, at the end of a long cord. They are also caught, we are assured, by means of a plank of softish wood, in which their teeth become engaged.

The negroes sometimes kill the crocodile by main force, and in the water, when they surprise him in a spot where he cannot support himself without swimming. They proceed to him, having the left arm guarded by a piece of ox-hide, and a bayonet in the right hand. They keep his mouth open by plunging the left arm into his gullet, and strike him with the bayonet in the throat. This information we also receive from Father Labat.

Adanson, the celebrated naturalist, returning from hunting on the island of Sor, found the nest in which a crocodile had just deposited its eggs, at about half a foot deep under the sand. There were thirty eggs in it. The negroes who accompanied him carried them away for the purposes of food.

It is said that the hippopotamus is one of the most formidable enemies of the crocodile; but the most dangerous, without question, is the ichneumon, which devours its eggs. It was even formerly pretended that this little animal was accustomed to enter the throat of this reptile while it was sleeping in the sun, and to tear its entrails: a fable which has long been utterly refuted. It was also asserted that the crocodile was the friend of the wren, and that this little bird used to perform for it the office of a dentist, cleansing its teeth from worms, which get between them, and from the flesh which happens to be there,—another fable still more ridiculous than the former.

The *double-crested crocodile* is the most common species in all the rivers which lead to the Indian ocean. It is found in Java. Peron has observed it at Timor, and the Sechelles islands. M. Delabillardiere informed M. Cuvier that it is a general opinion at Java that this animal never devours its prey on the spot, but that it buries it in the mud or sand, where it suffers it to remain untouched for three or four days.

In accounts of Macassar we read, that in the great river of that island there are crocodiles so ferocious that they do not confine themselves to making war on fish, but assemble in troops to watch the boats, and endeavour to overturn them, that they may devour the men who are in them.

It appears that this species is also to be met with in the rivers of Corea, and even in China.

The *crocodile of St. Domingo* was first published as a distinct species by M. Geoffroy St. Hilaire, on an individual sent to the Museum of Paris by General Rochambeau. Père Plumier had, however, described, drawn, and dissected this

DOUBLE-CRESTED CROCODILE.

C. BIPORCATUS. Cuv.

crocodile, but his observations remained in manuscript, excepting such of them as had been published by M. Schneider.

The males are much less numerous than the females. They fight together with great violence and inveteracy during the season of reproduction. The males are fit for generation at ten years of age, and the females at eight or nine. The fecundity of the latter seldom lasts above four or five years.

The female digs with the paws and muzzle a circular hole in the sand, on some slightly elevated mound, where she deposits twenty-eight eggs, moistened with a viscous liquor, ranged in beds separated by a little earth, and covered with earth well beaten down.

The laying takes place in March, April, and May, and the young ones issue from the egg at the end of a month. They are then only from nine to ten inches in length; the growth continues more than twenty years, and some individuals arrive to more than sixteen feet in length.

At the time of their disclosure from the egg, the female comes to scrape away the earth and let them out: she conducts, defends, and feeds them, (which latter operation is performed by disgorging to them her own food,) for about three months, a space of time during which the male would seek to devour them.

These crocodiles cannot eat in the water without running the risk of being suffocated. They dig holes in the bed of rivers, to drag in, and drown their victims, which they suffer to rot.

Such is the substance of the observations made personally at St. Domingo, by M. Descourtils. They are confirmed by a note from a physician of that island, who informed M. Parmentier, that these animals seek with especial avidity the flesh of negroes, and that of dogs, and that they never eat it until it is in a state of putrefaction.

It is reported, that to escape from this crocodile, the dogs bark, and horses strike the water, at a particular place, to attract the crocodile thither, and then hurry off, to drink at some greater distance.

The colonists and negroes give to this species the name of Cayman.

We cannot terminate the history of the crocodiles, without noticing the threrapeutic qualities, anciently attributed to them. Their blood was supposed to cure opthalmia, and to hinder the development of accidents caused by the bite of serpents. Persons labouring under fever were rubbed with their fat, and the ashes made from their skin when burned, and steeped in the lye of oil, were esteemed to be a powerful narcotic. Assertions such as these could only be opposed by facts, and, unfortunately, there is nothing of the kind to justify them. Another specimen of these absurdities, but of a different kind, was, that the crocodile itself furnished an antidote to its own bite. The reputation of these, and a multitude of other specifics, have been utterly destroyed by the progress of the philosophy of medicine.

The tribe of the CAYMANS, as far as it is known at present, is confined to the continent of America. But the word *Cayman* is generally employed by all the European colonists to designate the crocodiles which are most common around their habitations. Thus the Cayman of St. Domingo is a true crocodile. Authors are but little agreed on the origin of this name. Bontius will have it to be aboriginal to the East Indies, and Schauten is of the same opinion. Margrave tells us that it comes from Congo, and Rochefort that it was peculiar to the old inhabitants of the Antilles. M. De Tussac considers the assertion of Margrave to be the

PIKE-MUZZLED CAYMAN.

C. LUCIUS.

most correct. The slaves, on their arrival from Africa, at sight of a crocodile, give it immediately the name of cayman. It would appear from this, that it was the negroes who spread this name throughout America, where it is employed even in Mexico. This opinion is at least more probable than that it was transmitted from the East to the West Indies.

The Pike-muzzled Cayman (Crocodilus Lucius) inhabits North America. It proceeds pretty far towards the north. It ascends the Mississippi, as far as the Red River. Mr. Dunbar and Dr. Hunter have met with an individual of this species as high as 32° and a half north latitude, although in the month of December and during a very severe season. It was brought for the first time, from the Mississippi by M. Michaud, and afterwards Mr. Peale sent a very fine individual to the Paris Museum. Catesby seems to have given but an indifferent figure of this species. He informs us that these reptiles, in Carolina, conceal themselves in marshy places covered with woods, and live there in the midst of carnage. They spring upon domestic animals, such as pigs, sheep, and oxen, that are imprudent enough to penetrate into these vast solitudes, seize them with their powerful jaws, and drag them down to the bottom of the waters, where they are speedily devoured. The specimen above mentioned, sent to Paris by Mr. Peale, was only five feet in length—but Catesby has observed them as long as fourteen.

In the *Journal de Physique*, 1782, we are informed that these caymans, in Louisiana, take refuge in the mud of the marshes when the cold comes on, and fall there into a lethargic sleep, without being frozen. When the weather is very cold, they may be cut in pieces without awakening them. Catesby gives us pretty nearly the same information respecting those of Carolina, which on awakening from their state of rest, send forth horrible bellowings. According to the first of those observers whom we have just cited, they are reanimated by the warm days which occur even in winter. He also adds that the pike-muzzled cayman never eats in the water; but after having drowned his prey, withdraws to devour it. This reptile is said to prefer the flesh of the negroes to that of the whites, but constitutes itself no small portion of the food of many savages. Its voice resembles that of a bull. It is afraid of the shark and the great tortoise, and consequently avoids the neighbourhood of salt or brackish waters. Its mouth always remains shut when it is asleep.

Bertram appears to have spoken of this species when he relates that they assemble in numerous troops, in fishy places, and that the female deposits her eggs in layers, mixed with alternate beds of earth, so as to form little mounds of three or four feet in height. He adds, that she does not abandon them, and also keeps her young with her many months after their birth. He assures us that he has found some individuals in a stream of hot and vitriolic water.

According to the same traveller, the boldness of this reptile is equal to its strength. He says that his armed companions had to sustain a vigorous combat against one of them who came to attack their camp. At another time, on the banks of the river St. Jean in Florida, he saw two of these animals fight desperately together.

Its skin is of a thickness and hardness sufficiently great to resist musket balls. It is even said that the negroes form a kind of helmet out of it, that can resist the blow of a hatchet. To hit the animal at all with effect it must be aimed at under the belly, or towards the eyes.

The *Spectacled Cayman (Crocodilus Sclerops)* grows to a very considerable size. It inhabits South America, and is very common at Cayenne, and throughout all Guiana. Seba, who has given a tolerably good figure of it, was deceived in announcing the Island of Ceylon as its native country. It seems not improbable that the *jacarè* of Margrave, and the *yacarè* of M. d'Azara, which does not proceed beyond the fifty-second degree of south latitude, is the same animal. It has been very well described by M. Schneider.

In the great river of the environs of Surinam, some of these caymans have been seen, that attained to the length of twenty, and even twenty-four feet. The negroes sometimes eat their flesh, although it has a fetid and musky odour. Stedman assures us that they will not attack a man, as long as he remains in motion in the water. On land they do not possess one half of the swiftness of man, and but seldom attack him there, unless he approaches their eggs. These they defend with remarkable courage.

The female lays about sixty eggs in the sand; and covers them with straw or leaves, leaving them to be fecundated by the sun. According to Delaborde, they lay two and even three times in the year, within intervals of a few days, but this author asserts that the eggs are only twenty or twenty-four in number.

They always pass the night in the water and the day in the sun, sleeping on the sand. But they return to the water if they see a man or a dog.

M. d'Azara tells us that the inhabitants of Paraguay, to take the *yacarè*, have an arrow so constructed that, being shot into the flank, it leaves there the iron with which it is armed, and from which it detaches itself, but that nevertheless, those two parts remain connected by means of a long cord. The wood remains floating on the water, and indicates to the savages the place to which the animal has retired, whither they immediately repair and despatch him with their spears.

According to M. Correa de Serra, some Portuguese travellers are of opinion that the yacarès of the southern and temperate parts of Brazil are not the same as those of the north. Both lay their eggs in the sand confusedly, and not in layers or ranges. The place being recognised, the inhabitants seek out their eggs to pierce them with an iron-pointed spear. In the flat island of Marajo, or Johannes, at the mouth of the Amazons, these reptiles remain in summer in the marshes, and when these are dried, the little water which remains is so filled with them that the fluid ceases to be visible, and it seems probable that then the larger devour the smaller. They cannot re-ascend the river because the island is surrounded with salt water. According to Delaborde, they remain in Guiana, sometimes almost entirely dry, and it is at such times that they are considered as most dangerous.

The *Cayman with osseous eyelids (Palpebrosus)* certainly inhabits Cayenne, and is a very distinct species. But little is known of its habits, and the specimens are rare. The interval between its two hinder toes is less palmated, which, according to M. Cuvier, would render it more of a terrestrial animal than the preceding.

The sub-family of the Gavials have been observed only in the hottest countries of the ancient world. The first author who has spoken of these animals, was our countryman Edwards, who described an individual of them in 1756, in the forty-ninth volume of the Philosophical Transactions, which he announced as having come from the coast of Africa. In 1765, another was described, and Merck has noticed a third in 1785. But all these individuals were small, and the descriptions of them too brief to be of much utility. A very full and exact description, however, was subsequently published by the Count de Lacépède, of an individual of twelve feet in length, with all the dimensions and a figure. This naturalist was the first who gave to the species the Indian name of *Gavial*.

As yet we know but of two species of this subgenus, if indeed there be so many. The first is the *Great Gavial* (*C. Longirostris*). This gavial inhabits the Ganges, and probably some of the neighbouring rivers, such as the Burampooter. It feeds only on fish, and though it arrives to a gigantic size, is not dangerous to man.

M. de Lacépède observed in the collection of the Museum of Natural History in Paris, a portion of the jaw of a gavial of the East Indies, which must have been thirty feet ten inches in length, and whose dimensions were therefore considerably beyond the ordinary standard of these animals.

The same naturalist believes that it is to this species we must refer the crocodiles seen on the banks of the Ganges, by Tavernier, from Tontipour as far as the town of Acerat. But this great Asiatic river is also inhabited by a prodigious quantity of common crocodiles, a fact, which as our author has noticed, was not unknown to the ancients, and which has been verified by M. Fichtel, a skilful naturalist attached to the Museum of the Emperor of Austria.

As for the *Little Gavial* (*Tenuirostris*), we neither know to what size it may arrive, nor what country it inhabits, though it is suspected that it belongs to Africa.

These two reptiles have been very well figured by M. Faujas de St. Fond in his History of the Mountain of St. Pierre, pl. 46 and 48.

We now come to the great family of the LACERTIANS, which we shall treat of in detail, though as briefly as possible, confining ourselves to points of popular interest. We begin with the MONITORS proper, sometimes, but erroneously, called *Tupinambis.* These are easily to be distinguished from the crocodiles, which have the hinder feet palmated; from the dragons, which have angular plates upon the head; from the safe-guards, which have denticulated or serrated teeth; from the lizards and ameivas, which have the tail round; from the lophyri and basilisks, which have a crest upon the back; and in short from all the other saurian reptiles.

Until the expedition of the French into Egypt, Hasselquist, one of the disciples of Linnæus, was the only naturalist who had observed the *Monitor of the Nile*, or *Ouaran* of the Arabians, which Linnæus and Gmelin have regarded as a lizard, and Schneider has placed among the skinks. More circumstantial details concerning its conformation and manners were subsequently furnished by M. Geoffroy St. Hilaire; but as they chiefly relate to its anatomical conformation, we may be dispensed from their enumeration in this place. They are very fully detailed in the Baron's great work on the Fossil Bones.

The Egyptians pretend that the ouaran proceeds from the egg of a crocodile deposited in a dry soil. The ancients have frequently represented it on their monuments, perhaps because it is greedy after the eggs of the crocodile, and devours the young individuals of that species.

The *Tupinambis* or *Monitor* of Congo, *T. Ornatus* of Daudin, is about five or six feet in length, and devours all kinds of smaller reptiles and insects, which circumstance causes it to be held in high respect by the negroes, under the roofs of whose cottages it frequently pursues its prey.

The *Land Monitor* of Egypt is common in the deserts which border upon that country. The jugglers of Cairo employ it in the performance of tricks, after having drawn its teeth. It is the land crocodile mentioned by Herodotus.

The *Variegated Tupinambis* of New Holland is of a tint generally black, but varied with spots and stripes of different forms. There are several transverse ranges of round and yellow spots on the limbs, and the tail is covered, throughout its entire length, with numerous annulated bands alternately black and yellow. This reptile conceals itself at the bottom of the waters when it is pursued. Its total length is about three feet and a half. White has given the first notice of this species.

In the British Museum is a very beautiful species of

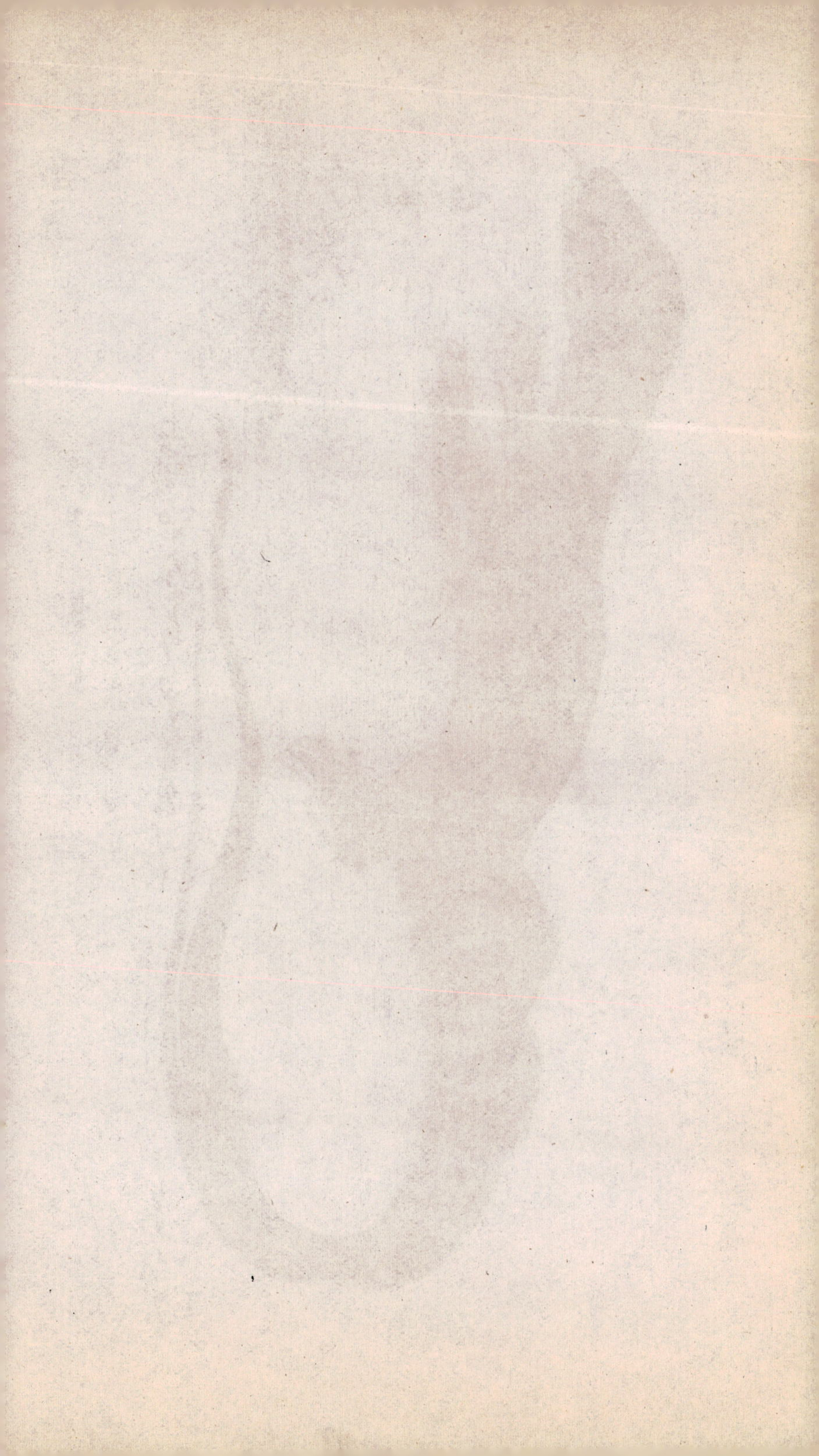

VARIED MONITOR.

MONITOR PULCHER. — *Leach.*

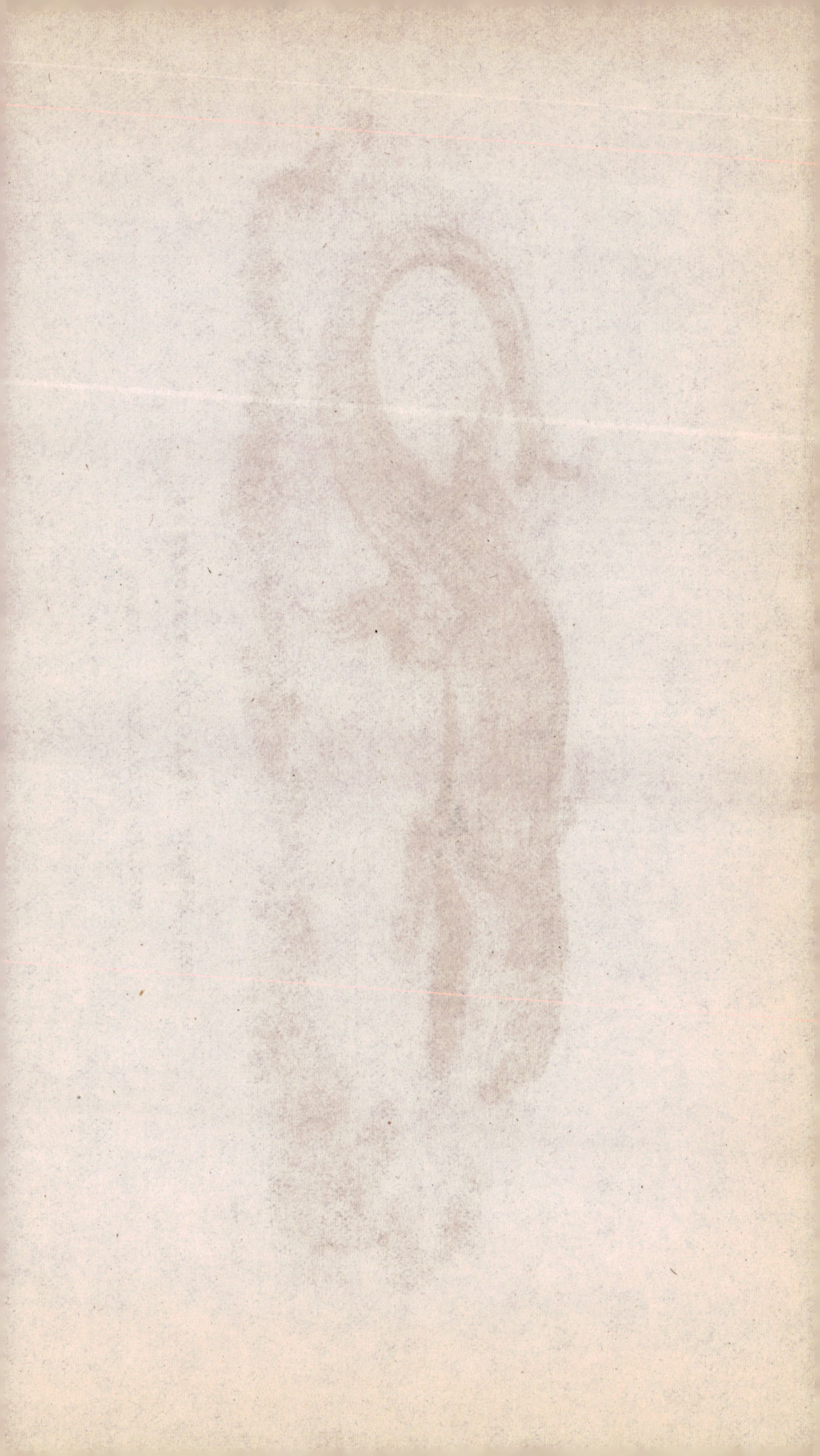

Mus. Brit.

THE GREAT DRAGON OF LACEPEDE.

MON. CROCODILINUS MER.

monitor, varied with white stripes, spots, and patches, on a black ground, to which Dr. Leach has given the epithet *pulcher;* the figure will sufficiently display the arrangements of the colours of the species and the general appearance of the genus.

The DRACÆNA of Lacépède, which must not be confounded with the dracæna of Linnæus, and still less with the dragon, constitutes a genus first established by the naturalist just mentioned. In addition to what is so briefly dwelt on in the text, we shall say that its distinctive characters are large scales, elevated with ridges like those of the crocodile, scattered over the back, and forming crests upon the tail. The teeth are conical, those in the bottom of the mouth being thick, and having the crowns rounded. The tail is round at the base, and compressed at the extremity. Between the principal scaly plates on the back and flanks, are several very small and rounded scales. The tongue is forked, and the tympanum apparent.

This saurian, the *Monitor Crocodilinus* of Merren, *Ada* of Gray, inhabits many regions of South America, Guiana more especially, where it is however comparatively rare. It resembles the crocodile as to form, but bears no similarity to it in its habits. It swims with some degree of difficulty, runs with tolerable swiftness, and climbs trees with great nimbleness and dexterity. It sometimes preys upon such animals as it meets with in the woods, frequents the inundated savannahs and marshy soils of its native regions, though it more generally sojourns on land and in the sun, than in the water. Our figure is from a specimen in the British Museum.

There is very considerable difficulty in taking it, because it conceals itself in burrows, and bites desperately. Its flesh is eaten, and reported to be no small delicacy. Its eggs are also in high estimation at Cayenne, and each female usually lays several dozens of them. The length of this reptile is generally from four to six feet.

M. Cuvier has given the denomination of Safe-guards to his third group of monitors. The following are their general characteristics.

Maxillary teeth serrated; no palatine teeth; tongue slender, extensible, and terminated by two long threads; body elongated; five toes on all the feet. All these toes armed with claws separated and unequal. Scales disposed by transverse bands under the belly, and around the tail. Those of the back small and without keel; one range of pores under each thigh.

The safe-guards are easily distinguishable from the last genus by the absence of ridged scales, from the monitors whose teeth are sharp and trenchant, from the lizards which have palatine teeth, from the iguanas whose tongue is neither extensible nor terminated by two threads, and from the crocodiles which have but four toes on the hinder feet.

This subgenus has been divided into two tribes; that of the *safe-guards* properly so called, which have the tail more or less compressed, and the scales of the belly more long than wide. They inhabit the banks of waters. The second division is that of the *ameivas*, whose tail is round, and furnished, as well as the belly, with transverse ranges of square scales.

Some naturalists place the *Lizardet* or *Lacerta Bicarivata* in this division: with the exception of its dorsal scales, it is pretty like the dracæna, although smaller. Its tail, long and compressed, is raised on the upper part by three or four keels of sharp scales. Its head is covered with broad polygonal plates. It is generally about a foot in length.

According to Linnæus, it is an inhabitant of the East Indies; while Gmelin, on the contrary, will have it to belong to the islands of South America.

The *Great American Safe-guard*, in addition to what is

1
THE AMEIVA OR SPOTTED LIZARD.

TEYUS AMEIVA.

2
THE SPURRED LIZARD.

CENTROPYX CALCARATUS.

mentioned in the text, is also distinguished by its large and projecting eyes.

This saurian is an inhabitant of Brazil and Guiana, and more especially of Surinam and Cayenne. It frequents the edges of waters and inundated situations, and arrives to the length of six feet.

The great safe-guard runs with rapidity on the ground, and throws itself into the water when it is pursued. It dives, but however does not swim, and never climbs trees. It lives on insects, reptiles, mollusca, the eggs of birds, and honey. It nestles in holes excavated by itself in the bosom of the sand.

Both the flesh and eggs of this reptile are eaten. The latter are of an oblong figure, and but few in number. According to Don Felix d'Azara, it is believed in Paraguay, that the rings of its tail are a preservative from palsy.

The *Ameiva* (*L. Ameiva*, *Lin.*) has an elongated head, compressed at the sides, and narrow above. The muzzle is pointed. The tail is longer than the body, and cylindrical. It is composed of at least one hundred and twenty verticillated scales, very slightly carinated. The size of this reptile is about a foot.

This saurian is commonly found in Guiana and the Great Antilles. It should not be confounded with the *Anolis* of Rochefort and Ray, but it appears to be the same animal as the *Anolis* of Surinam, described by Gronow, and as the large spotted lizard of Edwards.

The *Black-marked Ameiva*, which is enumerated in the notes of the text, is the *Lacerta Litterata* of Daudin. The tail is long, cylindrical, verticillated, and very pointed. The upper part of the body is of a very fine bluish green, rather deep, and varied with small black marks, numerous and irregular, disposed cross-wise over somewhat broad bands, ocellated here and there with small white rounded spots only on the flanks. The belly is of a very clear bluish green. The size is from eighteen to twenty inches.

Daudin has in a very unaccountable manner made this saurian a native of the temperate climates of Germany, Hungary, and Prussia. This is a mistake, for as our author has remarked, it is an inhabitant of America.

We must pass with brevity over many of the numerous sub-divisions of our author, and even over some of them in silence. The truth is, that when we come to treat in detail concerning the reptile tribes, there is an abundant scantiness of that sort of information concerning them which would prove interesting to the majority of readers. However useful and curious may be the investigation of their specific characters, it must be owned that to the many it is dry and repulsive. It is *caviare* to the general. It is, however, occasionally necessary for us to supply the deficiencies of the text in this way, which must be our apology for now and then deviating into matter merely of a descriptive kind. When the characters are of importance, and but superficially treated of in the "Règne Animal," the object of which work was to avoid amplification in this way, it would be unpardonable to omit all mention of them. This is peculiarly the case with generic characters, which are of the highest importance in classification. Nor can they always be clearly conceived, or properly impressed upon the mind, by a very brief description. Accordingly to those have we chiefly confined our attention, and to the notice of such species as are not described in the text, when they involve any interest or importance.

Returning from this, perhaps not wholly irrelevant digression, we shall now proceed to the consideration of the remaining genera of the saurian order, omitting to dwell on its minor subdivisions.

The Lizards proper form the second genus of the great Lacertian family. The following is an enumeration of their characters, rather more complete than what is to be found in the text.

The tongue is slender, extensible, and terminated by two long threads; the palate is armed with two ranges of teeth. There is a collar under the neck, formed by a transverse range of broad scales, separated from those of the belly, where there are only some small ones, as under the throat. The body is elongated, and wingless. No goitre; all the feet are furnished with five toes armed with claws, not opposable; separated, rounded, and unequal. Scales disposed by parallel and transverse bands under the belly, and around the tail, which is at least as long as the body, thick, cylindrical, and without crest or keel above. The anus is a transverse cleft. Part of the bones of the cranium advance over the temples and the orbits, so that all the upper part of the head is provided with an osseous buckler, or covered with large scales. The tympanum is on a level with the head, and membranous. The eyelid is of a single piece, cut longitudinally, and formed by a sphincter. Under each thigh is a range of small grains or tubercles formed of scales, rough to the touch and porous. There are transverse plates under the belly. The scales are carinated, but not imbricated, on the back.

The lizards are thus easily distinguished from the other saurians—from the *tachydromi*, which have not a range of pores under each thigh; from the *cameleons*, whose toes are opposable; from the *anolis* and *geckos*, which have the toes flat underneath; from the *agama*, which instead of plates have scales upon the head; from the *dragons*, whose sides are winged; from the *iguanas*, which have a denticulated goitre under the throat; from the *ameivas* and *safeguards*, which have no scaly collar; from the *monitors* and *dracæna*, which have the palate without teeth; from the *stelliones* and *cordyli*, which have the tail thorny; and from the *basilisks* and *lophyri*, which have a crest upon the tail.

The tail of the lizards is composed of articulations, which the slightest effort can separate, and it is susceptible of being reproduced when it has been broken by any external violence. The principle of life is very strong in them all, and they can pass a long time without eating ; and it also appears they can live a great number of years. None of them are venomous, but there are several that bite with great violence when attacked.

The lizards are very numerous, and inhabit various parts of both continents, seeming to delight equally in warm and temperate climates. Their movements are light and lively, but they fall into a lethargic state during the winter, at the bottom of their retreats. They are monoganous, and live only in pairs. They never go into the water, like many other reptiles, belonging as they do to the saurian order.

The genus of the lizards is far from containing at the present day all the species which Linnæus and the majority of systematic writers have thought proper to insert in it. Laurenti was the first who attempted, but without much success, to reform it. This undertaking has been better executed by more modern naturalists, such as Lacépède, Alexandre Brogniart, Cuvier, Daudin, Dumeril, &c.

Among several European species, confounded by Linnæus under the name *Lacerta Agilis*, is the *Great Green Lizard*, *(Lacerta Ocellata).* This reptile is one of the most brilliant and splendid of the whole saurian order. It is also the largest of all known lizards. It is found in the South of France, in Italy, and the other southern countries of Europe, in arid places among rocks exposed to the sun, and on the borders of woods. It may be frequently seen in the neighbourhood of Montpellier, frequenting the bushes and hedges, climbing on shrubs, and over large stones in search of insects. M. Poiret has met with it in Africa several times, towards the shores of the Mediterranean.

It would appear that it is not merely in warm climates that

this saurian is to be found. According to Ray and Linnæus, it also inhabits very northern countries, such as Sweden and Kamschatka. In the latter country it inspires terror, and is considered as an envoy of the infernal powers; a fact which Captain Cook ascertained during his residence in that remote and barbarous region.

We are assured that this reptile feeds not only on insects, but that it also swallows frogs, mice, shrews, and other small vertebrated animals. It seeks out worms, will swallow saliva, and take the eggs of passerine birds. M. Poiriet found in the stomach of a green lizard, which he dissected on the coasts of the ancient Numidia, a small lizard, completely entire.

According to M. de Lacépède this lizard is even known to attack some serpents, but he seldom comes off victorious in this sort of combat. He does not appear even much to dread the presence of man, and will bite with great violence and inveteracy the end of a stick with which any one may choose to torment him. This lizard not only runs with swiftness, but also leaps remarkably high, and being bolder than the grey lizard, he will defend himself against the dogs that attack him, fastening on their muzzle, and preferring to allow himself to be killed, than to let go his hold.

It is perfectly erroneous to regard the bite of the green lizard as venomous and mortal. On this subject, Laurenti has made a number of experiments perfectly conclusive and satisfactory.

If Gesner is to be believed, the Africans eat the flesh of those green lizards, which the majority of naturalists have regarded as a variety of the *lacerta agilis* of Linnæus. M.M. de Lacépède and Latreille were the first to distinguish them specifically.

The *spotted green lizard* (*Lacerta viridis*, Daudin) is also to be found in all the temperate climates of Europe. It frequents woods of little elevation, and exposed to the sun.

Laurenti has made a seps of this species under the name of *Seps varians*, and M. Latreille considers it as a variety of his green lizard.

The *green lizard* of Jamaica has been figured by George Edwards in his Natural History of Birds. It has the greatest possible relations with the ameiva, in the form of its head and body. It was seen alive in the metropolis by Edwards, and was brought here from Jamaica. In his work called *Gazophyllacium* it has been figured by Petiver, who, however, calls it the *lizard of Gibraltar*.

The *lacerta stirpium* of Daudin inhabits the woods of Germany and France, under the branches. It is very common in the Bois de Boulogne, and in that of Vincennes, near Paris. The upper part of the head, the back, and tail, are brown, and the sides and belly of a clear green; all the scales of the under part of the body and tail are marked with a black point.

It is very agile and by no means fearful, and glides off among the dry leaves when it is attempted to be taken. During the warm days of spring and summer, it quits its retreat, and goes to bask in the sunshine and give chase to the gnats, ants, and other small insects. These lizards usually live in pairs.

Almost all naturalists have regarded this lizard as a variety of the *lacerta agilis* of Linnæus, and M. Latreille has made of it a variety of the green lizard of Lacépède. It sufficiently appears that it is it which has been described by Seba under the names of *talectée* and *tamacolin* of New Spain.

M. Ruiz de Xelva found in the woods of Tuscany a variety of this reptile which does not differ from that of the environs of Paris, only by being somewhat of a larger size, and by the colour of its belly and flanks which are of a more lively green and destitute of black points.

Near Paris there also exists another variety, having sixteen callous tubercles under each thigh, the black of a bluish green, with longitudinal white lines and blackish spots.

Razaumowski, in his Natural History of the Jurat, has described a third which comes from Switzerland, and which has the under part of the tail a flesh-colour. The sides of the body are green, spotted with black. There is a band of brown spots along the back and tail.

Finally, Daudin, in the Bois de Boulogne, caught a fourth variety, the back of which is entirely of a reddish brown and without spots, and which evidently, according to him, is the same animal as the *Seps ruber* of Laurenti.

The *lacerta viridula* of Latreille resembles much in its form the *lacerta stirpium*. It was discovered by the Spanish naturalist Ruiz de Xelva, in that part of Mexico which is nearest the isthmus of Panama, where it lives in the clefts of rocks, and in the middle of heaps of stones in the neighbourhood of woods. The male may be distinguished by an orange-coloured spot, surrounded by blackish, which it bears upon the occiput and the neck.

The *lacerta tiliguerta*, Gmelin, is an animal of a shining green, relieved by black spots and stripes along the back. Its total length is from seven to eight inches.

This saurian has as yet been described, after nature, only by the naturalist Cetti. It is found at all seasons, in the lawns, fields, and on walls, in Sardinia, where it is known under the names of *tiliguerta* and *caliscertula*.

M. de Lacépède regards the tiliguerta rather as a simple variety of *ocellata* than as a distinct species, and our author, as may be seen by the notes on the text, thinks that it is only a mixture of an American ameiva with the green lizard of Sardinia, ill-described by Cetti.

The *nimble lizard, scaly lizard* of Pennant (*Lacerta agilis*, Linnæus), has the head triangular and depressed, the muzzle obtuse; the jaws armed with small and fine teeth, a little crooked, and turned towards the gullet. The neck is almost as thick as the body, and, like the latter, flatted on its four sides. The tail is cylindrical, prolonged into a point, and a little longer than the rest of the animal. The scales of the upper part, and the flanks, are very small, hexagonal, and not imbricated. There are seventeen porous tubercles under each thigh. The claws are recurved, and there are six ranks of plates under the belly. It is from five to six inches in length.

This species, called the grey lizard of the walls, is the most common saurian reptile in France, and in all the temperate climates of Europe. It inhabits sandy places and the walls of gardens, on which it climbs with a surprising degree of agility. It is also found in part of Asia and Africa. It lives on flies, ants, and other insects.

The vivacity of its motions, the grace of its rapid gait, its agreeable and slender form, cause it to be very generally remarked. It is susceptible of being tamed, and many persons consider it as peculiarly the friend of man.

It is so common in the environs of Vienna in Austria, that Laurenti declares that it might serve during the entire summer for the support of a great number of poor persons. Its flesh, wholesome, and productive of appetite, according to this observer, might be baked or fried, like that of small fishes.

Formerly, the properties of this same flesh were highly vaunted as a remedy against cutaneous and lymphatic complaints, against cancers, syphilis, &c. But its use is altogether abandoned at present for any medical purposes.

This animal passes the winter in a state of lethargy, at the bottom of its retreat, and begins to couple in the first fine days of spring. It is monogamous, and the individuals live only in pairs. The male and female remain in a perfect

2

3

1 GREEN LIZARD.

LAC AGILIS.

2 FIERCE ALGYRA.

ALGYRA BARBARICA.

3 EYED IACHYDROMUS.

IACHYDROMUS OCELLATUS.

union during many years, sharing the arrangement of their household, the care of excluding the young from their numerous eggs, of carrying them into the sun, of placing them in shelter from cold and humidity. These eggs are round, of the diameter of three or four lines, and covered with a calcareous envelope.

This lizard is subject to variations in its colours, according to age, sex, and particularly the climate which it inhabits, which is nothing surprising, considering that it is found at once both in the north and south of Europe.

We pass over the other species of lizards, particularly as most of them are doubtful, and but ill-ascertained, and nothing can be adduced concerning their manners or habits, that can compensate the reader for the trouble of attending to their descriptions. We also, for the same reasons, pass over the sub-genera of *algyra* and *tachydromus*, and proceed to the third great family of the saurians, the Iguanians.

The name of Stellio was given to the first genus of this family by Daudin, after the ancient Latin name of a not very well determined reptile.

This genus, which Linnæus had confounded with the lizards, is now generally admitted as distinct, and may be thus characterized.

Pointed tail, rounded, with spinous verticillæ; neck and paws distinct; the latter with free, unequal, unguiculated, and not opposable toes; no teeth in the palate; tongue fleshy, thick, not extensible, and merely emarginated at the end; head swelled behind by the muscles of the jaws; back and thighs bristled here and there with scales larger than the rest, and occasionally spinous. Ears surrounded with small groups of spines; thighs destitute of follicular spines.

The subdivision of *cordylus* we have passed over, as little of interest can be adduced concerning it. From it the stelliones proper are distinguished by not having under the thighs a line of very large pores. They are also easily distinguished from the cameleons, which have the toes opposable; from the agamæ, the iguanas, the lizards, the anolis, the dragons, and the geckos, none of which have the spiny tail.

Our author, as we have seen, admits of but one species among the stelliones proper. This is the *Stellio of the Levant* (*Stellio vulgaris*, Daudin. *Lacerta Stellio*, Linnæus. *Cordylus Stellio*, Laurenti). In form, this reptile has the head comparatively bulky; a little flatted, triangular, very wide, callous, and rough on the sides of the occiput. The tympanum is round, wide, and but little sunk in. The jaws are cleft as far as its level, and bordered with two or three parallel ranks of narrow, smooth, and nearly squared scales. The nostrils are round, and but moderately projecting. The eyes are behind over the cheeks; there are seventeen teeth on each side of the upper jaw; twenty-two on each side of the lower. The tail is longer than the body, and composed of seventy spinous verticillæ; the anus is transversal; there are five toes on all the feet; and the nails are small and crooked.

This saurian usually attains to the length of about a foot.

It is very common throughout all the Levant, and particularly in the islands of the Archipelago, in Egypt, and in Syria. It is also met with, they say, at the Cape of Good Hope. It appears to live in preference under the ruins of old edifices, amid heaps of stones, in the clefts of rocks, and in sorts of burrows, which it has the art and industry to excavate for itself.

It is extremely agile in all its movements, and feeds upon the insects which flutter over the sand.

The *Stellio of the Levant*, which the modern Greeks call κοσκορδυλὸς, and the Arabs *Hardim*, does not appear to be

THE UROMASTRIX OF EGYPT.

STELLIO SPINIPES.

the stellio of the Ancients, which is represented to us to be a spotted, venomous lizard, hostile to man, and extremely cunning. It may probably, as we have before mentioned, be the *tarentole*, or tuberculous gecko, of the south of Europe. Belon was the first who seems to have given rise to this false application of an ancient word.

In Egypt, according to the relation of this last traveller, and of some others, around the pyramids, and in the neighbourhood of the tombs of the Thebaïs, they collect, for the purposes of oriental pharmacy, the excrements of this reptile, which were also anciently employed in Europe as a cosmetic. The Turks still make some use of it, after the example of those coquettes of ancient Rome, of which Horace speaks in his Epodes:—

> "Nec illi
> Jam manet—colorque
> Stercore fucatus crocodili."

Be this, however, as it may, the Mahometans pursue and kill it, because, say they, it mocks them, by lowering its head, as they do when engaged in their devotions. Fortunate would it have been for the world, if superstition had never produced any worse result than the destruction of a comparatively harmless reptile!

Passing over *Doryphorus* and *Uromastix*, which latter genus includes the *Stellio spinipes*, which we have figured from a specimen in the British Museum, and which is described at p. 121, we shall say a word or two on the AGAMÆ.

The name of this genus is peculiar to the natives of Jamaica, and was employed by them to designate a species of lizard.

The characters of the agamæ are a body covered with tubercles, toes and tail rounded, a head bulky, cordiform, covered with scales, and a tongue very short and fleshy. Their toes are not opposable, which distinguishes them from the cameleons; their sides destitute of membranes, distinguish them from the dragons. Their tail without spines, the throat without goitre, and the head without plates, will not permit us to confound them with the stelliones, the iguanas, and the lizards.

In general, the true agamæ have the body thick, covered with a loose skin, which can be inflated at the will of the animal, and which is sown, throughout its whole extent, with small tuberculous scales, rounded, rhomboidal, or hexagonal, and more or less projecting. Their tongue is not extensible, the gullet is without teeth, the neck seems compressed, and on its sides, and behind the ears, are fasciculi of pointed tubercles. The tail is seldom longer than the body. The figure of *Agama spinosa*, so named by Mr. Gray, is from a specimen in the British Museum; the colour is uniformly yellowish green, and the species seems nearly allied to the *Agama Colonorum*, described at p. 123.

They seek out humid places in the warmest countries of the two continents, and never issue forth from their retreats until the evening.

The species which gives rise to our author's subdivision of the TAPAYES, or orbicular agamæ, namely, the *lacerta orbicularis*, of Linnæus, has some apparent relations with the toad. Seba has accordingly designated it under the name of the spinous toad of America. In Paraguay it is called cameleon, because it can swell its body and change colour a little when it is touched. Its body is nearly orbicular, broad, and inflated. Its head thick, short, widened, and swelled behind the eyes. The occiput is furnished with small spines, in the same manner as the back and the upper part of the tail. The tail is short, slender, and pointed. The feet have fine long and slender toes, armed with crooked nails. The colour is an ashen grey, more clear underneath, and shaded with brown spots more or

Mus. Brit.

SPINOUS AGAMA.

AGAMA SPINOSA. Gray.

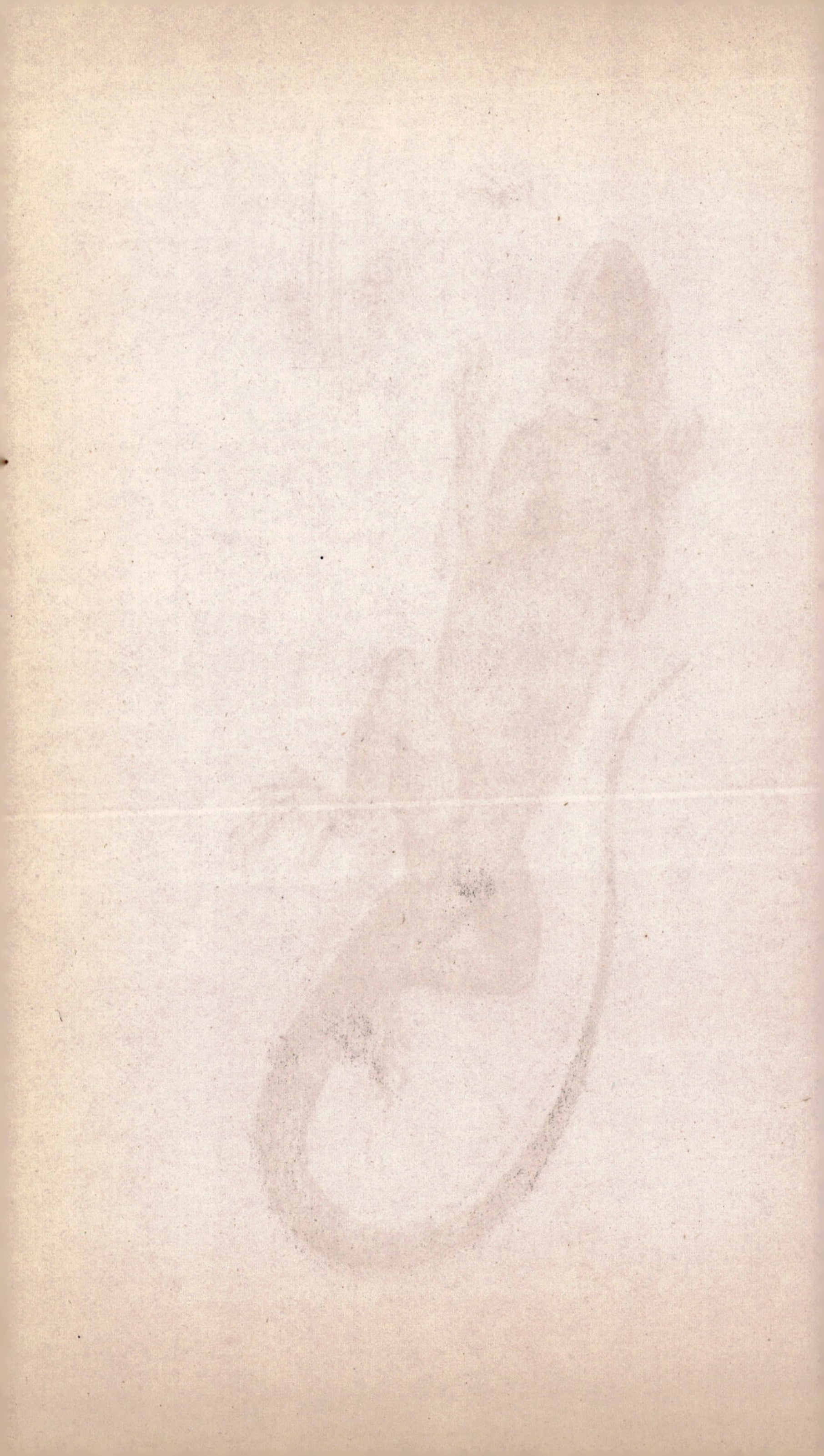

ORBICULAR LIZARD.

AGAMA CORNUTA. Harlan. L. ORBICULARIS.

Mus. Brit.

FRILLED LIZARD.

CLAMYDOSAURUS KINGII.

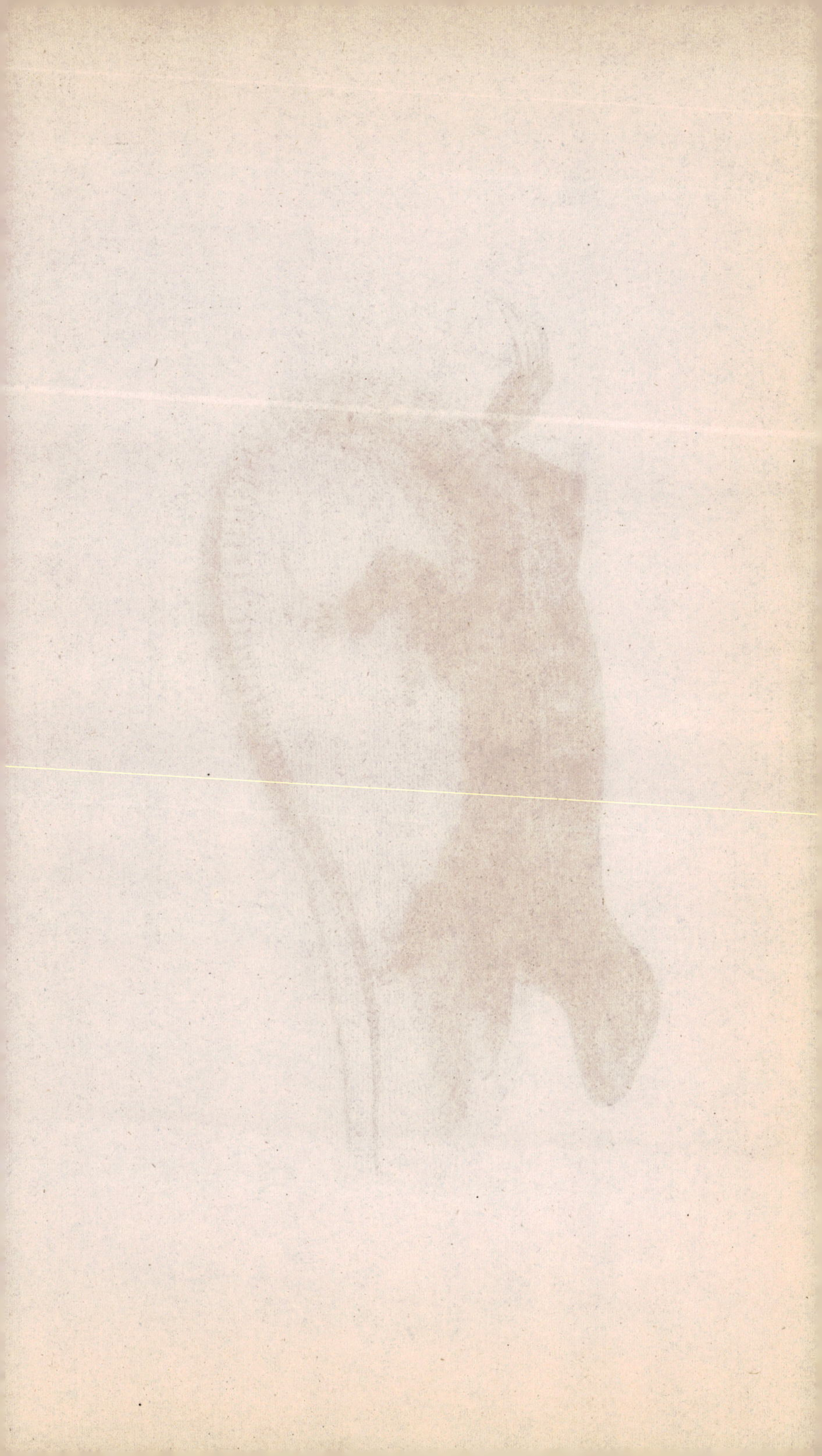

CLOUDED LIZARD.

L. NEBULOSA. Sn.

LOPHYRUS AGAMOIDES.

less deep. We have copied Dr. Harlan's figure of this animal, which he describes as specifically distinct, under the name *Agama cornuta*, but which Cuvier identifies with the orbicular lizard.

This animal most generally inhabits the mountains and rocks of Mexico and Terra Firma, where the inhabitants also name it *tapayaxin*, according to the traveller, Hernandez. Taking in the tail, its most general length is about six inches. Ray informs us that it is by no means dangerous, and that it can be tamed.

Allied to the agamæ, with femoral pores, is a species of lizard, brought by Captain King from Africa, and now in the British Museum, under the generic name Clamydosaurus. For a description of the species, as well as of the other figures of this order, we refer for the sake of brevity to the table.

Passing over all the intervening subdivisions, which our limits prevent our noticing, and of which we could say but little to interest our readers, we proceed to the division of the Dragons; premising merely that the *Lac. nebulosa* of Shaw belongs to Cuvier's subgenus Brachylophus, and that the *Lophyrus agamoides*, which we have figured from a specimen in the British Museum, is an inedited species belonging to that genus, which will be found described in the table.

To no word, perhaps, are attached ideas more extraordinary, and of greater antiquity, than to that of *dragon*. In all ages, and almost in all countries, the terrified imaginations of certain timid men, the fantastic notions emanating from disordered brains, or the interested efforts of charlatanism and superstition, have produced a belief in the existence of fabulous beings, of monstrous forms and redoubtable ferocity, of supernatural force and address, who were accustomed to carry trouble and devastation into entire provinces, to guard the entrance to consecrated places, or to watch over the security of hidden treasures which had been confided to their care. If we open the books, in which are preserved the traditions of the earlier ages of the world, if we survey the heroic history of Greece, or the Roman Fasti, if we consult that of the people who to the middle age covered the soil of Germany and Gaul, if we listen to the recitals of travellers, the same tales of mystery and marvel greet our eyes in every page, and echo in our ears at every instant.

We find the dragon, consecrated by the religion of the earliest people, become the object of their mythology. "Rendered celebrated," says the eloquent Lacépède, "by the songs of Greece and Rome, the principal ornament of pious fables imagined in more recent times, conquered by heroes, and even by youthful heroines, who were contending for a divine law—adopted by a second mythology which placed the fairies on the throne of the enchantresses of old, became the emblem of the splendid actions of valiant knights; he has enlivened modern, as he animated ancient poetry.

"Proclaimed by the severe voice of history, every where described, every where celebrated, every where dreaded; exhibited under all forms, always clothed with tremendous power, and immolating his victims by a single glance; transporting himself through the midst of the clouds with the rapidity of lightning; dissipating the darkness of night, by the terrific splendour of his glaring eyes; uniting the agility of the eagle, the strength of the lion, the magnitude of the giant serpent; sometimes presented under a human figure, endowed with an intelligence almost divine, and adored even in our own days in the great empires of the east—the dragon, in short has been all in all, and every where to be found except in nature."

Such were the dragons, some of which were winged, and vomited flames, while others were deprived of feet; such were those which Pliny has asserted to exist in Ethiopia and in the neighbourhood of Mount Atlas; which Strabo pointed out in Spain; which according to Herodotus copulated by the head; which Elian assures us were the sworn enemies of the eagle; which Aristotle informs us, poisoned the air with their breath; and respecting which, Gesner, Micander, Aldrovandus, Nieremberg, Jonston, Charles Owen, and a crowd of other writers have put forth so many lying fables. We are now forced to deny the reality of their existence, and leave them to the embellishment of the images of romantic poetry. In our days nothing of the kind is to be seen. The progress of intelligence banishing the phantoms, dissipating the clouds which disturbed the imagination, destroying without mercy the innumerable errors connected with philosophical absurdity and religious prejudice, has driven the dragons to take refuge among nations not yet visited by the light of civilization.

Were we desirous to unravel the confusion which involves every thing connected with our present subject, the task would be too mighty and nothing would be performed. Let us remember, however, that even to the present time, the cabinets of the curious, the shops of pharmacopolists, the laboratories of alchymists, and the itinerant exhibitions of mountebanks, have presented animals of this kind, perfectly well preserved in appearance, and of forms the most singular and hideous. Even philosophers have witnessed such things, and avowed that the illusion was complete. But such representations are the pure effect of art. All these dragons are fabricated with rays from which certain parts have been removed, with which the head is fashioned and the mouth is cut—with which the lips are exhibited, covered with a sort of Mosaic work; with which the genital appendages, in the males, are extended in the form of paws, the vast pectoral fins of which are dried, and raised into wings.

Thus to a certain extent may be explained the figures of hydras with seven heads, of crowned basilisks, &c. which may be found in the writers of preceding ages. Conrad Gesner, for example, has represented one of these monstrous animals, brought from Turkey to Venice in 1530, and sent from there to the King of France. Aldrovandus and Johnston have also published similar engravings. Seba has given one of an heptacephalous hydra, which was for a long time in Hamburgh, and which he considered not to be the production of art, but which has since been most evidently recognized as such, as Linnæus has informed us in his Systema Naturæ.

We must not forget neither, that, among the Greeks, the word δρακων designated, in general, a great serpent; that some of the ancients have made mention of dragons which bore a crest and beard, which can hardly be applied, says M. Cuvier, to any thing but the iguana; that Lucan was the first who spoke of flying dragons, in allusion, without doubt, to the pretended flying serpents, the history of which is related by Herodotus; and that St. Augustine and other later writers have constantly attributed wings to the dragon.

It is after the ideas generally formed of these fabulous beings, that modern naturalists have given the name of dragon (*Draco*), to a genus of saurian reptiles of that family, called *Eumerodes* by M. Dumeril, and Iguanians by our author. The animals comprehended in it are distinguishable at the first glance from all the other saurians, because their first six ribs, instead of turning round the abdomen, are extended in a right line, and support a production of the skin, which forms a sort of wing analogous to that of bats, but independent of the four limbs.

The characters of this genus of reptiles may be thus expressed:

Two membranous wings, supported by extended ribs; body covered with small imbricated scales; those of the tail and limbs carinated ; the tongue fleshy, scarcely extensible, and slightly emarginated ; a long pointed wattle under the neck, supported by the tail of the hyoid bone ; on the sides of this last, two others, supported by the horns of the same bone ; tail long ; thighs destitute of porous grains ; a slight indentation on the nape ; four small incisors in each jaw, and on each side a long and pointed canine, and a dozen of cheek teeth, large and trilobed ; toes free and unequal, six in number.

The wings are plicatile, and are developed, like a fan, at the will of the animal. In a state of repose they are horizontal. They support the dragon like a parachute, when he leaps from branch to branch, but they have not sufficient force to strike the air and elevate him like a bird. The goitre under the throat is a sort of dilatable sac, narrow, and capable of being folded in circular and concentric wrinkles.

All the dragons are very harmless animals, of a small size, living in the bosom of the forests which cover some of the burning regions of Africa, and a portion of the great islands of the Indian ocean, particularly Java and Sumatra. In these deserted places they pursue the insect tribes with dexterity and quickness, and may be almost said to take them on the wing. They rarely descend to the earth, on which they crawl with difficulty. They always couple on the branches of trees, and the females deposit their eggs in the hollows of trees exposed to the south. Such is the report made by Van Ernest, a Dutch naturalist, to M. Daudin.

It would appear, according to the observation of M. Palisat de Beauvois, that the dragons are amphibious reptiles. This philosopher remarked one of them, among several, in the Kingdom of Benin, which he was unable to procure, because the animal was swimming in a river.

These reptiles belong exclusively to Africa and Asia. Seba has led naturalists into an error by saying that they are to be found in South America. The contrary is now clearly proved.

In 1811, M. Tiedemann published at Nuremberg a German dissertation in quarto on the anatomy and natural history of the dragon.

Draco Lineatus, is a very rare reptile, which inhabits the great woods of the Island of Java. The wings of *Draco viridis* are membranous, adherent to the base of the thighs, very broad, and each of them remarkable for six large emarginations; the scales under the body, the lower face of the limbs and tail, are carinated. The colour is an uniform greenish, with the exception of the wings, which are of a very pale brown, and are each of them marked with four transverse brown bands, fringed at their edges with little white points.

This species is a little smaller, and more slender than *Draco lineatus*, but the wings are broader. Seba first described it under the name of *the winged dragon of America*, and subsequently figured it as the *flying dragon of Africa*. Bontius has published a tolerably exact sketch of it, and that ancient traveller informs us that this pretty reptile, which is common enough in the Island of Java, inflates its yellowish goitres when it flies, that it may be more light in the air, without, however, being able to traverse any great space. It only shoots from tree to tree, a distance of about thirty paces, and produces by the agitation of its wings, a slight noise. But he adds that it is neither venomous nor mischievous. The inhabitants of Java handle it without fear, and without danger, and it often becomes the prey of serpents.

Shaw, in his Miscellany, has given the figure of a flying dragon, which appears to be the same as this, except, indeed, that there are several sharp spines upon the neck. He says that it inhabits Africa, and proceeds from tree to tree, leaping or rather flying after the manner of the flying squirrels. He also believes that this animal fills its goitre with insects, preserving them there for some hours, and afterwards feeding upon them.

The *Draco fuscus, or brown dragon*, is so called from its colour. It is a little longer and more thick than the green. The wings are broader and the tail is less elongated.

Naturalists have given the name of IGUANA to a genus of the present order. It is thus characterized:

Rounded toes, separated one from the other, and not opposable; body and tail covered with small imbricated scales; a pectinated goitre, compressed and pendant under the throat; all along the back a range of spines or rather of erect scales, compressed and pointed. The head covered with plates; a range of porous tuburcles on the thighs; a range of compressed triangular teeth, with indented edges in each jaw; two small ranges of teeth at the posterior edge of the palate; the tail without spines; the tongue fleshy, and emarginated at the top.

Thus the iguanas are easily distinguished from the cameleons, which have the toes opposable and united as far as the claws; from the stilliones, which have the tail spinous; from the lizards and agamæ, which have no goitre under the throat; from the dragons, which have membranous wings; and from the anolis and geckos, whose toes are flatted underneath. The name *iguana* originated in St. Domingo.

The reptiles which most naturalists have hitherto regarded as belonging to the genus iguana are tolerably numerous; but modern naturalists, after having examined and compared them with more attention than their predecessors, have transferred many of them to the agamæ, and formed separate genera of the *basilisk* and the *marbled iguana (Molychrus)*. We shall slightly notice the principal species which have been left.

The *Common American Iguana (Lacerta iguana, Lin.;*

tuburculata, Laurenti) is from four to five feet long. It is very common in all the warm parts of America, where it remains in the woods, at the environs of rivers, and sources of spring-water. It passes most part of its time on trees, sometimes going to the water, and living on fruits, grain, and leaves. Without being either venomous or dangerous, its bite is exceedingly painful; and when it is angry, the goitre which it has under its neck becomes distended and expanded.

This reptile has great tenacity and endurance of life, and will resist the blows of a stick or cudgel very well. Accordingly, it is usually hunted with the bow or the gun.

The females are smaller than the males, but their colours are much more brilliant. They lay eggs in the sand, about as large as those of pigeons, but a little longer, and of equal thickness at both ends. The shell of these eggs is white, even, and soft. They are entirely filled by the yolk, and can hardly be said to have any albumen. They never harden by fire, but only become a little pasty. But their flavour is very agreeable, and they are constantly eaten in Surinam and Guiana. A single female will lay about six dozen.

The flesh of the iguana is considered as delicious, and is in great estimation throughout all the warm parts of America. It is white and delicate. Many persons, however, consider it as unwholesome, especially for those who are infected with syphilis, some symptoms of which, such as pains in the bones, &c. it is supposed to aggravate or cause the return of. At Paramaribo, it is sold extremely dear, and highly thought of by epicures.

Pison, and many other of the old travellers in America, have spoken in high terms of the virtues of the bezoar of the iguana, a kind of stone, found, say they, in the stomach or cranium of this reptile. But, at the present day, this substance is fallen into the most absolute disrepute among all medical practitioners.

The *slate-coloured Iguana, Ig. cerulia* is but three feet in length. It inhabits the same places as the former species, and may, as our author thinks, be merely a variety of it, in age or sex. Seba derives it from the Island of Formosa.

The horned Iguana of St. Domingo is about four feet long. It is frequently found in the hills of St. Domingo, between Artibonite and Gonaïves. It lives on fruits, insects, and small birds, which it seizes with marvellous agility, and during the day it couches on trees and rocks to watch for its prey. During the night, and the entire season of the great heats, it retires into the hollows of rocks, or into the holes of old trees, and it passes about five or six months of the year there in a state of lethargy.

This reptile is considered by the negroes as a delicious meat, and is accordingly sought after by them with great avidity. According to the report of the colonists, its flesh resembles in flavour that of the roebuck, and the maroon dogs make great slaughter among these reptiles. The colours of this iguana are not precisely known. M. de Lacépède was the first to describe it at the end of his Natural History of Serpents, and Bonaterre subsequently gave a good figure of it in the Dictionary of Erpetology, in the "Encyclopédie Méthodique."

Some authors place here the *iguana fasciata*. Its colour is deep blue, with transverse bands of a clearer tint. The goitre is moderate, and not denticulated. There is no large scale at the angle of the jaw.

This iguana belongs to the island of Java. It may probably be the reptile which Bontius has named *cameleon*. It is also probable, that to this species must be referred the very large iguanas which are found at Batavia, and which are sometimes as thick as a man's thigh. In his voyage with Captain Cook, Sir Joseph Bankes killed one of these, which was five feet in length.

The flesh of this iguana is eaten in the East Indies, as that of the common iguana is eaten in America. The eggs are likewise in great estimation.

Under the name of Basilisk is at present designated a genus of reptiles, of this saurian order, which exhibits many affinities with the iguanas and monitors.

No animal, perhaps, has been the subject of so great a number of prejudices as the one now under consideration. The most ancient authors have spoken of the basilisk, as of a serpent which had the power of striking its victim dead by a single glance. Others have pretended that it could not exercise this faculty, unless it first perceived the object of its vengeance before it was itself perceived by it. It was also most absurdly imagined to proceed from the eggs of old cocks. Aldrovandus, and several other writers have given figures of it. They have represented it with eight feet, a crown on the head, and a hooked and recurved beak. Pliny assures us that the serpent named basilisk has a voice so terrible, that it strikes terror into all other species, that it thus chases them from the spot which it inhabits, and of which it retains the sole and undisputed dominion. The name, indeed, of basilisk, Βασιλικος, signifies royal.

The fantastic forms, and fabulous properties thus attributed to an animal, which most probably never had any existence, rendered this name too celebrated for naturalists not to endeavour to apply it to another species, which accordingly they did. Seba has figured a species of lizard, whose head is surmounted with projecting lines, and the back furnished with a broad vertical crest, which extends as far as over the tail, and which that author believed to be intended for the purposes of flight. He has designated it under the name of basilisk, or dragon of America, a flying amphibious animal. This is the animal which has subsequently been described in all works of Natural History under the name of basilisk.

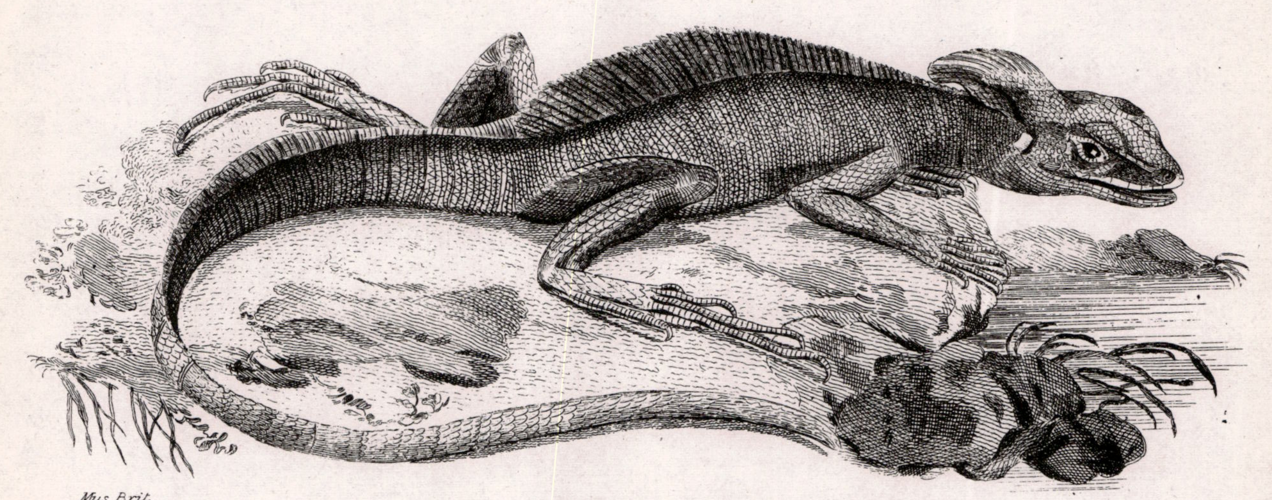

Mus. Brit.

THE MITRED BASILISK.

B. MITRATUS.

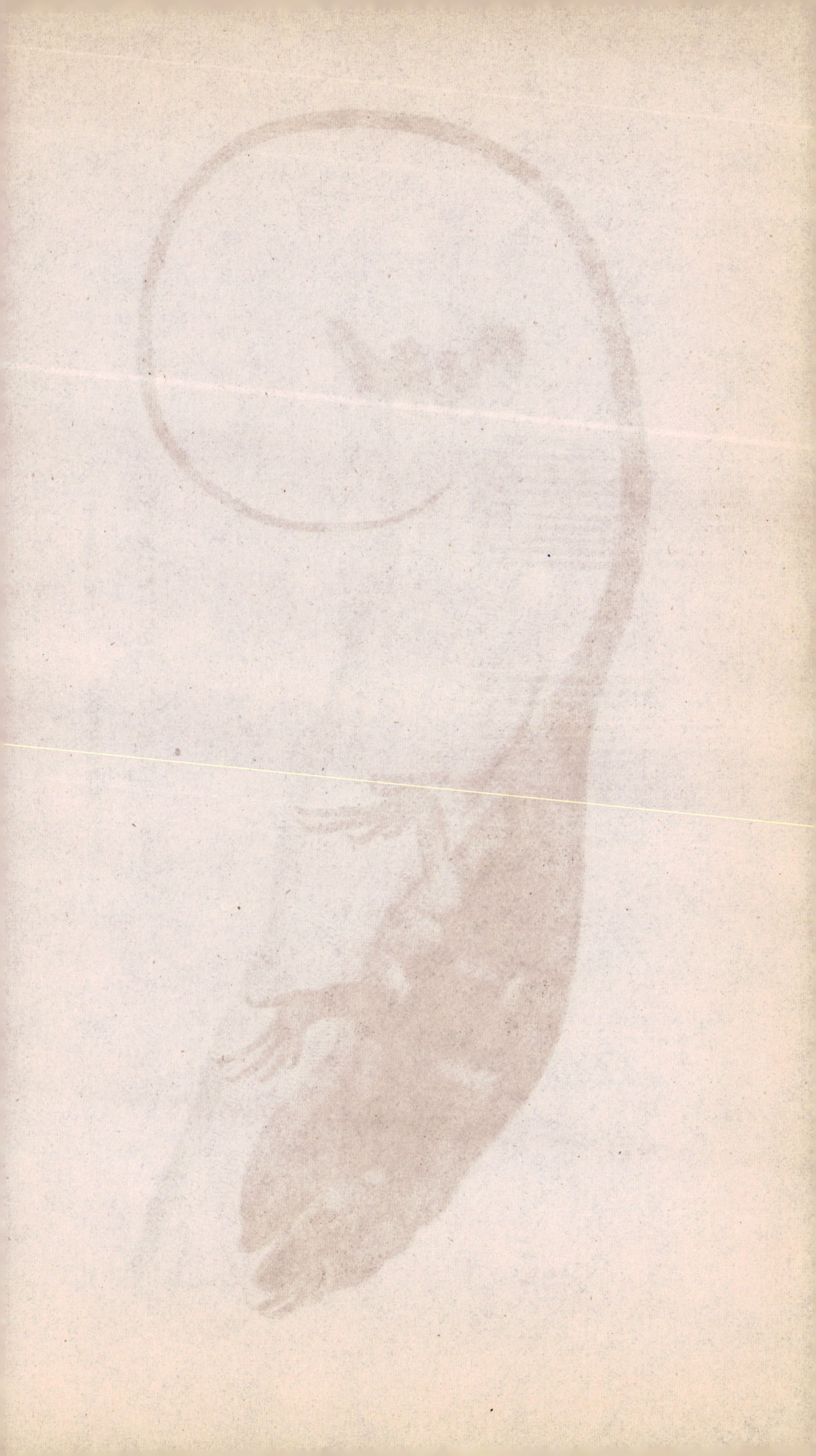

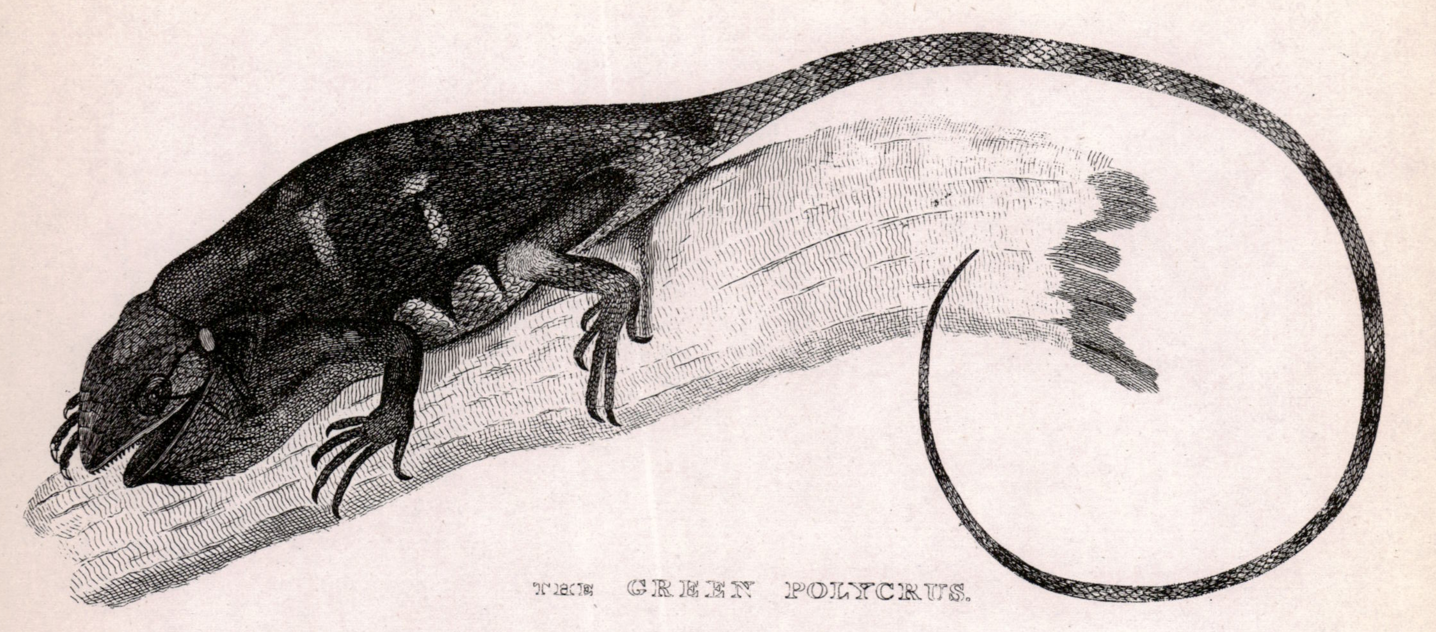

THE GREEN POLYCRUS.

P. VIRESCENS. — *Pr. von Neuwied.*

The individual described and figured by Seba, formed a part of the collection ceded by Holland to France, when France was the plunderer of nations; when, not contented with stripping them of their treasures, and forcing their children into her ranks, to combat for a perpetuity of bondage, emulous of the rapacity of ancient Rome, she also robbed them of the monuments of art, and the acquisitions of science. The specimen in question is at present deposited in the Museum of Natural History in Paris.

Laurenti was the first writer who considered this lizard as proper to form a separate genus. Linnæus arranged it among the stelliones.

This genus is very easily distinguished from almost all the other saurians, by the form of the tail, which is long, and compressed from right to left; from the crocodiles and the dracæna, because the scales which cover its back are pretty nearly similar to those of the rest of the body; from the monitors, by the crest which predominates along the entire back; and from the lophyri, because this crest is furnished with osseous radii.

The manners of the basilisks are very little known. The form of their tails sufficiently indicates that they live on the banks of waters, and that they can employ it for the purposes of swimming. It is probable that they live on snails and insects, like the majority of the lizards.

We have illustrated this genus, as well as that of POLYCRUS, or the basilisks without crest, by figures from specimens in the British Museum.

The genus of the ANOLIS was first established by Daudin. Our information concerning the manners of the species is rather limited. The one called *lacerta bullaris* is very much extended in the Antilles, in Mexico, and in Carolina, where M. Bosc has frequently had occasion to observe it. M. de Lacépède is of opinion that it approximates very much in conformation to the grey lizard. It never grows large, and attains to the length only of a few inches.

It lives generally in gardens; its agility is excessive, and it climbs with great facility, by the assistance of its crooked and long claws. It seeks out humid places, and runs among the stones. It holds the tail habitually raised above the back.

When it has run considerably, and is fatigued, it opens its mouth and lolls out the tongue just like a dog. This probably is the reason why it has been called in French *roquet* (a cur-dog). It destroys numbers of insects, and we are even assured that it feeds on the eggs of tortoises, and of other lizards. When it is angry, or violently agitated, it swells its throat, which becomes red, and utters a kind of dull sound, rather singular in its effect.

These lizards often fight together. When two of these animals attack each other, says M. de Lacépède, it is always with boldness. They appear to threaten, by rapidly agitating the head. The throat swells; their eyes sparkle, and they seize each other with fury, and fight with bitterness. They usually have, as spectators of their combats, and perhaps as witnesses of their power, the females who are to constitute the prize of victory. The weakest takes to flight, his enemy pursues him with vivacity, and devours, if he overtakes him. But sometimes he is only able to seize him by the tail, which breaks easily in his mouth, and which he swallows. Those which have been thus mutilated, grow timid, feeble, and languishing. Pere Nicolson has given the same details in his history of St. Domingo.

We have inserted the figure of an Anolis from a specimen in the British Museum. It is the *Anolis Edwardsii* of Merrem, from the West Indies.

We shall now consider the GECKOS, which our author has so considerably subdivided. This genus contains a great number of species, spread throughout the hot countries of both continents. Their singular conformation, their dull and

EDWARD'S ANOLIS.

ANOLIS EDWARDSII.

melancholy aspect, and their resemblance to toads and salamanders have caused them to be hated and feared, and have prevented naturalists from confounding them with other saurians.

The geckos are found in South America, in Africa, and in the Indies. Naturalists have divided them into many families. We shall consider the most interesting species, without much attention to order.

The *Smooth Gecko, Gecko Levis,* is so called because the skin is covered with a multitude of very small scales, which render it smooth, and, as it were, satiny, and which are rather more distinct and rounded under the belly and tail. The tail, as is mentioned in the text, is easily broken, and shoots forth again in the form of a little root, or radish. It inhabits Surinam.

The *Geitje* (*Lacerta Geitje*, Sparman) has no claws at all, no more than many others of the platydactyli. This gecko inhabits the Cape of Good Hope, where it is regarded as being exceedingly venomous. We are, in fact, assured, that its bite produces a terrible kind of leprosy, which invariably terminates in death, and that its effects do not become manifest until about the end of a year, or six months at soonest. Nevertheless, Sparman tells us, in his voyage to the Cape of Good Hope, that though this animal is frequently seen in spring near Gorée river, and in other places, the maladies caused by its bite are not frequently spoken of.

The inhabitants assured the Swedish traveller, that near Sitsikamma, this animal nestles usually in the empty shells of the *bulla achatina.*

The tail of this reptile will detach itself, and fall at the slightest shock. Its motions are slow.

The *Wall Gecko* (*Lacertus Facetatus*, Aldrovandi) is a most hideous reptile. It conceals itself in holes of walls, and heaps of stones, covering its body with dust and ordure. It inhabits all around the Mediterranean sea, even as far as Provence and Languedoc, where it is common, according to Olivier.

The Wall Gecko courts the heat, and avoids low and humid stations. Accordingly it is frequently found under the roofs of ruined houses, and old habitations of all kinds, where it passes the winter, without, hovever, falling into a perfect state of lethargy. In the first days of spring, it issues from its retreat, and proceeds to warm itself in the sun; but at the slightest noise, or when it commences to rain, it retires again into its hole.

This animal, which is tolerably agile, feeds on insects, and easily fastens itself to the walls, by the assistance of its crooked claws, and the scales with which the under part of the toes are furnished. Accordingly it is sometimes seen to walk in a backward position, along the ceilings of rooms, or remain a long time motionless under the vaulted roofs of churches, as Olivier has observed. It has been reported, but erroneously, to have been venomous. It utters no sound.

The *Lacerta Sputator*, which some naturalists have referred to the anolis, is placed by our author among the geckos, in the family *Spherodactyli*. This saurian, of a small size, is common in all the islands of America. Degur received one from the island of St. Eustache, in 1755. M. Bosc possessed one from St. Domingo. Aguilar says that it inhabits South America, in houses, and in timber, and that it is named in our language *Woodslave*. It climbs and runs with agility along the walls. Sparman, who first made it known, says, that if one approach too near, or disturb it, it shoots at the indiscreet intruder a black spittle, of a poisonous nature, which causes the part on which it falls to swell. This assertion does not appear to be worthy of much credit. It never issues from its hole but during day-light.

Daudin refers, provisionally, to his *Gecko Spinicauda*, a venomous gecko of the Indies, mentioned by Bontius and Valentyn, employed by the inhabitants of Java to poison their arrows. The first of these observers tells us that the bite of this hideous reptile is so dangerous, that if the part affected be not excided or burnt, death will ensue in a few hours. Its urine is also said to be one of the most corrosive poisons. Its blood and saliva, yellow and thick, are regarded as equally deadly.

The *House Gecko* (*Lacerta Gecko*, Hasselquist) is common in all the humid and gloomy parts of houses in the different countries which border the Mediterranean sea to the south and east, in Egypt, in Arabia, in Syria, and in Barbary, from whence it has subsequently spread, through the various countries of Southern Europe. At Cairo they name this Gecko, *Abou-burs* (father of the leprosy), because they pretend that it communicates this malady by poisoning with its feet provisions of all kinds, but more especially salted provisions, of which it is extremely fond. It produces redness and inflammation, by walking on the skin.

Its voice resembles the croaking of the frog, and the cry, it is said, may be expressed by the syllables *geck-o*.

Ancient writers, in speaking of this reptile, have attached too much importance to the fables related concerning it by the natives of the Levant. Bontius, for instance, was totally wrong in saying that the gecko could impress its teeth on the hardest bodies, even on steel. They are not even sufficiently strong to pierce the skin.

It is neither from its bite, its saliva, nor its urine, that this animal is hurtful. Hasselquist has remarked, that it is through the lobules of its toes that the poison exudes. This author, in 1750, saw at Cairo two women and a girl who were at the point of death, in consequence of having eaten some cheese over which this reptile had crawled. At another time he saw the hand of a man who would lay hold of a gecko, instantaneously covered with red pustules, inflamed, and accompanied with an itching equal to that produced by the stinging of a nettle.

We are told that the cats pursue the gecko, and feed upon

it. It is driven from the kitchens in Egypt, by keeping there a large quantity of garlick. It feeds on insects, and its eggs are about the size of a small nut.

The genus of the CAMELEON was confounded by Linnæus with the lizards. It was first separated from them by M. Alexandre Brogniart. As may be seen from the text, there are three peculiarities in these animals, any one of which would suffice for the establishment of a genus. These characters are derived from the conformation of the tongue, the toes, and the tail—but the text, in this place, has rendered it unnecessary for us to enter on merely descriptive particulars.

Authors are not agreed on the etymology of the word *cameleon*, which has come to us from the Greeks. Some maintain that it signifies *little lion*, others, that it corresponds to *camel-lion*, each supporting their argument with a pertinacity and violence, exactly proportioned to the absence of proof.

Without attaching too much consequence to the word, let us examine the thing. There are probably no animals whose names are more, or which have given rise to so many comparisons and allegories as the cameleons, the dragons, the basilisks, and the salamanders. A long list might easily be formed of the prejudices, errors, and falsehoods, which have prevailed and been published respecting them.

One species of this genus has been known from all antiquity, and has been long celebrated for its supposed faculty of living upon air, and changing colour according to the bodies to which it approximates. Observation, at the present day, has done justice to those fables of which this animal was the object. But in the language of the orator and the poet, the *cameleon* continues to be the emblem of those hypocrites who adopt the modes of thinking and acting of men in power, changing at will, whenever it is necessary for the purposes of their vile ambition, and of those flatterers, who, taking as it were the colour of circumstances, view with an attentive eye the person whom they fear, or from whom they entertain expectations; who never make a single step, which they do not believe to be advantageous to themselves, and who proceed like the cameleon, feeling their way with caution and servility. Such are stigmatized by La Fontaine as,

"Peuple caméléon, peuple singe du maître."

The cameleon has also furnished a comparison for those inconstant and capricious persons, whose inclinations are perpetually changing, not from interested motives, but from mere unsteadiness of mind, or those, who unstamped by the hand of nature with any original character, fall almost unconsciously into the tone of the society in which they are placed, and thus become "all things to all men."

It is only in the warmest parts of Africa and Asia, that the cameleons are to be found. It is very probable that there are none in America, though Seba talks of them as belonging to that quarter of the globe. Africa has been thought to be the region in which they most abound.

The disposition of their toes, gives to these animals, a very great facility in seizing the branches of trees, and remaining perched there like birds; while their long and prehensile tail, serves to give a still greater fixedness and solidity to their position.

The walk of the cameleons is very slow. They are sometimes to be seen for entire days on the same branch. It is only with great circumspection, and after having felt their way, and fixed themselves strongly with the tail, that they will venture on a few paces. This slowness of motion, and their total want of arms offensive and defensive, render them an easy prey to the numerous enemies who may choose to attack them. Accordingly an immense annual destruction of them takes place, and but for their amazing fecundity, it is probable that the species would shortly become extinct.

The cameleons live on insects, and principally upon flies. They seize them with vivacity by means of their long and gluey tongues, and bruise them between their jaws. Like other reptiles they can remain for months without eating, which has given rise to the opinion that they live on air. But to this power of fasting there are limits, and after a time they will succumb under the want of food. Golberry, who when in Senegal, instituted the most rigorous experiments to ascertain how long the cameleons could live without eating, has established four months as the maximum. They lay about a dozen eggs, which the female deposits in the sand, where the young are excluded by the simple operation of the solar heat.

The exact duration of the life of the cameleon is unknown —but we may presume, that but few individuals arrive at the term fixed by nature to their existence, or in other words die a natural death, since, as we have just said, they cannot, but by the greatest chance, escape the numerous enemies that wage perpetual war against them; so that to use the words of a French writer, " un *caméléon* aperçu est un *caméléon* perdu." In countries where cold sometimes prevails a little, such as Lower Egypt, and the coasts of Barbary, they conceal themselves in winter in holes under heaps of stones, where they remain in a state of perfect immobility, but without being asleep, or in a lethargy, like most other reptiles.

The Indians and Africans consider the cameleons as useful animals, and love to see them around their houses destroying those insects by which they are tormented. They never do them any injury, and are fond of caressing them. The cameleon on its part, is extremely gentle. One may take it in his hand, and put his finger into its mouth without any apprehension or danger of being bitten. Some assert that it cannot utter any thing like a genuine cry—others, that it sends forth a slight hissing noise when it is surprised and laid hold of.

"But," says M. de Lacépède, "whether the cameleon climbs along the trees, whether, concealed under the leaves, it waits patiently for the insects which constitute its food, or whether, in fine, it walks upon the ground, in every situation it appears to equal disadvantage. It presents neither agreeable proportions in its form, nor grace, nor lightness in its gait. It moves with the utmost circumspection. As it cannot grasp the branches on which it is desirous to climb, it assures itself, at every step, that its claws have well entered into the fissures of the bark. When on the ground, it gropes along, never raising one foot until it is certain of the position of the other three. All these precautions impart a sort of gravity to its gait, at once ludicrous and awkward."

The cameleon then would never have attracted the attention of those who confine their observation to the most prominent objects of the animal kingdom, if the faculty of presenting according to its different states, colours more or less varied, had not rendered it celebrated for so long a period.

These colours, in fact, change with equal frequency and rapidity, but it is by no means true, that they are determined by those of surrounding objects. Their shades depend on the volition of the animal, on the state of its feelings, on its good or bad health, and are besides, subordinate to climate, to age, and to sex.

It was believed in the time of Pliny, that no animal was so timid as the cameleon, and, in fact, not having as we have observed, any means of defence, and being unable to secure its safety by flight, it must frequently experience internal fears and agitations more or less considerable. Its epidermis is transparent; its skin is yellow, and its blood of a very lively violet blue. From this it results, that when any passion or impression causes a greater quantity of blood to pass from the heart to the surface of the skin, and to the extremities, the mixture of blue, violet, and yellow, produces, more or less, a number of different shades. Accordingly, in its natural state, when it is free and experiences no inquietude, its colour is a fine green, with the exception of some parts, which present a shade of reddish brown or greyish white. When in anger its colour passes to a deep blue green, to a yellow green, and to a grey more or less blackish. If it is unwell, its colour becomes yellowish grey, or that sort of yellow which we see in dead leaves. Such is the colour of almost all the cameleons which are brought into cold countries, and all of which very speedily die. In general, the colours of the cameleons are so much the more lively and variable as the weather is warmer, and as the sun shines with greater brilliancy. All these colours grow weaker during the night. Such are the observations made by Opsonville and Golberry, and which have been repeatedly verified on an animal of the same family, but of a different genus, by M. Bosc. This was the *Lacerta Bullaris*, which is equally of a clear green in its natural state in warm weather, and which changes at will, and very rapidly, to a black green, to a yellow green, to grey, and to brown, according as it is affected by strange objects which have the power of agitating it. In cold weather it is of a grey colour, shaded with brown in some parts, and it has no longer the faculty of varying its tints, because its blood can no longer come to the surface of its skin to modify the yellow by which it is coloured. During the winter, in this country and in France, the same is positively the case with the cameleons.

The cameleon possesses another property which merits a

THE LOBED CAMELEON

CHAM. DILEPIS. Leach.

Mus. Brit.

THE FORKED-NOSE CAMELEON. *CHAM. BIFURCUS.*

particular examination. It can inflate at will the different parts of its body, so as considerably to increase its entire volume. This, in all probability, with its colour resembling the leaves among which it dwells, are the only feeble means of defence vouchsafed to it by nature, who appears in all else to have been a step-dame to this harmless reptile.

It is by slow and irregular motions, and not by progressive oscillations, that the cameleon swells itself out. It fills itself with air so as to double its diameter. This inflation extends even into its feet and tail. It remains in this state sometimes during two hours, gradually diminishing in bulk from time to time. Its dilatation is always more sudden than its compression. It is more than probable that this phenomenon takes place by the introduction of the air from the lungs between the epidermis and the skin; but there are no positive observations on this subject so very worthy of the researches of travellers. It is certain, at least, that these animals can also very considerably inflate their lungs; for those who have dissected them are very far from being in accordance respecting the volume of that organ, some pronouncing it to be very small, and others extremely bulky.

We have illustrated this genus by two of its species from specimens in the British Museum. The *C. Bifurcus* is described at page 61. *C. Dilepis* is so named by Dr. Leach; the specimen was brought by Mr. Bowdich from Africa.

Among the various species of the Skinks we shall slightly notice the following:—

The *skink* of pharmacopolists, as it is termed by our author, *the common skink* (*scincus officinalis*), does not exceed the length of six or eight inches, and is a native of Nubia, of Abyssinia, Syria, and Egypt, from which, by the way of Alexandria, commerce has extended it throughout all Europe, but more especially throughout Asia. It also appears to frequent the coasts of Barbary, and, perhaps, even

Sicily, some of the islands of the Archipelago, the environs of Smyrna, and some provinces of India. When it is menaced, it digs for itself a hole in the sand with so much promptitude that, according to Bruce, one would think that it rather found the opportunity of disappearing in a retreat already existing, than the means of preparing one for itself. It is fond of stretching itself in the sun, and runs in a creeping pace. In Arabia it is named *el adda*, and *dhab* in Abyssinia.

For a long time the skink has been regarded by the Arabian physicians and their followers as a sovereign remedy against certain maladies. Before this it was extolled by Pliny as a specific for the wounds caused by poisoned arrows; subsequently it has been vaunted as an aphrodisiac, and quackery or ignorance has placed it in the rank of those medicaments which merit the distinguished honour of being employed to reanimate the exhausted powers, and to rekindle the fires of love when extinguished by the frosts of age, or the excesses of debauchery. Its flesh has been administered as depurative, excitant, anthelmintic, analeptic, anti-cancerous, sialagogue, and anti-siphyphilitic. Notwithstanding that this confused mass of medical properties, thus put together without discrimination, as if to form the *vade mecum* of some empiric, now appears completely ridiculous, yet even at the present day, in many countries, fables are still published respecting the success of this remedy. In spite, however, of the discredit into which it has fallen among the faculty in general, it does not appear to be totally devoid of efficacy in some complaints.

The oriental physicians still continue to recommend it against elephantiasis, and all cutaneous maladies; also against ophthalmia, and even cataract.

After so many virtues, real or pretended, it is not at all astonishing that in the south of Egypt the skink should be

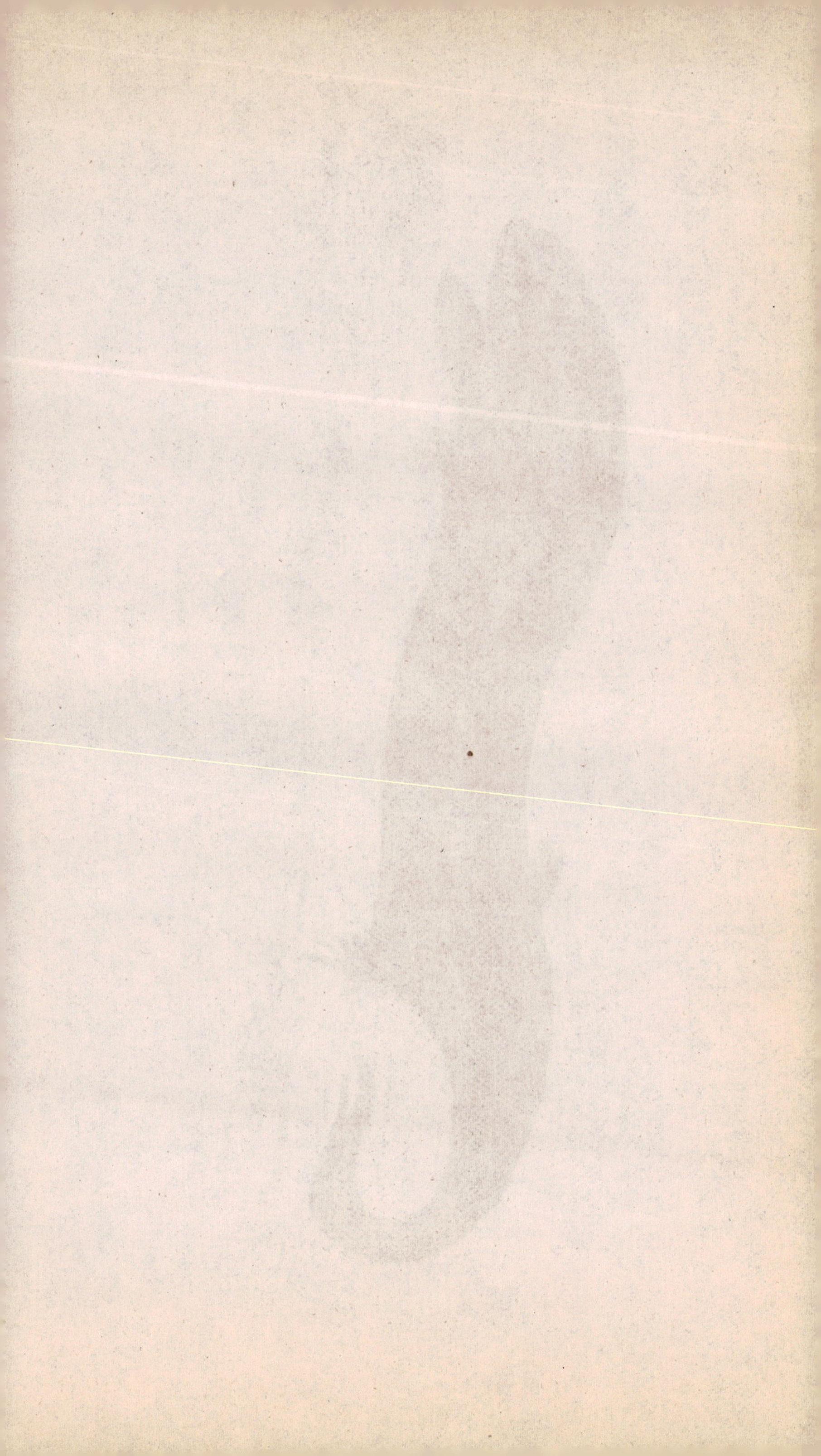

Mus. Brit.

THE TILIQUA OF MALTA.

SCINCUS TILIQUA.

hunted with a sort of fury. The inhabitants of the deserts cause it to be dried, and send it thus to Grand Cairo and Alexandria, where the pharmacopolists of Europe and Asia come to provide themselves with it.

White, under the denomination of the *scincoïd*, has spoken of a reptile of New Holland, which has the greatest relations of form, of size, and even of colour, with the Egyptian skink, and which Daudin believes to be only a variety of it.

There is a species called *galley-wasp*, in the West Indies (*Lacerta Occidua*, Shaw), which is placed among the skinks. It is considerably larger than the last, being more than a foot in length. It frequents the Antilles and Jamaica in particular, where, according to Sloane, it is amphibious, and is attached to marshy situations. It is at least partial to humidity, and conceals itself under the rocks. Like many other saurians it has also the name of *mabouia*, which the negro slaves bestow upon all hideous and maleficant beings.

In Jamaica, it is believed, but without adequate evidence, that the bite of this large skink is extremely venomous, and causes immediate death.

There is another skink, also called *mabouia*, but, as our author remarks, improperly so called, in the West Indies, and which the negroes in like manner consider as maleficent. It climbs trees with dexterity, and runs rapidly over the cabins of the natives and slaves ; but its most usual retreat is in the holes of old rotten trees, from which it seldom issues but during the very hot weather.

Among the skinks destitute of palatal teeth, is the species we have figured, under the name of *Scincus tiliqua*, from a specimen brought from Malta by the Rev. Mr. Heunah.

The word Seps is employed to designate an animal which some have considered as a lizard, and others as a serpent. It comes from the Greek word σηπεῖν, to corrupt. Anatomy approximates the seps to the skinks, but the habits, movements, and mode of living, of these reptiles, indicate the strongest affinity to anguis. Among them,

The *Anguis Quadrupes* is a native of Africa. It is of a shining and greyish colour, and crawls after the manner of serpents. In Barbary its bite is regarded as very dangerous, but without foundation.

The *Cicigna (seps tridactylus)* inhabits the southern parts of France, Provence, Italy, Sardinia, and many countries of Africa. Ray found it on the sand of the shore not far from Leghorn. Nicander mentions its existence in Lybia, in Syria, and in the island of Cyprus, and Imperati in the marshy meadows of Campania. The ancients believed that its bite was mortal, particularly to mares, and this prejudice exists among the Sardinians at the present day. Sauvages and Cetti have proved that it was an animal altogether innocent. According to the report of Columna and Imperati, it appears to be viviparous. It burrows in the earth on the approach of winter, and falls into a lethargic state. There is a new species in the British Museum, described in the table under the name *Cicigna Burnettii.*

The Bipeds still marked the gradation to the ophidian order, if, indeed, both they and the bimana may not properly be regarded as belonging to it; for, as Latreille has well observed, these pretended feet, which have no claws, and are extremely short, may rather be considered as a prominence, or even as the organs of generation, than as feet. At all events, our judgment must remain suspended respecting those animals, until repeated observations on many individuals shall enable us to fix it.

The Chalcides, although they have four very short feet, and sometimes claws, are yet extremely remote from the rest of the saurians, and are closely allied to the ophidians, especially to the amphisbenæ. Of the manners and habits of this genus, or of the bipedes and bimana, we have nothing to communicate.

THE THIRD ORDER OF REPTILES.

The Ophidians, or Serpents,

Are reptiles without feet, and, consequently, those which best merit the appellation of reptiles. Their body, greatly elongated, moves by means of the folds or windings, which it makes on the ground.

They must be divided into three families.

Those of the first, or

Anguis,

Have still the osseous head, the teeth and tongue resembling those of seps. The eye is furnished with three lids, &c. They may, in fact, be considered as seps without feet.

They all enter into the genus of the

Snakes (Anguis, Linnæus),

Characterized externally by imbricated scales, which cover them completely. They are formed into four subgenera, the first three of which have under the skin, the bones of the shoulder and pelvis.

Scheltopusik (Pseudopus,* Merem.)

Have the tympanum visible externally, and on each side of the anus a small prominence, in which there is a little bone analogous to the femur, and appertaining to a true pelvis, concealed under the skin. As to the anterior extremity, it can hardly be said to be visible, there being only a fold, not easily remarked, and no interior humerus. One of the lungs is a quarter less than the other. The scales are square, thick, semi-imbricated, and between those of the back and belly, are some smaller ones, which produce a longitudinal furrow on each side

Pallas has made known a species of the South of Russia, which is also found in Hungary and Dalmatia (*P. Pallasii*, Nob. *Lacerta Apoda*. Pall. Nov. Com. Petrop. xix. pl. ix. f. 1.), one or two feet in length. The scales of the back are smooth; those of the tail are carinated.

M. Durville has discovered another in the Archi-

* *Pseudopus* (*false foot*). I have not been able to perceive, more than M. Schneider, any division at the extremity of this little vestige of foot.

pelago, in which the scales of the back are rough, and carinated like those of the tail. *Ps. Durvillii*, Nob.

A neighbouring subgenus, that of

OPHISAURUS, Daud.

Does not differ from the Sheltopusiks, but in having no appearance of posterior extremities. But the tympanum is still visible, and the scales also have a fold on each side of the trunk. The smaller lung makes one-third of the greater

The species most anciently known (*Oph. Ventralis, Ang- Ventralis*, Lin. Catesb. II. lix.) is common in the South of the United States. Its colour is a yellowish green, spotted with black above. Its tail is longer than the body. It breaks so easily, that this animal has been termed the *glass serpent*.*

THE SNAKES, proper, (Anguis, Cuv.)

Have also no appearance of extremity visible without. Their tympanum is even concealed under the skin. Their maxillary teeth are compressed and crooked, and they have none in the palate. Their body is surrounded with imbricated scales, without any fold on the side. One of the lungs is one-half less than the other.

* Add *Ophis Punctatus*, *Ophis Striatulus*, Nob., two new species.

We have one species of them very common in all Europe, (*Anguis Fragilis*, L. Lacep. II. xix. 1.) with very smooth, shining scales; silvery yellow above, blackish underneath, three black threads along the back, which change with age into divers series of points, and end by disappearing totally. Its tail is of the length of the body. This animal reaches to the size of a foot and some inches. It lives on worms and insects, and is viviparous.*

These three subgenera have still an imperfect pelvis, a small sternum, an omoplate, and a clavicle, concealed under the skin.

The absence of all these osseous parts, also necessitates the separation from the snakes of the subgenus which I shall name.

Acontias,†

And which resembles them by the structure of the head and the eyelids, but which has no sternum nor vestige of shoulder or pelvis. Their anterior ribs are united, one to the other under the trunk, by cartilaginous prolongations. I have found in them but one moderately sized lung, and one very

* The *Anguis Erix*, L. is only a young common snake in which the dorsal lines are still well marked; and the *Anguis Clivicus*, of which Daudin makes an erix without any obvious reason, is an old common snake, with truncated tail. It is only mentioned after Gronow, who quotes the coluber of Gesner, which is the old common snake.

† *Acontias* (Javelin), the Greek name of a serpent, which was supposed to dart like an arrow on the passengers. (ακοντιζω, *jaculor.*)

small. Their teeth are small and conical. I think I have perceived a few in the palate. These are easily to be recognized by their muzzle which is enclosed as it were in a kind of mask.

The species well known, (*Anguis Meleagris*, L.) Seb. II. xxi. 1,* comes from the Cape of Good Hope. It resembles our snake, but its obtuse tail is much shorter. Upon its back are eight longitudinal ranges of brown spots. The same country produces other species, one of which is entirely blind. (*Ac. Cæcus*, Cuv.)

The second family, or that of

THE TRUE SERPENTS,

Which is by far the most numerous, comprehends the genera without sternum, or vestige of shoulder, but whose ribs surround a great part of the circumference of the trunk, and in which the bodies of the vertebra are articulated, each by a convex facet, into a concave facet of the following one. They are destitute of the third eyelid, and of the tympanum, but the osselet of the ear exists under the skin, and its handle passes behind the tympanic bone. Many of them have under the skin, a vestige of posterior

* Daudin has also made an *erix* of the *Anguis Meleagris*, but without any reason, for its lower scales are not larger than the others. I have ascertained, by dissection, that this serpent has not the sternum which M. Oppel has attributed to it.

member, the extremity of which even appears in some externally, in the form of a little crook.*

We subdivide them into two tribes.

That of the Double Walkers has the lower jaw supported, as in all preceding reptiles, by a tympanic bone immediately articulated to the cranium—the two branches of this jaw are soldered in front, and those of the upper jaw are fixed to the intermaxillary bone, which prevents their mouth from dilating as in the following tribe, and makes the head all of a piece with the rest of the body. This form enables them to move either backwards or forwards. The osseous framework of the orbit is incomplete behind, and the eye very small. As for the rest, their body is covered with scales, the anus situated very near its extremity, the trachea long, the heart very far back. None are known to be venomous.

There are two genera, one of which approximates to the chalcides and bimana, the other to the snakes, and acontias.

Amphisbœna,† L.,

Have the entire body surrounded by circular ranges

* See the German dissertation of M. Meyer, on the posterior extremities of the ophidians, in the *xiith vol. Curieux de la Nature* of Bonn.

† From αμφις and βαινειν, walking in both directions. The ancients believed that it had two heads. This name has been erroneously applied to serpents of America, unknown to the ancients.

of quadrangular scales, like the chalcides and bimana among the saurians, a range of pores in front of the anus, teeth not numerous, conical, in the jaws only, and not in the palate. There is but a single lung.

Two species have been for a long time known. (*Amph. Alba.* Lacep. II. xxi. 1, and *Amph. Fuliginosa,* L.) Seb. II. xviii. C. 3, and lxxiii. 4. Both of South America. They live on insects, and often remain near ant-hills. This gives rise to the popular belief that they live on the large ants. These amphisbœna are oviparous.*

There is one of Martinique entirely blind. (*Amp. Cœca.* Cuv.†)

The Leposternons, Spix, are amphisbenæ, the anterior part of whose trunk has underneath a union of some plates which interrupt the rings. They have no pores in front of the anus. Their head is short. Their muzzle a little advanced.‡

Typhlops, Schn.,||

Have the body covered with small imbricated scales,

* *Amp. Flavescens,* Pr. Max, 9th book.

† May not this be *A. Vermicularis?* Spix, xxv. 2. He says *oculi vix conspicui.* I can discover none: he uses the same expression for his *A. Oxyura.*

‡ *Lep. Microcephalus,* Spix; or *Amphis Punctata,* Pr. Max.

|| Τυφλοψ τυφλινη (*blind*), were the names of the snake among the Greeks. Spix has changed this name into Stenostona.

like the snakes, with which they were for a long time classified, the muzzle advanced and furnished with plates, the tongue rather long and forked, the eye a sort of point, hardly visible through the skin, and the anus almost entirely at the extremity of the body. One lung is four times larger than the other. These are small serpents, which, on a careless glance, resemble earth-worms. Their species are to be found in the hot climates of both continents.*

There are some whose head is of one piece with the body, and obtuse. They resemble the ends of slender packthread.†

The majority have the muzzle depressed and obtuse, and furnished with several plates in front.‡

In some, the front of the muzzle is covered with a single large plate, with an interior edge, a little trenchant.||

Finally, there is one whose muzzle is terminated by a small conical point, and which is entirely blind.

* I have not been able to discover any teeth in those which I have examined.

† *T. Braminus*, Nob. or *Rondas Talaloopam*, Russ. Serp. Corom. xliii. or *Eryx Braminus*, Daud. or *Tortrix Russelii*, Merr.

‡ *Ang. Reticulatus*, Schn. Phys. Sacr. pl. DCCXXLVII. 4; *Typhlops Septemstriatus*, Schn. *T. Undecemstriatus*, Nob. *T. Cinereius*, Sch. *T. Crocotatus*, id. *T. Leucorhous*, Oppel, &c. Seb. I. vi. 4. is a species of this division.

|| *Anguis Lumbricalis*, Lacep. II. pl. xx. Brown. Jam. xliv. 1. Seba, I. lxxxvi. 2. *P. Albifrons*, Opp. As in all genera where the species are very similar, authors have not well determined the different typhlops, and a monograph of this genus would be desirable. We are acquainted with about twenty species.

Its posterior extremity is enveloped in an oval and corneous buckler.*

The other tribe, or that of the SERPENTS, properly so called, has the tympanic bone, or pedicle of the lower jaw, mobile, and almost always suspended to another bone, analogous to the mastoidian, attached on the cranium by muscles and ligaments which permit its mobility. The branches, likewise, of this jaw, are united to each other, and those of the upper jaw, to the intermaxillary only by ligaments, so that they can be separated more or less, which gives to these animals the faculty of dilating their throat so as to be enabled to swallow bodies larger than themselves.

Their palatine arches participate in this mobility, and are armed with sharp teeth curved backwards, the most marked and constant character of this tribe. Their trachea is very long. The heart is placed very far back. The majority have but one large lung, with a small vestige of a second.

These serpents are divided into venomous and non-venomous, and the latter subdivided into venomous with many maxillary teeth, and venomous with isolated fangs.

In the non-venomous, the branches of the upper jaw are furnished all along, as well as those of the lower jaw, and the palatine branches with teeth

* *Typhlops Philippinus*, Nob. From the Philippine Islands; eight inches long, and entirely black. *Typhlops Oxyrhyncus*, Schn. must approximate very closely to it.

fixed and not pierced. There are, therefore, four ranges, pretty nearly equal of these teeth in the upper part of the mouth, and two in the under.*

Those among them which have the mastoidian bones comprehended in the cranium, the orbit incomplete behind, the tongue thick and short, resemble the double walkers, in the cylindrical form of the head and body, and have been formerly united with the snakes, in consequence of their small scales.

These are,

Tortrix, Oppel.†

They are distinguished from the snakes, even externally, because the scales of the range which predominates along the belly and under the tail are a little larger than the others, and because their tail is extremely short. They have but a single lung.

Those which are known belong to America. The most common must be

* The common opinion is that none of the serpents without fangs pierced in front of the jaws is venomous, but I have some reason to doubt this. They all possess a maxillary gland, often very bulky; their back molars frequently exhibit a furrow, which may well answer the purpose of conducting some fluid. It is quite certain that many of these species, in which the back teeth are very large, are considered excessively venomous in the countries which they inhabit, and the experiments of Lalande and Leschenauld seem to confirm this opinion. A repetition of such experiments would be desirable.

† These are also the Anilius of Oken, the Torquatrix of Gray, and the Ilysia of Hemprich and Fitzinger.

Anguis Scytale, L. Seba, II. xx. 3.

Two feet long, painted with irregular rings of black and white.*

The Uropeltis, Cuvier, are a new genus, approximating to tortrix, whose tail, still shorter, is obliquely truncated above, and the truncature is flat and rough with small grains. The head is very small. The muzzle pointed. Under the belly is a range of scales, a little larger than the others, and under the stump of the tail a double range.†

Such of the non-venomous serpents as have, on the contrary, the mastoidian bones detached, and whose jaws are capable of considerable dilatation, have the occiput more or less swelled, and the tongue forked, and very extensible.

They have been formed long since into two principal genera, the *boas* and the *adders*, distinguished by simple or double plates under the tail.

The Boas‡ (Boa, Lin.)

Formerly comprehended all serpents, venomous or

* Add *Ang. Corallinus*, Seb. II. lxxiii. 2, 1. 3. which is, perhaps, but a variety of *Scytale; Ang. Ater*, id. xxv. 1, and vii. 3; *Tortr. Rufa*, Merr. which appears but a variety of *Ater; Ang. Maculatus* et *Tesselatus*, Seb. II. c. 2; F. *Latta*, N. Seba, II. xxx. 3. Russel. xliv.; *Tortr. Punctata*, Nob. Seba, II. ii. 1, 2, 3, 4. and VI. i. 4.

† *Uropeltis Ceylanicus*, Nob.; *Uropeltis Philippinus*, two new species.

‡ *Boa*, the name of certain large serpents of Italy, probably of the

not, in which the upper part of the body and of the tail is furnished with transverse scaly bands of a single piece, and which have neither spur nor rattle at the end of the tail. As they are very numerous, independently of the subtraction of the venomous species, the others have been again subdivided.

The Boas, more especially so named, have a crook on each side of the anus. The body is compressed, more bulky towards the middle, the tail prehensile, and the scales small, at least, on the posterior part of the head. Amongst them are found the largest of all the serpent tribe. Certain species attain the length of thirty and forty feet or more, and can swallow dogs, deer, and even oxen, according to the report of some travellers, after having crushed them between their folds, covered them with their saliva, and enormously dilated the jaws and throat. This operation is very long. A remarkable circumstance in their anatomy is that their small lung is only one-half shorter than the other.

These serpents are again subdivided, according to the teguments of their head and jaws.

1. Some have the head covered as far as the end of the muzzle, with small scales similar to those of

adder with four stripes, or of the serpent of Epidaurus, among the Latins. Pliny says, that they were thus named because they sucked the udder of cows. The *boa*, one hundred and twenty feet in length, which it was pretended was killed in Africa by the army of Regulus, was probably a python.

the body, and the plates with which their jaws are furnished, are not hollowed into fossets.

Such is,

The Boa Constrictor, Lin. Lacep. II. xvi. 1, Seb. I. xxxvi. 5, liii. III. lxxxviii. 5, xcix. 1, CI. *Devin.*, or *Boa Empereur* of Daud.*

To be recognized by a broad chain, formed alternately of large blackish spots, irregularly hexagonal, and of pale oval spots emarginated at both ends, which predominates along its back, and constitutes a very elegant drawing.

2. Others have scaly plates from the eyes to the end of the muzzle, and are destitute of fossets to the jaws.

The Anacondo. (*Boa Scytale* et *Murina.* L.) Seb. II. xxiii. 1, and xxix. 1. *Boa Aquatica.* Pr. Max. 2d liv.

Brown, a double series of round black spots along the back, and eye-like spots on the sides.

* Daudin believed that this boa was found in the old continent, but it certainly belongs to Guiana—M. Levaillant and Humboldt have brought it thence. Prince Maximilian has found it in Brazil; M. Levaillant also brought from Surinam the two following species, and every body knows that the *bojobi* is from Brazil. I do not believe that the ancient continent possesses any true boas of a large size. The very large serpents of India and Africa are pythons; the name of *devin* (conjuror) is derived from an erroneous attribution to this serpent of what is related of certain large adders, of which the negroes of India make their fetishes.

3. Others again have scaly plates on the muzzle, and fossets to the plates on the sides of the jaws.

The Aboma. (*Boa Cenchris*, Lin.) *Aboma* and *Porte-anneau* of Daudin. Seb. I. lvi. 4, II. xxxviii. 2, and xcviii. *Boa Cenchrya*, Pr. Max. 6th liv.

Fawn colour, with a series of large brown rings along the back, and variable spots on the sides.

These three species, which attain nearly to an equal length, remain in the marshy situations of the warm parts of America. Adhering by the tail to some aquatic tree, they suffer their body to fluctuate, for the purpose of seizing the quadrupeds that come to drink.

4. There are some which have plates on the muzzle, and the sides of the jaw are hollowed with a foss, in the form of a cleft, under the eye and farther back.*

There are some, in fine, which want fossets, and have the muzzle furnished with plates a little prominent, and cut obliquely from back to front, and truncated at the end so as to terminate in a corner.

* *B. Hortulana*, L., Seb. II. lxxxiv. 1; and *L'elegant*, Daud. V. lxiii. 1, which do not differ. The *bojobi*, *B Canina*, L. Seb. II. lxxxi. and xcvi. 2, or *Xiphosoma Araramboja*, Spix, xvi. The *B. Hipnale*, Seb. II. xxxiv. 1, 2, and Lacep. II. xvi. 2, appears to be only a young bojobi. The *B. Merremii*, Schn. Meri. Beytr. II. ii., or *Xiphosoma Dorsuale*, Spix, XV. of which Daudin has made his genus Coralle, on the character, probably accidental and individual, of the first two plates being double under the neck.

Their body is very much compressed. The back carinated. These come from the East Indies, and may give rise to a distinct subgenus.*

Schneider has separated from the boa,

SCYTALE, Merr. (PSEUDO-BOA),

Which have plates not only on the muzzle, but on the cranium, like the adders, no fossets, the body round, the head of a piece with the trunk, as in tortrix.†

Daudin has also separated the

ERIX,

Which differ in having a very short obtuse tail, and ventral plates more narrow. Their head is short, pretty nearly of a piece with the body, and these characters would approximate them to tortrix, if the conformation of their jaws did remove them from that genus. Their head, moreover, is covered only with small scales. They have no crooks at the arms.

To them we may approximate

* The *B. Carinata*, Schn., or *Ocellata*, Opp *B. Viperina*, Sh. Russel, pl. iv. N. B. These two subdivisions form the genus XIPHOSOMA of Fitzinger. CENCHRIS of Gray.

† *Scytale Coronata*, Merr. Seb. II. xli. 1, Pr. Max. 7. liv. N. B. We must not confound the *scytalæ* of Merrem with those of Daudin, which are the *echis* of Merrem.

The Erpetons. Lacep.

Very remarkable for two soft prominences, covered with scales which they have at the end of the muzzle. Their head is furnished with large plates. Those which predominate under the belly are not very broad, and those of the under part of the tail, scarcely differ from the other scales. But this tail is rather long and pointed.*

The Adders (Coluber. L.)

Comprehended all the serpents, venomous or not, in which the plates of the under part of the tail are divided into two, that is to say, ranged by pairs.

Independently of the separation of venomous species, their number is so enormous that recourse has been had to all kinds of characters to subdivide them.

We may, at first, separate from them the

Pythons,

Which have crooks near the anus, and the ventral plates narrow like the boas, from which they differ only in having double plates únder the tail. Their

* *Erpeton Tentacule*, Lacep. Ann. Mus. II. l., the name given to this genus by M. de Lacépède, who was the first describer. It is from the Greek ἑρπετος, a serpent. Merrem has changed it into Rhinopirus.

head has plates on the end of the muzzle, and there are fossets to their lips.

There are some species as large as any boa. Such is the *Ular-sawa*, or *Great Adder of the Sunda Islands*, (*Colub. Javanicus*, Sh.) which attains to more than thirty feet in length. Seba, I. lxii. II. xix. 1, xxviii. 1, xcix. 2.*

Some of these pythons have the first, others the last plates of their tail simple. Perhaps it may be sometimes but an accidental variety.†

Cerberus, Cuv.

Have like the pythons almost the entire head covered with small scales, and plates only between and in front of the eyes. But they have no crooks at the anus. They have also sometimes simple plates at the basis of the tail.‡

* This *Ular-sawa*, or *Python Amethyste*, Daud.; *Boa Amethystina*, Sch., of which we have a large skeleton and some skins brought from Java by M. Leschenault, approximates at least very closely to the *Pedda-poda* of Bengal (*Python Togre*, Daud.), Russel, xxii. xxiii. xxiv. *Col. Boæformis*, Sh., *Boa Castanea et Albicans*, Schn., and it appears to us that, in general, all the pretended boas of the old continent are pythons. *Ular-sawa* signifies, in Malay, serpent of the river.

The *Boa Reticulata, Ordinata, Rhombeata*, Schn., belong to the pythons.

† The *Bora*, Russ. xxxix. (*Boa Orbiculata*, Sh.)

‡ We have seen these simple plates in one individual, while others of the same species had them all double; a sufficient proof of the small importance of this character. To this group belong the *Col. Cerberus*, Daud. Russ., pl. xvii; *Homolopsis Obtusatus*, Reinw., and some neighbouring species.

XENOPELTIS, Reinw.,

Have behind the eyes large triangular plates, and imbricated; so that they are confounded with the scales that follow, and which are merely smaller.*

HETERODON, Beauvois,

Have the usual plates of the adders, but the end of their muzzle is of one piece, short, and formed like a three sided pyramid, a little raised, and one ridge of which is above, which has caused them to be named serpents, with the snout of a hog.†

The HURRIA,‡ Daud.,

Are adders of the East Indies, in which the plates of the basis of the tail are constantly simple and

* *Xenopeltis Concolor*, Reinw.

† *Heterodon Noiratre*, Beauv., *Heterodon* of Daud., and *Heterodon Tacheté* (*Cenchris Mokeson*, Daud.), belong to this genus; but Beauvois has established it on a character which is found in a great number of adders, of having the hinder maxillary teeth larger, and Daudin appears to have been acquainted with his *Mokeson* only by a drawing. We understand by it the *Hognose* of Catesby, II. pl. lvi. which Daudin himself has cited. It has sometimes a portion of the plates of its tail entire; but at its basis, and not at the end, as Daudin tells us. Linnæus indicated this serpent very well in his tenth edition, under the name of *Coluber Constrictor*; we cannot tell wherefore he should have changed it in his twelfth edition to *Boa Contortrix*.

‡ *Hurriah* is a barbarous name derived from that given in Bengal to the species represented. Russ. xl. copied. Daud. V. lxvi. 2. Another, Merrem, II. iv.

those of the point are double. But these trifling anomalies merit very little regard.

The DIPSAS of Laurenti (BUNGARUS Oppel)

Have the body compressed, much less broad than the head, and the scales of the range, which predominates over the spine of the back, are larger than the others, which we shall see again in the bongares.

Such is,

Dipsas Indica, Nob. (*Colub. Bucephalus*, Sh.) Seb. I. xliii.

Black, ringed with white.*

DENDROPHIS, Fitzinger (AHÆTULLA, Gray)

Have, like the dipsas, a line of scales, more broad along the back, and more narrow scales along the sides. But their head is not broader than the body which is very slender and elongated. Their muzzle is obtuse.†

* *Dipsas*, a Greek name of a species of serpent which was believed to cause a mortal thirst by its bite, from διψα (thirst). The figure given by Conrad Gesner at the word *dipsas*, is precisely of this sub-genus.

The *dipsas indica* is entirely different from the *vipera atrox*, Mus. Add. Fred. xxii. 2, with which Linnæus, Laurenti, and Daudin, have confounded it.

† *Col. Ahætulla; Col. Decorus*, Shaw; *Col. Caracaras*, id. (*Bungarus*

DRYINUS, Merr. (PASSERITA, Gray),

Have the body as long and narrow as the preceding; but at the end of their muzzle is a small slender and pointed appendage.*

DRYOPHIS, Fitzinger,

Have still the elongated form of a thread or cord. Their muzzle is pointed, but without any appendage, and their scales equal.†

We may further distinguish

OLIGDONON, Boié,

Small adders, with obtuse head, short and narrow, which are destitute of palatine teeth.‡

Filiformis, Oppel.) I join to them the SIBONS, Fitz., at least in the *Col. Catenulatus*, Russ. pl. xv. The dorsal scales are rhombridal, and larger, as in *Col. Ahœtulla*.

* *Coluber Nalusus*, Russ. Sup. pl. xii. and xiii.

† *Coluber Fulgidus*, Daud. VI. lxxx. Seba, II. liii. *Dryinus Veneus*, Spix, III.

‡ By these I particularly mean the *Tyria*, *Malpolon*, *Psammophis*, *Coronella*, *Xenodon*, and *Pseudoelaps* of Fitzinger; at the most we can only adopt his DUBERIA, in which the head is short, obtuse, and of a piece with the body, as in elaps, and his HOMALOPSIS, in which the eyes are more vertical than in the other adders. Mark, that I have withdrawn the *Cerberi* from them. Laurenti had already attempted to divide the adders into *Coluber* and *Coronella*; these last were those which have the scales on the sides of the temporal plates, sufficiently large to be themselves reckoned as additional plates, but the passages from one group to another are almost insensible.

But the other subgenera, dismembered from that of the adders, by various authors, appear to us not capable of being continued. They are founded on slight differences of the proportions of the head, bulk of the trunk, &c.

Even after all these separations, the adders continue to be the genus of serpents the most numerous in species.

There are many in France, as

The Collared Adder. (*Coluber Natrix*, L.) Lac. II. vi. 2.

Very common in meadows, and dormant waters; ash-coloured, with black spots along the sides, and three white spots forming a collar on the nape. The scales are carinated, or raised in ridges. It lives on insects, frogs, &c. It is eaten in many provinces.

There is in Sicily a species greatly approximating to this, but much larger, and with a black collar. (*Col. Siculus*, Nob.)

The Viperine Adder. (*Col. Viperinus*, Lath.)

Brownish gray, a series of black spots forming a zigzag along the back, and another of smaller eye-like spots along the sides, colours which give it a remblance to the viper; the under part spotted, like a draught-board with black and grayish. The scales are carinated.

Col. Austriacus, Gm.
Lacep. II. ii. 2.

Brown-red; marbled with steel-colour underneath; two ranks of small blackish spots along the back, the scales smooth, having each of them a small brown point towards the edge.

The Green and Yellow Adder. Col. Atro-virens,
Lacep. II. vi. 1.

Of our woods, spotted with black and yellow above, altogether yellowish green beneath; smooth scales.

These four species are to be found in the neighbourhood of Paris.

The South of France and Italy produce the

Bordelaise Adder. (*Col. Girondicus*, Daud.)

Almost of the same colours as the Viperine, but with smooth scales, the spots of the back smaller, and more separated.

The Four-striped Adder. (*Col. Elaphis.* Sh.)
Lacep. II. vii. 1.

Fawn-coloured, with four brown or black lines upon

the back. It is the largest of our European serpents. It sometimes exceeds six feet in length. We may believe that this is the *Boa* of Pliny.

The Esculapian Serpent. (*Col. Esculapii*, Sh.)

More bulky, and not so long as the last. Brown above; strawy yellow, beneath, and on the sides; scales of the back nearly smooth. Of Italy, Hungary, and Illyria. It is that which the ancients have represented in their statues of Esculapius, and it is probable that the serpent of Epidaurus was of this species.—N.B. The *Col. Esculapii* of Linnæus is a totally different species, and belongs to America.

The foreign adders are innumerable, some are remarkable for the vivacity of their colours—others for the regularity of their distribution. Many are uniform enough in their tints. There are but few which arrive to a very great size.*

* The adders presenting but few interesting varieties of structure, I have not thought it necessary to give the long catalogue of them here. It will be found in the works of Gemlin, of Daudin, of Shaw, and of Merrem, but their enumerations must be consulted with caution and critical attention. They are full of repetitions of the same species under different names, and of transpositions of synonyms.

For example, the *Col. Viridissimus* and *Col. Janthinus*, Merr. I. xii. differ only in consequence of the action of spirits of wine; *Col. Honidus*, Daudin, Merr. II. x. (*Col. Viperinus*, Sh.), is the same as the *Demi-collier* of Lacep. II. viii. 2; the *Coul. Violette*, Lacep. II viii. 1, and *Col. Reginæ*, Mus. ad. fr. xiii. 2, differ only by the action of the spirituous fluid. We should regard as the same *Col. Lineatus*, Seba, XII. 3. Mus. ad. fr. XII. 1. XX. 1; *Col. Jaculatrix*, Seb. I. 9. ix. 2. Scheuz. dccxv. 2. *Col.*

ACROCHORDUS, Hornstedt,

Are easily distinguished in this family, by the small uniform scales which cover the body and head, above and underneath.

The species known, *Oular Caron* of Java, (*Acrochordus Javensis*, Lac. II. xi. 1, 2. *Anguis Granulatus*, Schn.) has its scales raised, each with three small ridges. It grows very large. Hornstedt has advanced, but erroneously, that it lives on fruits, which would be very extraordinary in a serpent.*

The true venomous serpents, or those with isolated fangs, have a very peculiar structure in their organs of manducation.

Their upper maxillary bones are very small, and supported on a long pedicle, analogous to the external pterygoid apophysis of the sphenoid, and are very mobile. There is fixed in them a sharp tooth, pierced by a small canal, which gives issue to a liquor secreted by a considerable gland situated

Atratus, Seb. I. 9. ix. 2., and even *Terlineatus*, Lac II. xiii. 1; *Col. Sibilans*, Seb I. ix. 1. II. lvi. 4; and *Col. Chapelet*, Lac. II. xii. 1, appear equally identical, as well as the *Col. Esculapii*, Jacq., and *Flavescens*, Scopol, &c. &c. As to the transposition of synonyms, they are innumerable.

N.B. The ENHYDRES of Daudin should be the non-venomous adders with compressed tail; but the only species which he cites, *Anguis Xyphura*, Herm. Aff. An. p. 269, and Obs. Zool. p. 288, is evidently an hydrophis, or a pelamides.

* We have seen nothing which resembles the peculiar bone which M. Oppel says that he has observed in the *Acrochordus*, and which is a substitute for the poisonous fangs, and we are moreover assured, by the testimony of M. Leschenault, that the achrocordus is not venomous.

under the eye. It is this fluid, poured into the wound by the tooth, which carries destruction into the bodies of animals, and produces effects more or less fatal, according to the species of the serpent from which it comes. This tooth is concealed in a fold of the gum, when the serpent does not choose to make use of it, and there are behind it several germs destined to replace it, if it should be broken in a wound. Naturalists have named these venomous teeth mobile fangs, but it is, properly speaking, the maxillary bone which moves. It bears no other teeth whatsoever, so that in this kind of maleficent serpents only the two ranges of palatine teeth are visible in the top of the mouth.

All these venomous species, whose mode of reproduction is well known, bring forth their young alive, because the eggs disclose before they have been laid. This is what has caused them to receive the general name of *vipers*, which is a contraction of viviparous.

The venomous serpents with isolated fangs, present external characters pretty nearly of the same nature as those of the preceding; but the great majority have the jaws very dilatable, and the tongue remarkably extensible. Their head, large behind, has, in general, a ferocious aspect, answering in some manner their natural character. There are particularly two great genera of them, the *crotali*, and the *vipers*, the second of which has suffered various dismemberments.

The Crotali, (Crotalus, Lin.) vulgarly, *Rattle-snakes,*

Are celebrated above all other serpents for the fatal subtlety of their poison. Like the boa, they have simple transverse plates under the body and tail. But what best distinguishes them is the noisy instrument which they carry under the tail, and which is formed of many scaly cornets embossed loosely in each other, which move and resound when the animal moves his tail. It appears that the number of these cornets increases with age, and that there remains an additional one at each moulting. The muzzle of these serpents is hollowed with a small round fosset behind each nostril. All the species, whose country is well known, come from America. They are more dangerous in proportion to the heat of the climate or season. But their natural disposition is in general tranquil, and rather lethargic.

The rattle-snake crawls slowly, does not bite except when provoked, or for the purpose of killing its prey.

Though it does not climb on trees, it nevertheless makes its principal food of birds, squirrels, &c. It was for a long time believed that it had the power of torpifying them by its breath, and even of fascinating, that is, of forcing them by its glance alone to precipitate themselves into its mouth. It appears, however, that it is enabled to seize them, only during those irregular movements which the fear of its aspect causes them to make.*

Most part of the species have scales on their heads similar to those on the back.

The species most common in the United States, (*Crotalus Horridus,* L.) Catesb. II. xli. is brown, with irregular blackish transverse bands.

That of Guiana, (*Crotalus Durissus,*) Lacep. II. xiii. 2, has lozenge-formed spots bordered with black, and four black lines along the upper part of the neck. Both are equally redoubtable, and can cause destruction in a few minutes. They generally attain to the length of six feet.†

Some species have the head furnished with large plates.‡

We should approximate to the *Crotali,*

TRIGONOCEPHALUS, Oppel; BATHROPS, Spix; COPHIAS, Merrem;

Which are distinguished by the absence of the noisy apparatus, but have the same fossets behind the nostrils, and are at least equal to the crotali in the violence of their poison.

* See BARTON, "*A Memoir on the Faculty of Fascination attributed to the Rattle-Snake.*" Philad. 1796.

† These two names of *durissus* and *horridus* have been variously exchanged between the two species by naturalists.

‡ It is of this subdivision that Mr. Gray has made his genus CROTALOPHORUS, and Mr. Fitzinger his genus CAUDISONA. The *Crotalus Miliaris,* L. Catesb. II. 42, appertains to it.

There are some whose subcaudal plates are simple, as in the crotali, and whose head is furnished with plates as far as behind the eyes. Their tail is terminated by a sting.*

Such is,

The Brown Viper of Carolina. (*Colub. Tisiphone*, Shaw. Catesb. II. xliii., and xliv.)

Brown, with cloudy spots of a deeper brown.

Others have the subcaudal scales double, and the head furnished with scales similar to those of the back.†

Such is, among others,

The Yellow Trigonocephalus ; Yellow Serpent of the Antilles ; *Vipere Fer-de-lance*, Lacep. II. v. 1. (*Trig. Lanceolatus*,) Opp.‡

The most dangerous reptile of our sugar islands. It is yellowish or grayish, more or less varied with brownish, and attains to six or seven feet in length. It lives in cane plantations, and subsists chiefly on rats. It causes many fatal accidents to the negroes.||

* These are the TISIPHONE of Fitzinger.

† CRASPEDOCEPHALUS, Fitzinger. All the Bothrops of Spix belong to it.

‡ This species also inhabits Brazil, and doubtless other parts of the continent of South America. I would even believe that it is the one which Spix names *Sourancou*, pl. xxiii., and regards as the *Crotalus Mutus*, or *Lachesis*.

|| Here comes the *Trimesure Vert.* of Lacépède. *An. Mus.* IV. lvi. 2, or

GREEN COPHIAS. BOODROPAM. *Rufsel.*

COPHIAS VIRIDIS. Merrem.

Some of these trigonocephali with double plates under the tail have the head furnished with plates.*

Others with small scales on the head, have double plates under the tail, excepting the little end, which is furnished both below and above with small imbricated scales, and is terminated by a small sting.†

Of this number is,

The Lozenged Trigonocephalus. (*Crotalus Mutus.* Lin. *Colub. Alecto,* Sh.) Seb. II. lxxvi. 1. *Lachesis Rhombeata,* Pr. Max. 5. liv.

Yellowish, the back marked with large brown, or black lozenges. Its scales are raised in the middle. It reaches six or seven feet, and is not less formidable than the rattle-snakes.

THE VIPERS (VIPERA, Daud.),

Confounded for the most part with the adders, by Linnæus, as having also the plates of the under part of the tail double, ought to have been separated

Boodroopam, Russ. Serp. Corom. ix., which has sometimes two or three entire plates under the origin of the tail, but this is only an individual accident. Add, *Cophias Bilineatus,* pr. Max. 5. liv. *C. Atrox; C. Jacaraca.*

* M. Fitzinger reserves the name of TRIGONOCEPHALI to this subdivision.

† It is the genus LUCHENIS of Daudin, adopted by Fitzinger, but badly characterized. Its subcaudal plates are certainly double as far as near the end, where there are only small scales. The Prince of Wied represents it perfectly.

from them, in consequence of their poisonous fangs; and some serpents are naturally united with them, which have the under part of the tail simple, either in the whole, or in part.

They are all distinguished from the crotali and trigonocephali, because they have no fossets behind the nostrils.

Some have on the head only imbricated and carinated scales, like those of the back.*

Such is

The Short Tailed Viper. (*Vip. Brachyura,* Nob.) Seb. II. xxx. 1.

One of the most terrible from its poison.†

Others have the head covered with small granulated scales.

Such is

The Common Viper. (Col. Berus. L.)

Brown; a double range of transverse spots on the

* This division and the following form together the subgenus *Echidna* of Merrem, which, with his *Echis,* of which we shall speak farther on, compose his genus VIPERA. Fitzinger portions out our first three divisions into three genera, which he names VIPERA, COBRA, and ASPIS.

† Add, the *Aspic* of Lacep. II. ii. 1. (*Vip. Ocellata,* Latreille), a large foreign species neighbouring to *Atropos,* Lin. Mus. ad. fred. XIII., but very different from the *Aspis* of Linnæus, which is only a variety of the common species; *Vip. Coltro,* Seb. II. xciii. 1; *Vip. Lachesis,* id. xciv. 2; *La Daboré,* Lacep. II. xiii., or *La Brazilienne,* id. iv. 1.; *La Vipere Elegante,* Daud. Russel, vii., &c.

SHORT-TAILED VIPER.

V. BRACHYURA Cuv. *COL. LACHESIS.* Sh.

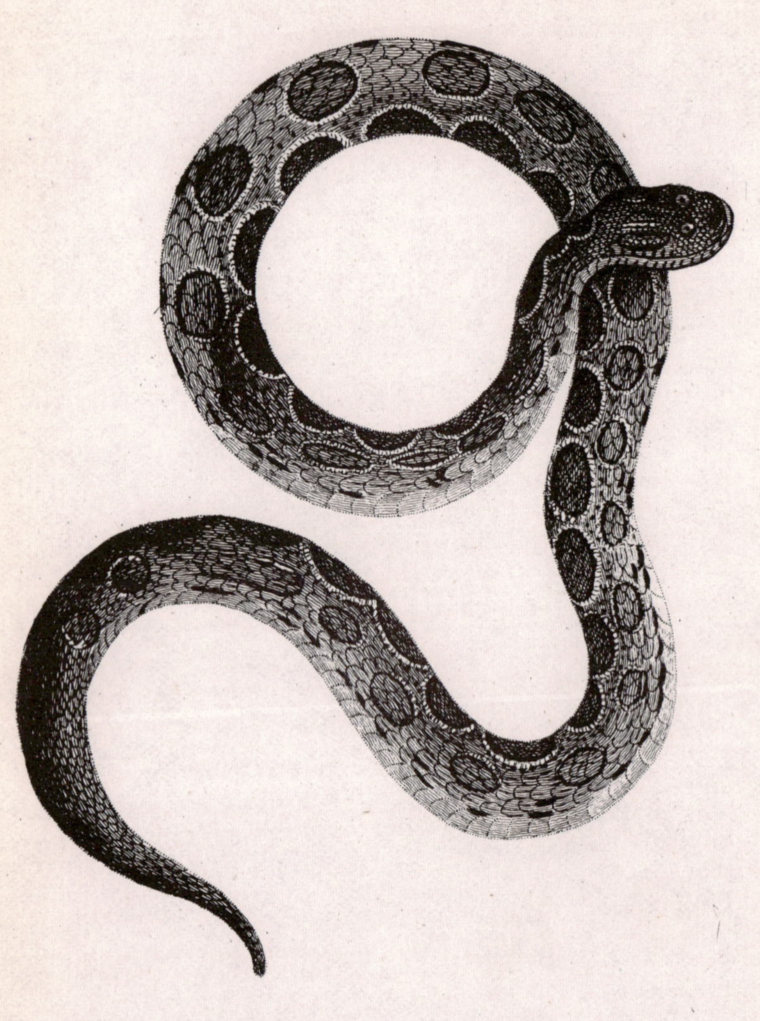

THE CHUCKEERA-BORA OR BAD-RINGED, OR RUSSELIAN SNAKE.

back; a range of black or blackish spots on each side. Sometimes the spots of the back unite in transverse bands; at others, they only form altogether one longitudinal band, turned zig-zag, and then it is the *Coluber Aspis*. Lin., which is sometimes named *aspic*, in our neighbourhood. It is this variety which was multiplied some years ago in the forest of Fontainbleau. There are also individuals almost entirely black.*

The Viper with Horned Muzzle. (Col. *Ammodytes.*) Jacquin. Collect. IV. pl. xxiv. and xxv. *Vip. Illyrica*, Aldrov. 169.

Nearly similar to the common, but eminently distinguished by a small soft bone, covered with scales, which it bears on the end of the muzzle. It is found in Dalmatia, in Hungary, &c.

The Cerastes, or *Horned Viper*, (*Col. Cerastes*, Lin.) Lacep. II. i. 2.

Is remarkable for a small pointed bone over each

* *Aspis* is the name of an Egyptian serpent of which there were many species, and one of which, from what is said respecting the expansibility of its neck, must have been the *Haje*.

Berus is the name of a serpent employed only by the authors of the middle ages, such as Albert, Vincent de Beauvois, and for an aquatic species, probably the collared adder. The *Viper of Charas*, of which Laurenti was also desirous to make a species, and which is the *Col. Aspis* of Gmelin, poes not differ from this common viper, which, in my opinion, is the true eyebrow. It is greyish, and keeps itself concealed in the sand, in Egypt, Lybia, &c. The ancients have often mentioned it.

The Plumed Viper, Vip. *Lophophris*, Nob. (Voyage of Paterson,)

Has on each eyebrow, instead of a horn, a small group of short and corneous threads. It lives in the environs of the Cape.

Other vipers, in other respects very similar to the preceding, have on the middle of the top of their head three plates, a little larger than the scales which surround them.*

The Little Viper (*Col. Chersea*, Lin.), *Col. Berus* of Laurenti and Daud.

Is nearly similar to the common viper, and is especially distinguished by the three plates in question. It is more rare, and smaller in size. It is also pretended that it is more venomous.†

There are some individuals almost entirely black,

Berus of Linnæus, inasmuch as he only quotes on this subject Aldrovandus, 115, which is this species.

* Merrem has made of this division his genus Pelias.

‡ It is the Æsping of the Swedes (*æsping*, a corruption of *aspis*), clearly represented in the Mem. de Stockholm for 1749, pl. vi. Nevertheless, Saurenti, Spec. Med. p. 97, has transformed to it the name of *Berus*. It is also the *Pelias Berus* of Merrem, *Vip. Berus* of Fitzinger.

which have been named the *black viper*. (*Colub. Prester*, Lin.) Laurenti, pl. iv. fig. 1.*

Next in order come the vipers which have the head furnished with plates almost like the adders.

In this number there are some, distinguished by nothing excepting these plates from the most common vipers.†

Such is,

Col. Hæmachates, L. (Seb. II. lviii. 1. 3),

A serpent of the Cape, of a red-brown marbled with white, with the muzzle cut obliquely underneath.

The Naia,

Are such of these vipers with the head furnished with plates, whose anterior ribs have a power of being raised and drawn forward so as to dilate this part of the trunk into a dish of greater or less width.

The most celebrated species is,

* *Prester*, πρεσθὴρ, the Greek name of a serpent, which many authors pronounce to be the same as the *Dipsas*, from πρήθειν, to burn.

† Of this subdivision Merrem has formed his genus Sepedon. Add. *Col. V. Nigrum*, Scheuz. Phys. Sac. IV. dccxvii.

N.B. The *Ophis*, Spix, Serp. xvii., would seem to be a venomous serpent like to these Sepedon, but in which the poisonous tooth is preceded by some small, simple teeth. Not having seen the species, I fear it may be one of those adders with the larger hinder maxillary teeth of which we have already spoken, and many of which appear to us to be at least liable to the suspicion of being poisonous.

The *Cobra Capello* of the Portuguese of the East Indies, (*Colub. Naia*, Lin., *Naia Tripudians*, Merr.) Seb. II. 85, 1. 89, 1, 4, &c. Lacep. II. iii. 1:

Named *Spectacled Serpent*, from a black line drawn on the widened part of its disk in the form of spectacles. It is extremely venomous, but it is pretended that the root of the *Ophiorhyza Mungos* is a specific against its bite. The Indian jugglers tame some of these serpents, and teach them to play tricks and dance to astonish the people, after having taken care, however, to pull out their poisonous teeth.

The same use is made in Egypt of another species,

The *Haje*, (*Coluber Haje*, Linn.) Geoffr. Egypt, Rept. pl. vii., and Savigny, same work, Supl. pl. iii.:

In which the neck widens somewhat less, and which is greenish, bordered with brown. The jugglers of the country, by pressing its nape with the finger, know how to throw this serpent into a kind of catalepsy, which renders it stiff and immoveable, *thus seeming to change it into a rod or stick*. The habit which the haje has of raising itself upright, when approached, made the ancient Egyptians believe that it guarded the fields which it inhabited. They made it the emblem of the protecting Divinity of the world, and sculptured it on the portals of all their temples, on the two sides of a globe. It is incontestably the serpent which the ancients have described under the name of *Aspic of Egypt*, of *Cleopatra*, &c.

The ELAPS (ELAPS, Schn., in part),

Are vipers, with the head furnished with plates, of an organization quite opposite to that of the Naia. They are not only unable to dilate their ribs, but even their jaws can hardly separate behind, in consequence of the shortness of their tympanic bones, and of their mastoidean bones especially, from which it results that their head, like that of tortrix and amphisbœna, is altogether of a piece with the body.*

The most common species is

Elaps Lemniscatus, (*Col. Lemniscatus*, L.) Seba. I. x. ult. et II. lxxxvi. 3,

Are marked with black rings, approximated to each other, three by three, on a white ground. The

* M. Schneider comprehended among his elaps all the serpents which he supposed to be destitute of a separate mastoidean bone; but of this he judged only externally, by the small degree of swelling in the occiput. Accordingly, this character is not found except in the tortrix of Oppel, or the Ilysia: he also paid no regard to the scales or the venom. Ἔλαψ, Ἔλοψ, are Greek names of a non-venomous serpent.

end of its muzzle is black. It belongs to Guiana, where it is very much dreaded, and where it also causes to be feared, the *Tortrix Scytale* and *Coluber Esculapii*, because, though perfectly harmless, they resemble it in form, size, and colours. There are, besides, many elaps, in both continents, whose colours are distributed pretty nearly in the same manner.*

The MICRURI, Wagler,

Are elaps, with the tail very short.

The PLATURI, Latreille,

Have also the head enveloped with plates, and double plates under the tail; but this tail is compressed in the form of an oar, which constitutes them aquatic serpents.†

Finally, at the sequel of the vipers should be placed some serpents which do not differ from them, but because their subcaudal plates are simple, in whole, or in part. They are distinguished from

* Such are *Elaps Anguiformis*, Schn.; *Vipère Psyche*, Daud. VIII. c. 1.; *Col. Lacteus*, Lin. Mus. ad. Fr. xvii. 1; and better, Seb. II. xxxv. 2; *El. Nob. Surinamensis*, Seb. II. vi. 2 and lxxxvi. 1; *Col. Latonius*, Merr. I. ii.; and Seb. II. xxxiv. 4: and xliii. 3, the same as *Col. Lubricus; Col. Flavius*, &c.

† *Col. Laticandatus*, L., or *Hydrus Colubrinus*, Sh. Daud. VII. lxxxv.

tisiphone, because they have no fossets behind the nostrils.

Sometimes the plates of the base of the tail are entire; they are

The TRIMERESURI, Lacep.,

Which have large plates on the head, a part of their plates double, and the others simple.*

Others,

The OPLOCEPHALI, Cuv.,

Have large plates upon the head, and all the subcaudal plates simple.†

Others again,

The ACANTHOPHIS of Daudin, or the OPHRIAS of Merrem,

Have plates on the front of the cranium and the head. Their tail terminates in a crook. Almost all its plates are simple. Sometimes it has some double ones under its extremity.‡

* *Trimeresure Petite Tête*, Lacep. An. Mus. IV. lvi. 1.

† The species are new.

‡ *Acanthophis Cerastinus*, Daud. V. lxxvii., and Merrem; Beytr., II. ix., or *Boa Palbebrosa*, Sh.; *Ac. Brorouii*, Leach, Zool. Miscell. I. iii., the most venomous reptile in the environs of Port Jackson.

ECHIS, Merr., or SCYTALE, Daud.,

Have the head covered with small scales, and all the subcaudal plates simple.*

We may also place here,

The LANGAHA, Bruguieres,

Which have the head covered with plates, the muzzle projecting and pointed, the anterior part of the tail enveloped with entire rings which completely surround it, and the posterior furnished below as well as above with small imbricated scales.†

Besides these two tribes, anciently observed, of serpents properly so called, a third has been recognized in these latter times, in which the jaws are organized, and armed pretty nearly as in the non-venomous serpents, but which have the first of their maxillary teeth larger than the others, and pierced for the purpose of conducting the poison, as in the venomous serpents with isolated fangs, of which we have just spoken.

These serpents form two genera, distinguished, like those of the two neighbouring families, by the vesture of their belly and of the under part of their tail.

* *Horatta,* Pam. Russel, II. pl. 2, or *Boa Horatta,* Sh., or *Pseudoboa,* Carinata, Schn., or *Scytale Lizonata,* Daud. V. lxx.; *Pseudoboa Krait,* Schn., or *Scytale Krait,* Daud.

† The *Langaha* of Madagascar, Lacep. I. xxii., a serpent which is only known by the figure given by Brugnieres.

The BUNGARI, Daud., in part, (PSEUDOBOA, Oppel,)

Have, like the boa, the crotali, the echis, simple plates under the belly and under the tail. Their head is short, covered with large plates, and their occiput is swelled. What best characterizes them is, that their back, very much carinated, is furnished with a longitudinal range of scales, wider than the lateral scales, as in the dipsas.*

These serpents come from the East Indies, where they are called rock-serpents. There is one species which attains to the length of seven or eight feet.†

The HYDRAS,‡ (HYDRUS, Schn., in part, *Hydrophis* and *Pelamides*, Daud.)

Have the posterior part of the body and tail, very much compressed, and greatly raised in the vertical direction, which giving them the facility of swimming, constitutes them aquatic animals. They are very common in certain latitudes of the Indian seas. Linnæus arranged those with which he was acquainted with the snakes, in consequence of having

* *Bungarus*, a barbarous name derived from that of *Bungarum Pamma*, which is given to the largest species in Bengal.

† The *Bongare à Anneaux*, Daud. V. lxv. *Boa Fasciata*, Schn., copied from Russel, III. Add, *Bong-bleu, Boa Lineata,* Sh. Russ. I.

‡ *Hydrus*, Greek name of an aquatic serpent, perhaps our common adder, but the *Marine Hydras* of Ælian are precisely of this genus.

almost all their scales small. Daudin has subdivided them as follows—

THE HYDROPHIS, HYDRUS, Schn. in par.,

Have under the belly, like tortrix and erpeton, a range of scales a little larger than the others. Their head is small, not swelled, obtuse, and furnished with large plates. Some species have been found in the salt-water canals of Bengal, and others farther on in the Indian sea.*

The PELAMIDES

Have also large plates upon the head, but their occiput is swelled in consequence of the length of the pedicles of their lower jaw, which is very dilatable, and all the scales of their body are equal, small, and disposed like hexagonal pavement.

The species most known, (*Anguis Platurus*, Lin.; *Hydrus Bicolor*, Schn.) Seb. II. lxxvii. 2, Russel, xli.,

* *Hydrophis*, water serpent. See the hydrophis of Russel, serpents of Corom. pl. xliv. and II. part, pl. vi. x. Add, *Hydrus Curtus*, Sh., *Hydrus Spiralis*, id. pl. 125. *Leyoselasme* and *Disteyre*, Lacep. Ann. Mus. IV., also come in the subgenus hydrophis; I even believe that this last is the *Hydrus Major*, Sh. pl. 124. There are also serpents of the Indian sea venomous, and with several maxillary teeth.

N.B I do not find, like M. Fitzinger, that the *Pelamides* and *Disteyres* are harmless. I am assured, on the contrary, that their poisonous gland and fangs are constituted like as in the hydras and bungari. As to the *Aispysure*, Lacep. An. Mus. IV., I have never met with it, nor can I verify what it is.

FASCIATED ACROCHORDUS.

ACROCHORDUS FASCIATUS.

is black above, yellow underneath. Although very venomous, it is eaten at Otaheite.

I have added to these two subgenera, that of

Chersydrus, Cuv.,

Whose head and entire body are equally covered with small scales.

Such is,

The *Oular-limpé*. (*Acrochordus Fasciatus*, Sh.) Rept. pl. cxxx.

A very venomous serpent which inhabits the bottom of the rivers of Java.

The third and last family of the Ophidians, or

The Naked Serpents,

Comprehends but one very singular genus, and which most naturalists think should be referred to the Batracians, although it is not known whether it be subject to metamorphosis. It is that of

Cœcilia.

Thus named, because its eyes, excessively small, are pretty nearly concealed under the skin, and are sometimes wanting. The skin is smooth, viscous, and furrowed with folds, or annular wrinkles. It appears naked, but when dissected, exhibits in its thickness some scales, altogether formed, though slender and regularly disposed, on many transverse ranges between the wrinkles of the skin. The head of the cœcilia is depressed, the anus round, and pretty nearly at the end of the body. Their ribs are much too short to surround the trunk. The articulation of the bodies of the vertebræ is made by facets like a hollowed cone, filled with gelatinous cartilage as in the fish, and some of the last of the batracians, and their cranium is united to the first vertebra by two tubercles, also like the batracians. Amongst the Ophidians the amphisbœnæ alone approximate to this structure. The maxillary bones cover the orbit, which is only pierced with a very small hole; and those of the temples cover the temporal foss, so that the head presents above nothing but a continued osseous buckler. Their hyoïd bone, composed of three pairs of arches, might lead to the belief that in their early age they had gills. Their maxillary and palatine teeth are ranged on two concentric lines, as in the proteus, but are often sharp, and curved backwards, like those of the serpents, properly so called. Their nostrils open at the back part of the palate, and their lower jaw has no mobile pedicle, inasmuch as the tympanic bone is enchased with the other bones in the buckler of the cranium.

The auricle of the heart in these animals is not divided sufficiently deep, to be regarded as double, but their second lung is equally small as in the other serpents. Their liver is divided into a great number of transverse foliations. Vegetable matters, mould, and sand, are found in their intestines. The only osselet of the ear is a small plate over the fenestra oralis, as in the salamanders.

Some have the muzzle obtuse, the skin loose, the folds very marked, and two small hairs near the nostrils.

Such is,

The Annulated Cœcilia, (*Cœc. Annulata*, Spix, xxvii.)

Blackish, with eighty and some more folds, marked with white circles; teeth conical.

It lives in Brazil, remaining many feet under ground in a marshy soil.

The Tentaculated Cœcilia, (*C. Tentaculata*, Lin.), Amen. Acad. I. xvii. 1,

With one hundred and some folds. Which, two by two, especially near the tail, do not surround the entire body. It is black, with white marblings under the belly.*

* This cœcilia is not more tentaculated than others of its subdivision. Add, *Cœcilia Albiventris*, Daud. VII. xcii. 1, if it be not the same as the tentaculated; *Cœc. Interrupta*, Nob., in which the white line of the rings

Others have the folds much more multiplied, or rather serrated, transverse striæ.

The Glutinous Cœcilia (*Cœc. Glutinosa*, Lin.), Seb. XXV. 2, and Mus. Ann. Fred. iv. 1,

Is of this number. It has three hundred and fifty folds which unite underneath in an acute angle, and is blackish, with a yellowish longitudinal band along each side. It is found in Ceylon.*

There are some, in fine, in which the folds are almost effaced. Their body is slender, and very long. Their muzzle projecting. One species is entirely blind, (*Cœcilia Lumbricoides*, Daud. VIII. xcii. 2,) blackish, two feet long, and about as thick as the tube of a pen.†

does not correspond below; *Cœc. Rostrata*, Nob., with the muzzle a little pointed, and without white edges to the rings.

N.B. It is impossible to tell why Spix has attributed to his *Cœc. Annulata* two hundred and odd folds: his figure exhibits no more than eighty.

* It is truly of Ceylon, though Daudin asserts that it is of America. M. Leschenault brought it to us from Ceylon, but it is true that there is in America a closely approximating species. *Cœc. Bivittata*, Nob.

† Linnæus gives it, Mus. ad. Fred. V. 2, but confounds it with tentaculata.

We have a skeleton of a cœcilia more than six feet long, and with two hundred and twenty-five vertebræ; but the external characters of which we are not acquainted with.

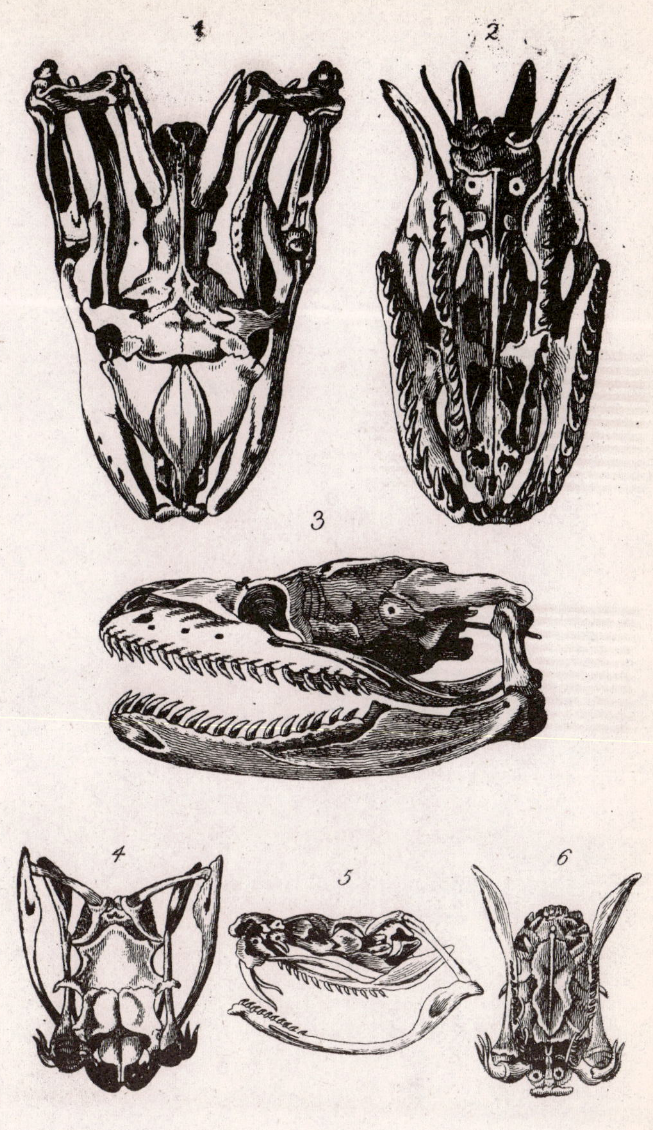

1.2.3. Head of Great Python. _Col. Javanicus. Sh.

4.5.6. Head of the Rattle Snake.

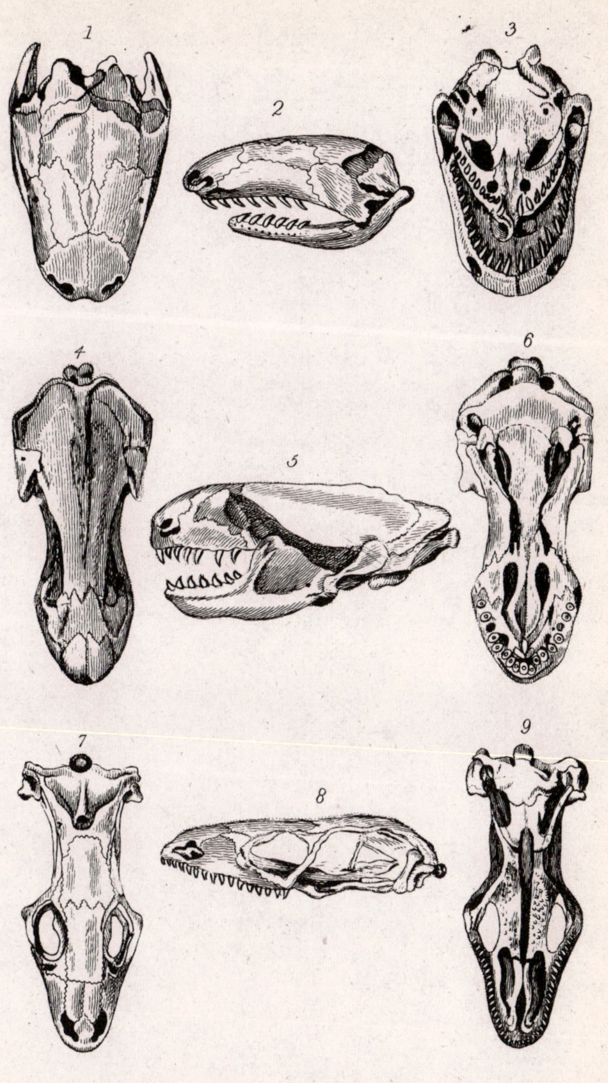

1. 2. 3. Head of Genus Cœcilia.

4. 5. 6. Head of Genus Amphisbœna.

7. 8. 9. Head of Ophisaurus.

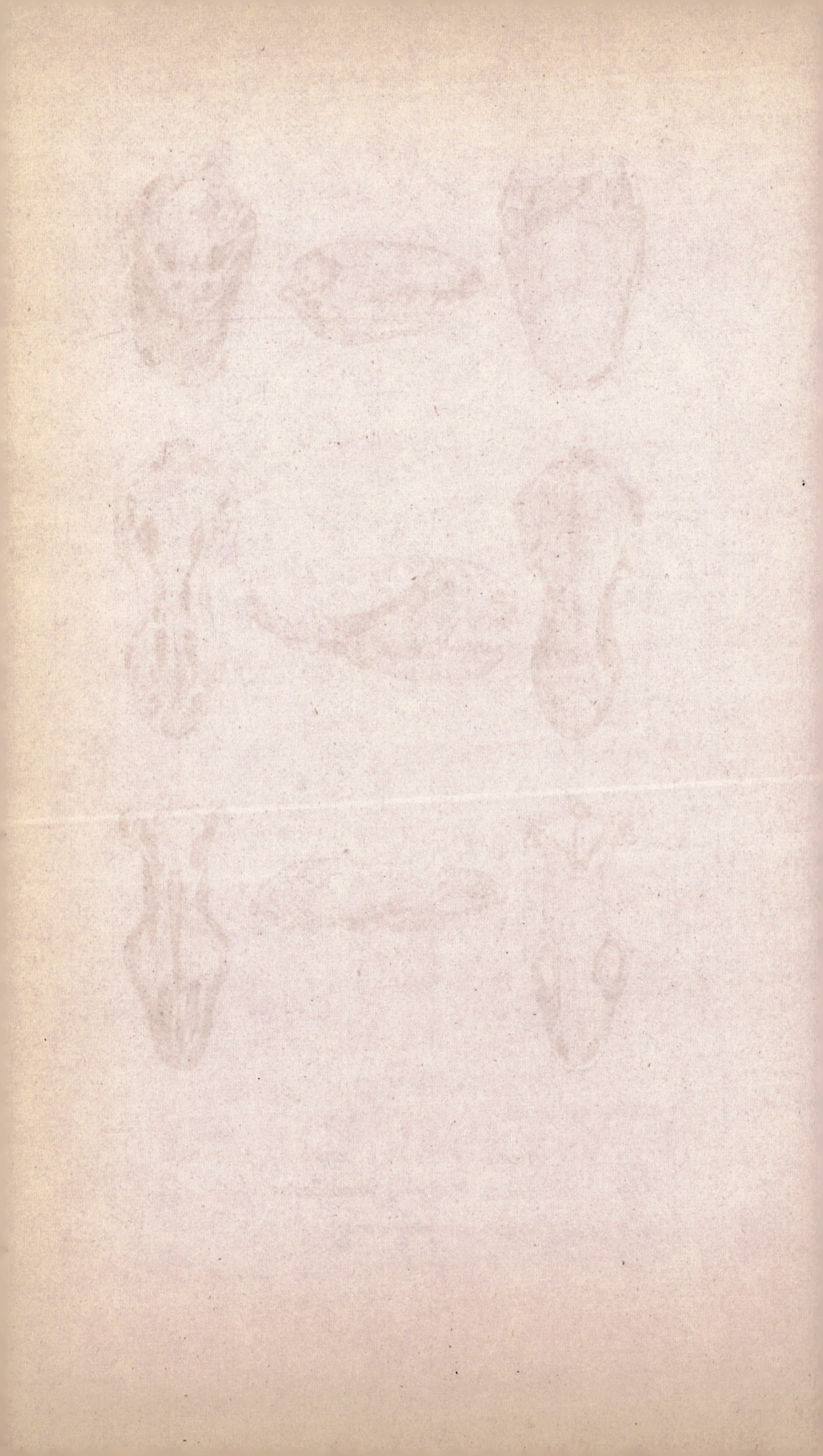

SUPPLEMENT ON THE OPHIDIANS.

This name was first given by M. Alexandre Brongniart to this order, which comprehends the animals designated by Linnæus under the collective appellation of *Amphibia Serpentes*, and which, in the vertebrated class, are, beyond all contradiction, the most easily to be distinguished by unequivocal characters, and common signs derived from their conformation and habits.

The name is of Greek origin. It is derived from ὄφις, the word by which Aristotle and all his compatriots designated a serpent, and εἶδος, which signifies form or figure.

The order ophidia, adopted by MM. Cuvier and Dumeril, as well as by the majority of modern erpetologists, is very natural. Its generic characters are as follow; the

Body, elongated, narrow, without paws or fins; no mobile lids, or, properly speaking, distinct tympanum: some teeth in the jaws; teguments formed by scales, or by an annulated coriaculary or granular skin.

Considering only their external characters, the animals of this order have been grouped by some naturalists, for the sake of convenience, into two distinct families. The one has the skin naked, or equally scaly underneath. These are called Homodermata. The other has the skin covered above with small scales, and below with large corneous plates. This is the family called Heterodermata. The first are harmless, and of small size in general, and feed on insects, and other very small animals. Among the second are many venomous species, and many of very formidable dimensions.

A superficial examination is sufficient, generally speaking, for the purpose of immediately distinguishing an ophidian from every other reptile. Nevertheless, there are some of them in which, without a certain degree of attention, relations might be discovered to species of genera more or less remote from them. If they are distinguished from the chelonians by the absence of limbs and the presence of two penes; from the batracians, by undergoing no metamorphosis; from the saurians, by the want of eyelids, and from the fish by that of gills, yet do they approximate, in some points, to all these animals. Thus are they connected to the first by the *emys longicollis* of Shaw; to the second by the *cœcilia;* to the third by *anguis;* and by *hydrophis* and *pelamedis* to the last.

There is nothing more wonderful and admirable in nature than this sort of connection between the classes, orders, groups, and genera of the animal kingdom. It is not a regular gradation of being, like the steps of a ladder, according to the Platonic system, nor do we think that it can be very easily reduced to any definite plan, notwithstanding the very ingenious and laudable attempts in this way, of some recent naturalists. But we find in every class, and every order of animals, connecting links with all the other classes, and all the other orders. Somewhere or other, we are sure to find the existing bond of affinity. Thus we have flying mammalia, and walking birds—swimming birds, and flying fishes, —in short, some out of each borrow the characters of others, and lose some of those peculiar to their own division.

In consequence of such affinities, which are for the most part but external, it is incumbent on every person, who is desirous of fathoming the history of the animals on which we are now writing, to study attentively their internal organization, and by means of it, to establish the points of comparison which throw light on the theory of their classification. We shall therefore now briefly touch upon their different organs, beginning with those of locomotion.

Progression takes place in these animals by means of sinuosities and springs executed in the water, or on the ground, by a true *reptation*, as well as by the faculty which many of them possess of contorting themselves, and consequently climbing around the branches of trees.

The ophidians, then, are eminently *reptiles*, and their reptation consists in an impulse of the body forwards or backwards by an alternate movement of one or several of its under parts against the ground. This movement takes place in various ways; sometimes by vertical waves or swellings, as in the adder of Esculapius. Sometimes these windings are horizontal, an example of which is the collared adder. Sometimes the posterior region of the body alone contributes to the motion, while its anterior part is raised vertically. Sometimes it is performed by a series of gentle undulations, owing to the alternate approximation and separation of the transverse plates of the abdomen and the tail, or by an analogous action of the rings of the body, as in the amphisbœnæ.

When they repose on the earth, they form with their body several circles placed one above the other, or around the others, and surmounted by the head. It is by the sudden deploying of all these circles, or a part of them only, that, although deprived of feet, they are enabled to spring and shoot forward.

The species which, like the collared adder, and the pelamides, can sustain themselves in the water, swim on the surface of this fluid by vertical undulations, and breathe externally.

The muscles of the ophidians are endued with a contractile force, which is truly prodigious. The boa constrictor, by entwining itself round them, can suffocate almost the largest quadrupeds between its folds, which may be compared to tightened knots. The same is done to the large squirrels of North America, by the *coluber constrictor*, which, moreover, according to Catesby, runs with the most inconceivable agility over the roofs of the houses in Carolina. This muscular power partly explains to us why the ancients, in their mythological traditions, so often founded on exact observations, have made strength the attribute of the serpent, and why they supposed that Achelöus, about to combat with Hercules, clothed himself with the form of this reptile. It is also, without doubt, its agility, and the promptitude of its movements, which caused it to be selected from the very origin of civilization among the Egyptians and the Greeks as a symbol of the swiftness of time, and the rapidity with which the years roll on in succession, one after another. It was also chosen as the emblem of Saturn, and as that of eternity, without beginning and without end, like the perfect circle formed by this animal in biting its tail.

The principal pieces of the skeleton of the ophidians, present modifications not found in the other vertebrated animals. They are destitute, for instance, of sternum and of pelvian bones, not to mention those of the limbs. The spine is composed of vertebræ, which have pretty nearly the same form, from the head to the tail.

Among these vertebræ, the number of which is considerable, some support ribs, and in certain species are more than two hundred in number. While those which do not, and which may be about one hundred and twenty, belong to the tail, and have no connection with the rest of the skeleton.

Nevertheless, in each of these bones, whose assemblage constitutes almost the entire skeleton, we can distinguish a body, spinous, articular, and transverse apophyses.

The first of these apophyses, which predominate all along the back, are separated from each other in the boas, while in the crotali they are so broad, that they appear to touch by their next edges. On the vertebræ of the tail, they are replaced by tubercles, in all the ophidians in general.

The articular apophyses are imbricated, and cover each other after the manner of tiles.

The upper face of the body, thin, bears a very sharp spine directed towards the tail, and which limits the movement only when it might produce luxation, without otherwise constraining it. But its anterior face presents an hemispherical tubercle, which is received in a corresponding cavity of the preceding vertebra, so that each vertebra is articulated by enarthrosis, with that which follows and with that which precedes. In fact, in plainer language, this articulation resembles that of the knee, and easily explains the peculiar nature of the motions executed by the ophidians.

Although articulated by a projecting condyle, with three facettes disposed trefoil-like, yet the head, in the ophidians, is not more moveable on the atlas, than the other vertebræ are upon each other.

Like that of all reptiles the sensibility of these animals is obtuse, and this remarkable attribute of the vital power may be apparently destroyed in them, often during a long period, as in winter, when they fall into an absolute lethargy. But on the other hand, their irritability is truly astonishing; their heart will still palpitate a long time after it has been plucked out, and they will open and shut their mouth after the head has been separated from the body during several hours. Redi and Boyle have seen serpents still exhibit some signs of the preservation of this faculty after having remained about four-and-twenty hours in vacuo. The rattle-snake which Tyson had occasion to dissect so long ago, appeared to live many days after the skin had been torn and most part of the viscera removed.

These facts induce us to believe that the ophidians derive their sensibility less from the brain than the nerves.

The head, though in many species very voluminous, is formed only in small part by the cranium, which closely embraces the encephalon. It lodges, morever, the organs of the senses, and gives attachment to the muscles destined to move both the jaws and itself on the spinal column.

The cranium of the ophidians advances between the orbits, as in the frogs. It presents two frontals almost square, and one parietal; the occipital presents an apophysis directed backwards, and bearing a particular bone, which is mobile, and articulates with the lower jaw, and with the arches which form the upper. The sub-sphenoidal foss is a little sunk, but it is not limited by the clinoid apophyses.

Their brain weighs little more than the seven hundreth or eight hundreth part of the rest of the body. Its two hemispheres form together a mass more broad than long. The optic lids, hollowed each by a ventricle, are almost globular, and placed behind the hemispheres, which do not cover them, and have twice their volume; all the regions of the brain are smooth and without circumvolutions.

The cerebellum, very small and flatted, has the figure of the segment of a circle.

The olfactory nerve presents no sensible bulb, and proceeds from the anterior extremity of the hemisphere. The origin of the other nerves exhibits no peculiarity. But all those of the organs which depend on the cerebro-spinal system are, as in the chelonians and batracians, very thick, relatively to the brain. For the rest they have nothing very remarkable.

All the ophidians have two eyes placed laterally at the right and left of the head. These eyes in appearance are without lids, a slight edging formed by the skin seems to be their only protection. This fact has been remarked at all times, for Aristotle, in his immortal writings, expressly notices this pretended want of mobile and protecting veils of the organ of vision in the serpents, and his opinion has been partaken at divers periods by anatomists and zoologists, and even by M. Cuvier. Nevertheless, recent researches undertaken by M. Cloquet, and verified by our author and by M. Dumeril, have demonstrated that the eye of the ophidians is covered by a single lid, very large and immoveable, and which appears as it were enchased in a projecting frame, which forms around the orbit a variable number of scales, but most usually from seven to eight.

There is a circular *cul-de-sac* of no great depth, between this frame and the eyelid, which is itself composed of three membranous leaves placed over each other.

The first of these leaves is an epidermic plate, elastic, thicker towards the centre than at the circumference, which continues insensibly with the cuticle of the scaly edging of the orbit. This alone is detached, and falls with the rest of the epidermis at the epocha of moulting.

The second is very fine, soft, and perfectly transparent in the centre.

The third is formed by the conjunctive membrane or *albuginea*, which represents a great sac, without external aperture, as in those individuals of the human species in whom is observed that peculiarity of structure denominated by pathologists, *ankyloblepharon*.

This conjunctive membrane invests the two anterior thirds of the globe of the eye, to which it adheres closely, and a part of the motive muscles of the organ, as well as the lachrymal gland, whose conduits appear to traverse it behind. In front and below, it is pierced with a hole or rounded pore, with a single lachrymal point, which is continued along with a membranous conduit, very slender and transparent. This last is engaged in an infundibuliform aperture, which the os unguis presents to it, passes into the external paries of the nasal fosses, and opens at the exterior part of a large anfractuous pouch, which receives the tears and transmits them into the mouth.

As for the lachrymal gland, the existence of which in the ophidians has been generally denied, until latterly, it is voluminous in many species, and lodged in the orbit behind the globe of the eye. Its form is triangular. Its external face is covered by the skin, which adheres to it but little. The anterior face sends slender and transparent threads to the conjunctive membrane, which appear to be the excretory conduits of the organ. It is enveloped by a cellular membrane, very slender, and composed of a multitude of rounded granulations, whitish, tolerably voluminous, and united together by means of vessels and nerves which penetrate it through its external face.

In the majority of serpents with venomous fangs, the viæ lachrymales exhibit a remarkable modification, inasmuch as the lachrymal canals, immediately pour the tears into the nasal fosses, without depositing them in the sac or inter-maxillary reservoir which we have described.

In all of them, generally, in spite of the existence of the fluid secreted by the apparatus in question, the eye, otherwise constantly fixed, is always dry at its surface.

As well as the other reptiles, the ophidians have an organ of hearing, composed of a vestibulary sac, a vestige of cochlea, and three semi-circular canals; but none of them present any external aperture, nor pavilion for the ear. The box of the tympanum itself seems to be wanting, as well as the membrane which closes it. The only osselet which is there observed touches, by its exterior extremity, the bone which supports the lower jaw, is surrounded by the flesh, and is applied to the fenestra by a small concave plate whose edges are irregular.

The apparatus, therefore, destined to the perception of sounds is far from being perfect in these animals, and accordingly they do not appear to have the sense of hearing very fine.

It is the same with the sense of smelling, the organs of which, in the ophidians, appear to be still more incomplete.

Their nostrils are short, very little developed, usually simple, and situated at the extremity or on the sides of the muzzle. In some species, as the ammodytes, and coluber nasica, they are prolonged so as to represent a kind of nose. Their nasal fosses exhibit nothing which can be compared to the sinuses which are hollowed in the bones of the head in mammalia and birds. The projecting laminæ which divide the interior of these cavities, have not been described in a satisfactory manner. The pituitory membrane is furnished with a net-work of blackish vessels.

The crotali, and some other venomous serpents, have underneath, and behind each nostril, a hole tolerably deep, but the use of which is unknown.

The sense of taste in these animals whose general history we are thus sketching, is very feeble, and perhaps still less developed than that of olfaction.

The tongue of the ophidians is, in fact, very extensible, and is terminated by two long points, which though extremely mobile, are semi-cartilaginous and corneous. Its surface is smooth, though soft and humid. This organ appears rather destined to seize the aliments than to perceive savours. It serves more for deglutition than gustation, which should naturally be the case with animals which do not masticate. In a state of repose it is most usually shut up within a membranous sheath.

The sensation of touch in these reptiles exists of course in all the parts of the body which can embrace objects; but it is blunted by the scales and by the corneous epidermis which embraces them in all parts. This epidermis is removed at least once a year, carrying off with it even the most superficial membranous leaf of the eyelid, and the animal gets rid of it in the form of a sort of sheath, or like a glove turned inside out; that part which was inside when the body was covered with it, being now outside.

The mucous body which exists under the scales has very lively and very varied colours in the ophidians, the teguments of which, in other respects, present no appearance of the papillary tissue which constitutes a part of the skin of man and so many other vertebrated animals. But it has for its basis a very strong and thick dermis placed under the scales, or in other words, under certain compartments of the skin, between which the epidermis sinks and moulds itself, and the figure and disposition of which varies greatly according to the species.

The ophidians, as may be seen from the text, have in many species certain little appendages from the nostrils or muzzle. Thus one cœcilia has two little barbles dependent from the nostrils. The erpeton of Lacépède has two tentaculæ on the muzzle. The ammodytes, a small fleshy eminence on the nose. The cerastes a mobile horn above each eye. It is a question whether all these appendages are to be considered as organs of tact.

These reptiles feed on living flesh and insects, worms, and mollusca, never drink, and cannot suck, digest slowly, and eat seldom, especially in the cold season. One repast will suffice them for many weeks; and it is reported that both adders and vipers have been kept for more than six months without any aliment whatever, and without losing a particle of their energy and activity. Notwithstanding this, when they find the opportunity, they will swallow at a time an enormous mass of food. It is very common to see the collared adder swallow toads and frogs whose body is of considerably greater diameter than its own, and also mice, rats, and field-mice.

In the Dutch colonies of the East Indies, André Cleyer purchased of the hunters of the country an enormous serpent, in the body of which he found a deer of middle age, altogether entire with its skin and limbs. In another individual of the same species also examined by this traveller, a wild he-goat was found, with its horns, and a third had evidently swallowed a porcupine with its quills. He adds that a pregnant woman also became the prey of a reptile of the same genus in the island of Amboyna, and that this kind is sometimes kept for the purpose of attacking the buffaloes in the kingdom of Aracan on the frontiers of Bengal. We need hardly be astonished at this, when Prince Maurice of Nassau-Siegen, one of the governors of Brazil, in the seventeenth century, assures us that he himself was an eye-witness of stags, and other equally voluminous mammifera, and even of a Dutch woman, being devoured in this manner in that region of South America where he commanded. Father Gumilla, in his history of the Orinoco, recounts analogous facts concerning an ophidian which he calls *bajo*, and a multitude of others of the same kind, may be read in the works of travellers and naturalists, who also inform us that serpents have been seen to employ many days in swallowing a large prey. So that the part which had got into the stomach was already digested before the rest was entered upon.

We shall soon see sufficient reasons to explain such astonishing peculiarities, in our examination of the organs of deglutition in the ophidians.

In some homodermatous serpents, such as the acrochordus, typhlops and amphisbæna, the two branches of the lower jaw are soldered together, so that they cannot move either forward or externally. They are short, and articulated to the condyle by their most posterior point. These reptiles subsist only on prey of a small size.

But in all the heterodermatous serpents, the branches of the lower jaw are simply united one to another by a ligamentous apparatus, which renders them mobile, and capable of being approximated to, or removed from each other, at the will of the animal, and the articulation of this jaw takes place pretty nearly in the same manner as in birds; for there is no maxillary condyle, and at the posterior extremity of the bone, is hollowed an articular facet, to receive an eminence which has much analogy with the *os quadratum*, and from which it differs only by not being so mobile and so free.

From this disposition it happens, that the lower jaw on each side can not only be raised, and lowered, and open and close the mouth, but also it can move outwards. Now, it would have been difficult that the branches of this lower jaw should have been thus made capable of separation, without permitting the upper at the same time to widen. This, in fact, is what does take place in the majority of cases. The upper jaw is as it were suspended, distinct from the cranium and subordinate to the movements of the lower jaw, which by the separation of its posterior extremities, obliges the pterygoidian arches to separate. This movement, by the approximation of their anterior extremities, simultaneously draws outwards the posterior extremities of the palatine and maxillary arches, while if, on the contrary, the articular extremities of this jaw tend to approximate, the anterior extremities of the same arches proceed outwards, and are removed from each other.

In the heterodermatous serpents which are not venomous, such as the boas and adders, all the bones of the upper jaw are thus mobile on the cranium.

The upper maxillary bones represent two long osseous branches, in which the teeth are implanted. They form the external edge of the foss of the palate. They are, like a lever of the first order, articulated towards their middle part, on a little bone analogous to the jugal, and which forms the anterior edge of the orbit. Pretty nearly towards this point, they have an apophysis which bears upon and glides over the palatine arch. This double arthrosis, gives them the faculty of performing a see-saw motion, and so much the more so, as their anterior extremity is free, and the posterior receives the extremity of a peculiar bone, which serves to unite it to the palatine arches.

These last are two osseous, interior branches, formed of two parts. The one, *anterior*, is free in front, and articulated behind, with an osseous stem, which proceeds towards the articulation of the lower jaw, outwards, along with the peculiar bone which unites it to the maxillary arch, and above, over the base of the cranium, in front of the orbits. The other, *posterior*, is analogous to the pterygoidian lamina, and unites in front, with the posterior extremity of the first portion behind, with the lower jaw on the internal side, and outwards, with the bone which joins it to the maxillary arch.

The incisive bones do not always carry teeth, and sometimes even, as is the case with the boas, they do not unite the upper maxillary bones.

Moreover, a final palato-maxillary bone, pretty nearly cylindrical in its middle, and flatted at its extremities, is articulated outwards with the posterior extremity of the maxillary arch, and inwards with the middle and external part of the pterygoidian region of the palatine arch.

The heterodermatous serpents with venomous fangs, present a new modification, because in them, not only can the jaws separate, but also because their upper maxillary bones are capable of being moved forwards. The palatine arches are very short, entirely directed forward, and carry none but venomous teeth. An intermediate bone proceeding above the upper maxillary, which is itself articulated in front of the orbit, over the short and mobile cheek-bone, unites them to the pterygoidian arches, so that by the movement of the lower jaw in front, the palatine arch, drawn in this direction, drives before it the bone which unites it to the maxillary, which last, being extremely mobile, rises immediately, and proceeds forward, being put in play over the bone of the cheek.

It is clearly to the conformation we have now described, that the majority of the ophidians are indebted for the singular faculty which they possess of dilating their throat so as to swallow bodies larger than themselves, as we have already mentioned.

The muscles which produce this dilatation deserve to be known, and present numerous peculiarities.

All those of the lower jaw are concealed in the thickness of the lips, and form, on each side, the circumference of the mouth. One of them, which appears to be the substitute of the masseter, is stronger, and constituting the anterior edge of the commissure of the lips, proceeds, through a great extent, to terminate, at the upper edge of the submaxillary branch, after having taken its origin by a strong aponeurosis on the tendinous purse, which encloses the poison bladder. We find immediately behind it the analogue of the temporal muscle, which is only a small fleshy band, which is confounded below with the preceding, and descends by an emargination behind the orbit. Still farther back, over all the lower part of the *os quadratum*, there is a peculiar muscle, which is an accessory of the temporal and the masseter, while the analogue of the digastric occupies the entire length of the lower part of the same *os quadratum*, and terminates at the hindermost apophysis of the branch of the jaw, just beyond its articulation.

It is easy to conceive that the first two of these muscles tend to approximate the jaws to each other, and to close the throat.

Those which act upon the upper jaw are more numerous. One of them, very fleshy, springs from the capsule, which surrounds the articulation of the jaw along with the *os quadratum*, and spreads over the purse of the venomous teeth, and over the posterior apophysis of the maxillary bone, so that in contracting it must carry down the fangs when they are upright. Two others, directed in an inverse way, are situated between the middle line of the base of the cranium, and the palatine arches. The first, subcutaneous, occasions the protraction of the maxillary bone, or the raising of the fangs and the narrowing of the mouth, by the approximation of the two interior arches. The second, more slender, and situated above it, is destined to draw back the entire mass of the upper jaw, producing at the same time the approximation of the two branches which form it. Thus, in biting any body, the serpents can twist the mouth, at the same time that they dilate it beyond measure.

All the ophidians have the mouth furnished with teeth, but these teeth do not serve for the purposes of mastication. They are only fit for the retention of prey. The muscles intended for moving the osseous frame-work which supports them, cannot perform the operation of grinding. The only faculties they possess, are those of raising, of lowering, of separating, of approximating, and of carrying backwards and forwards.

The tissue of the teeth presents nothing very peculiar. The osseous portion is hard and compact. The enamel is of no great thickness, and no cement is ever observed in their composition. In the non-venomous species they are conical, curved, very pointed, directed backwards, and implanted all along each of the maxillary, palatine, and mandibular arches, on four ranks in the upper jaw, and two only in the lower. Their number, though always considerable, varies very much.

But in the venomous species, the maxillary branch bears at its extremity but one hollow tooth, very long, and traversed by a canal for the transmission of an empoisoned fluid, which we shall treat of hereafter. Farther back this branch contains several germs of analogous fangs, concealed in a large pouch, which constitutes the gum. These are destined to replace the visible tooth when it shall have fallen. We find then, in the greatest part of the mouth but the two ranges of palatine teeth, and the two ranges of the lower jaw, and the fang itself, when the serpent does not think proper to make use of it, remains concealed in a fold of the gum.

By a natural consequence of the defect of mastication, the salivary glands constitute a less important apparatus in the organization of the ophidians than in that of the mammalia. They are not, however, altogether wanting. It is even observed, in some genera, as in those of the adders and boas, that there are underneath the skin, along the external face of the branches of the lower jaw, two elongated granulous glands, the fluid from which is poured to the external side of the correspondent teeth, and which, in the amphisbœnæ, are lodged immediately under the tongue between the genio-glossal, and genio-hyoidean muscles.

As for the glands which secrete the poison in most of the heterodermatous serpents, they are found on the sides of each branch of the upper jaw, behind the orbit, and almost below the skin. Their tissue is granulated, like that of the salivary glands, and two muscles, destined to raise the fangs, traverse them from front to back, one outwards, the other underneath, so that they cannot act without compressing the gland, and impelling the poison into its excretory canal, which conducts it to the base of the fangs, where it penetrates by a cleft, which prevails throughout their whole extent, and opens towards the point obliquely, like the cut of a pen.

When the irritated animal bites its victim, the fangs are raised upright, they penetrate into the flesh, and deposit there the fatal poison, the certain germ of death and destruction. But they do not deserve, properly speaking, the appellation of *mobile fangs*, by which certain naturalists have designated them. It is not the teeth which raise themselves upright, but, as we have already observed, it is the maxillary bone which moves them.

The ophidians have no epiglottis, and their pharynx, only a little wider than the œsophagus, has no muscle destined to move it, or cause it to change form. The mucous membrane by which it is lined, presents a number of longitudinal folds.

Their œsophagus is very dilatable, preserves pretty nearly the same diameter throughout its whole extent, and is not very precisely distinguished from the stomach, so that it is very difficult to indicate, in a very exact manner, the situation of the cardia. The fleshy membrane of this conduit is also very little marked.

Their stomach has simply the form of a gut a little wider than the rest, and without curvature. When its parietes are contracted, its internal membrane constitutes longitudinal folds. The pylorus is only marked by a slight contraction and by a greater thickness of the parietes.

In consequence of the kind of aliments with which these reptiles are supported, their intestinal canal is very short, and in the collared adder, for example, it is to the total length of the body in the relation of one to one and a half. It is long and slender, in the first part of its course, to which succeeds an intestine, thick and short, into the interior of which its extremity is prolonged like a circular valvule, but with no appendage to mark the place of their division.

The parietes of the thicker intestine are always more strong and thick than those of the smaller. It proceeds in a serpentine course as far as the rectum, but without turning aside, and preserves, pretty nearly, the same diameter, through its whole extent.

The mucous tunic forms, in the narrow intestine, broad longitudinal leaves, folded like ruffles. It is marked with rugosities, and constitutes thick and irregular folds in the rectum, whose extremity is dilated into a rounded cloaca.

In the majority of the species, the anus is only a transverse cleft, placed under the origin of the tail, and which conducts into the cloaca, a kind of common reservoir of the fluids, or the products of generation, of urine, and of solid excrements. This orifice has two lips, one of which moves against the other, and closes the aperture, like a covercle with hinges.

The liver, long and cylindrical, has but a single lobe. Its general colour is yellowish. In many species the common trunk of the hepatic canals is usually separated from the cystic, and is not inserted with this last in the intestinal canal.

The gall-bladder is absolutely separated from the liver. It is situated beside the stomach, in the neighbourhood of the pylorus, and a little behind it. Its figure is, in general, ovoïd. The gall which it contains is, usually, very green, very acrid, and extremely bitter.

The pancreas is very irregular, and situated at the right of the origin of the intestinal canal. The spleen adheres to the commencement of this canal, and is elongated.

The peritoneum appears to be confounded with the pleura, even in virtue of the union of the cavities of the thorax and the abdomen, in consequence of the want of diaphragm.

The mesentery forms a very narrow fold, which does not proceed immediately from the vertebral column, and between the laminæ of which the blood-vessels creep without being divided.

There are no epiploa, properly speaking, in the ophidians. Many of them, however, have appendages under the intestinal canal loaded with fat.

Lymphatic vessels have also been recognized in these animals: but the ganglia which appertain to the system of these vessels have not been yet discovered.

The kidneys are extremely elongated, and formed of a great number of lobes, separated, and as it were enchained, one before the other. From each of these lobes comes the urine, by a particular branch, into a common conduit, which follows the internal edge of the organ, and constitutes the ureter, which, when arrived above the cloaca, dilates into a small oval vesicule, before it is terminated by a separate orifice. In consequence of this last disposition, the urinary bladder is wanting.

The growth of the ophidians is rather slow. They live a long time, and the lethargy to which they are subject in the winter season, appears to suspend their existence. There are some species which, in the course of time, attain to the prodigious length of thirty or forty feet. The giant-serpent observed by Adanson, in Senegal, must have been one of this kind; as also that against which Regulus was forced to bring his machines of war to bear, if indeed it be possible to credit the report of Valerius Maximus, that it was one hundred and twenty feet in length.

Circulation, which always goes on slowly in these reptiles, is yet subordinate to the act of respiration, to the temperature of the atmosphere, and to the development of the passions. The heart has but one ventricle and two auricles, of which the right, which receives the blood of the body, is the largest. The parietes of these two cavities are slender, and appear transparent in the intervals of the fleshy bundles which consolidate them, and the inter-crossing of which is irregular. A membranous partition separates one from the other. They open, each at the side of the other, by a mouth covered with a semicircular membranous valve, into the ventricle, which has the figure of an elongated and not very regular cone, surmounted by an appendage on the left side of its base, and divided internally into two lodges, one upper and one lower, which are separated but in part by an incomplete horizontal partition, composed of fleshy bundles between which the blood can pass. The interior of these lodges is traversed in all directions by a multitude of muscular columns, which strengthen its parietes, and concur to produce a more intimate mixture of the blood which comes from the lung, with that which arrives from the rest of the body.

The orifice of the pulmonary artery corresponds to the lower lodge. The left aorta springs from the same lodge immediately below the right, which commences in the upper lodge, and which thus receives a portion of the blood of the lungs and body before its passage into the lower lodge, from which it is expelled into the left aorta and the pulmonary artery.

In consequence of the absence of sternum and diaphragm, the mechanism of respiration is altogether different in the ophidians from what it is in mammalia and birds. They have, moreover, but one lung, which is prolonged above the œsophagus, the stomach and the liver, and considerably beyond the two last. The trachea consequently is not divided into branchiæ, and when arrived at the single lung, it terminates abruptly in the cavity of the viscus. Its parietes are very membranous, for there are no fibro-cartilaginous portions found in it, except in the lower third of its circumference.

The parietes of the lung, or rather of the sort of sac or vesicle, which it represents, are lined by polygonal cellules, themselves edged with a firm, white, opake net-work, formed of cords of a tendinous nature, which divide the interior of these cells into smaller arcolæ, resembling a net-work with loose and very fine meshes.

There is no epiglottis in the ophidians, and they are equally destitute, as we have already mentioned, of the veil of the palate. Their larynx is formed only by a lower plate and two lateral pieces, contracting a little the edges of the glottis, and accordingly they can utter no other sound than a hissing noise, or rather a powerful sort of breathing.

"Sibila lambebant linguis vibrantibus ora."

The organs of generation are double. The testes in the males are situated in front of the kidneys in the abdomen, and on each side of the vertebral column. The epididymis, of small volume, soon changes into a deferential canal, very flexuous, and opening into the cloaca, by the middle of a papilla, sometimes improperly described as a penis; there are no seminal or accessory vessels. The females have two ovaries, in which the eggs are ranged in chaplets, and not agglomerated in mass as in the batracians. The oviducts are folded, very long, and terminate in the cloaca. The eggs, which are agglutinated by a mucous matter, are rounded, ovoid, and enveloped by a soft membrane, not porous, and slightly encrusted with a calcareous substance; the yoke is orange-coloured and oily. The albumen is greenish, and scarcely coagulable, as in the chelonians. There is no incubation, but sometimes the eggs exclude the young in the interior of the body, and the latter are thus born alive. Such is the case with the viper, which owes its name to this peculiarity.

The females often take care of their little ones in their

early age. Some have been observed in the moment of danger to receive their family into their œsophagus, and give it back again as soon as the alarm was over.

The first family, that of Anguis, which we shall call the *snakes* (though English writers have used this latter term with much greater latitude, and not a little inexactness), are not comprehended among the ophidians by many modern naturalists, but are attached to the last order; in fact, they properly form a passage from one to the other. Their characters are sufficiently given in the text, nor is it necessary for us to dwell in detail on their subgenera and species. We shall content ourselves with speaking of the *anguis fragilis*, *(common slow-worm, Sh.)*, the most common of these animals.

It generally attains to the length of from eight to ten inches; sometimes it is eighteen inches long, and according to some naturalists even three feet. It lives on insects, larvæ, small mollusca, &c. &c.

By the aid of its muzzle it excavates holes in the earth, three or four feet in depth, and conduits describing different circuits and having several issues. It conceals itself during rain, for a part of the day and night, especially when threatened by any danger, and during the season of frost. This animal is viviparous, and seems to produce twice a year, in spring and autumn. It does not cast its old skin until towards the middle of the month of July.

It appears more capable of resisting cold, than the majority of the serpents with which it has been confounded, for it is to be met with in Europe, in very northern latitudes, in Russia, in Sweden, in Poland, in Prussia, and in Germany, as commonly as in France and Italy; but it is never seen in Africa. In the environs of Paris these animals are commonly found under stones, the bark of old trees, in the grass, and under moss. When seized it stiffens itself with such violence, that according to Laurenti and other natu-

DURVILLE'S PSEUDOPUS. *PSEUDOPUS DURVILLII.*

ralists, it sometimes breaks in two, and this peculiarity, united to the great fragility of its tail, causes it in many countries, as well as the *ophisaurus ventralis*, to receive the name of *glass serpent*. It is an extremely mild and perfectly innoxious animal, and no more merits the reputation of being poisonous than that of being blind, which the vulgar in some countries have persevered in calling it, in defiance of the evidence of their senses. It not unfrequently becomes the prey of hens, ducks, geese, swans, hedgehogs, adders, frogs, and large toads.

It has already appeared by the text of Cuvier, that the first family or division of this order, called by him *Anguis*, includes, in fact, but one genus, designated by the same name, but divided into four sub-genera, by the characters stated, namely, *Pseudopus*, *Ophisaurus*, *Anguis*, and *Acontias*. Having nothing of interest to offer on the manners or habits of these animals, we shall merely refer to their specific characters in the text and table, and to the figures inserted to illustrate them.

The family of the true SERPENTS are more properly and particularly ophidians, and to them our preceding remarks on the order in general may be more especially applied. We have there stated all that relates to the nature and organization of these animals, which in all times have inspired both men and the majority of other animated beings with fears justly founded, and an almost insurmountable horror. This horror appears to be so innate, and so preconceived, that even the species which are unacquainted with the danger, or have but little cause to dread it, are affrighted at the sight of these creeping beasts, just as the rodentia in general will fly at the aspect of the wolf, and the mouse will tremble on perceiving the cat. Such is the effect of that mysterious instinct which puts every animated being on its guard for the preservation of its existence, by marking out its natural enemies, an instinct which leads it to study their manner, their character, their general means of injury, and in short, all their habits and peculiarities.

Under this point of view we shall now examine the serpents, all of which are not dangerous, and which often, without terror, as well as without peril, are contemplated by the naturalist, who knows how to appreciate their power and their arms, even when with sparkling eye, inflamed mouth, and tooth upraised for death, they rear themselves with dreadful hissings, and astound the vulgar with their terrific glance.

The sources of the evil to be dreaded from the serpent race, we have already pretty well explained. We shall now endeavour to point out the remedy, or rather to teach the prevention of such evil, by an examination of the habits of these animals. By knowledge of this kind, the curious traveller may cross without risk, the desert waste and the marshy wilderness, the entrance to which appears to be interdicted by these formidable reptiles; may repose himself under the shade of forests which they appear to have unpeopled; may catch by surprise the secrets of nature, and may possess himself of her riches even in the profound caverns which she seems to have confided to the care of these redoubtable guardians.

We shall, therefore, as far as existing materials will allow, describe their general habits, notice the species which merit our avoidance, point out the accidents which they occasion, and the means, as far as they are known, of counteracting the terrible effects resulting from the wounds which they inflict.

Independently of the emblems which we have already mentioned to have been furnished to the ancients by the serpent tribe, there were some others, worthy of remark, as shewing how accurately their habits and peculiarities were observed by those whom modern conceit would term children in science. The insinuating winding progress of the serpent, was considered as a lively image of that soft and persuasive, but too often hypocritical eloquence, which "can make the worse appear the better reason." Thus this animal was placed on the caduceus of Mercury, in the Heathen, and in the Christian mythology, became the symbol of the grand original tempter of mankind. As remorse penetrates and glides into the heart of the criminal to torment him, the Greeks by another ingenious fable have changed into serpents the hair, and the scourges, of the revenging Eumenides. The serpent Python, born from the slime of the Deluge of Deucalion, and killed by Apollo, is an allegory of the contagion, which is developed from marshes and stagnant waters, and again annihilated by the heat of the sun, as in Egypt the pestilence is observed to cease during the summer solstice. The serpent which lacerates the heart of Envy, or arms the bloody hand of Discord, is in the same manner only an ingenious truth masked beneath the charm of fiction.

Similar mysterious applications which appear to have preceded the heroic ages, which have furnished to poetry so many brilliant metaphors to enrich the literature of Greece and Italy, though altered by ignorance, embellished by imagination, and falsified by superstition and by fear, prove that the ancients were very well acquainted with the manners of the serpent tribe. Thus they surrounded the wand of Esculapius with one of these reptiles, and the god himself was adored at Epidaurus under the form of a serpent. Another was consecrated to Hygeia, the goddess of health, to show that temperance was the source of lengthened life; and the mirror of the divinity of Prudence was adorned with serpents, as symbolical of intelligence, foresight, and divination.

Subsequently, from motives of a very different nature, and doubtless from the extreme terror inspired by these animals, many nations have elevated them to the rank of gods.

> "Esse deos fecit timor, quâ nempe remotâ
> Templa ruent."

The serpent became an object of veneration among several of the rude and uncivilized hordes of Africa and America. In the kingdom of Whidah, it had its priests, its temples, and its victims. Every year in that superstitious country, the finest girls, rich offerings, silken stuffs, jewels, delicate meats, and even flocks, are consecrated to the reptile divinities, which of course (as is the case with the observances of superstition, under all its various forms throughout the globe) turn an immense revenue into the coffers of the priests, who are always sure to luxuriate on the credulity and besotted ignorance of the people. Little, therefore, is it to be wondered at, that they should invariably seek to prolong the reign of ignorance and credulity, and pertinaciously oppose themselves, even in the most civilized countries, to all that, by enlightening the human mind, must destroy the sources of their wealth and the foundations of their power.

All the serpents live on animal substances, and digest slowly in consequence of the weakness of their membranous stomach. Accordingly they eat but seldom, especially during the season of cold; one repast suffices them for many weeks, and they never drink, for their thick and scaly skin permits transpiration with great difficulty.

In our European climates they pass the winter in a state of lethargy. In the rigorous months, while overwhelmed in this death-like sleep, they remain concealed in holes in the earth, coiled up, and many of them entwined together, until they are awakened by the genial temperature of the returning spring, and restored to perfect vitality by the reanimating influence of the sun.

At this time they change their epidermis, for these animals undergo a moulting every year, from the effect of which the most external of their teguments dries up, splits, detaches into strips, or even comes off in a single piece, preserving the form of the body.

The serpents very seldom attack man without provocation ; on the contrary, they usually appear to dread his presence. Athough cunning, they are timid and fearful, apparently mild in their manners, and patient or quiescent to excess.

Their spontaneous movement from one place to another is rather slow, in consequence of their complete want of limbs ; but by rolling on themselves, the head being elevated above the ground, and the body let go suddenly, after the manner of a spring, they can dart occasionally a considerable distance and with much force, from the place which they occupied with their circumvolutions.

Twisted round a tree, the boa, or the python, of enormous length and prodigious force, awaits in ambuscade the arrival of its fated victim, which it immediately envelopes in its tortuous folds, and strangles in its murderous embrace. The smaller serpents climb up trees in search of birds which they devour even on the nest.

It has been almost universally believed that by certain special emanations, by the fear which they inspire, or even by a sort of magnetic or magic power, the serpents can stupify and fascinate the prey which they are desirous to obtain. Pliny attributed this kind of *asphyxia*, to a nauseous vapour proceeding from these animals ; an opinion which seems to receive confirmation from the facility with which, by the assistance of smell alone, the negroes and native Indians can discover serpents in the savannahs of America. Count de Lacépède seems inclined to adopt this notion in his history of serpents.

P. Kalm assures us that being fixedly regarded by a serpent hissing, and darting its forked tongue out of its mouth, the squirrels are, as it were, constraïned to fall from the summit of the trees into the mouth of the reptile, which swallows them up. According to the report of many travellers, one would think that by the effect of some charm, the durissus and boïquira, those redoubtable rulers of the steppes of America, possess the power of forcing their prey to fall into their mouths. At their aspect, it is said that hares, rats, frogs and other reptiles seem petrified with terror, and far from attempting to fly, will precipitate themselves upon the fate which awaits them. Even at a sufficient distance for escape, they are paralyzed by the sight of their tremendous foe, and deprived of all their faculties in a manner that appears wholly supernatural.

But this fact, which is so interesting in animal physiology, is not only far from being clearly explained, but even far enough from being sufficiently demonstrated. Notwithstanding the ingenious conjectures of Sir Hans Sloane on this subject, the observations of Kalm, whose assertions were implicitly received by Linnæus; those of Lawson, Catesby, Brickel, Colden, Beverley, Bancroft, and Bartram; notwithstanding a work published *ex professo* on the matter by Doctor Burton of Philadelphia, and notwithstanding some recent accounts by Major Garden of this stupifying power in the serpents, which he attributes both to the terror which they inspire, and to certain narcotic emanations from their bodies at particular times, it must be confessed that this subject is still liable to controversy, and still involved in a considerable degree of obscurity.

On the other hand, as the look of the dog stops the progress of the partridge, so we might imagine that the presence of man has a considerable influence over the faculties of some very justly dreaded serpents, and obliges them to obedience, by, as it were, a certain kind of fascination. From the most ancient time, certain hordes of Arabia, such as the Psylli and the Marsi, were acquainted with some art of charming and taming these reptiles Kæmpfer, and many other travellers, have left us accounts of the dance which the Indians make the *naia* perform. We also know, beyond any doubt, that the Egyptian jugglers cause the asp of the ancients, the *hajé* of the modern Arabs, to play a variety of tricks at the word of command, and that they seem to imitate the magicians of Pharaoh, who pretended to turn their rods into serpents. It is also a remarkable fact, that music has a very considerable influence on these animals, to which we cannot otherwise attribute any very large portion of sensibility. The Viscount de Chateaubriand relates, that in 1791, in the month of July, in Upper Canada, on the banks of the Genesee, he saw a native appease the anger of a rattle-snake, and even cause it to follow him, merely by the music of his flute, without having recourse to any other method.

The coral serpent, which is found in Florida, is extremely gentle, capable of a certain degree of domestication, and is worn around the neck as a collar by the women of that country. Even in Europe women have been known to tame the collared adder (*coluber natrix*), which reptile, as well as others of the race, has shown itself susceptible of some degree of attachment to those who take care of it.

It is but rarely, as we have already remarked, that the serpents will attack man without being highly provoked to do so, and we may observe here, that their poison is more subtile and active in proportion to the heat of the climate which they inhabit. The hot and humid steppes and savannahs of Asia and America, and the burning sky of the African deserts, seem by far the best suited to the multiplication and developement of these reptiles. Only fifteen or sixteen of their species inhabit Europe, while Russel has described forty-three merely for the coasts of Bengal and Coromandel. Equatorial America, scorched by the burning rays of the sun, and incessantly watered by those immense rivers which roll the tribute of their waves towards its eastern boundaries, furnishes, of itself alone, according to the observation of M. de Humboldt, one hundred and fifteen species, out of three hundred and twenty which have been described in the ophidian order. In the provinces which it contains, the earth, peculiarly lavish in the support of poisonous weeds and hurtful animals, has peopled with impure and dangerous reptiles the inundated morasses, and yet untrodden forests of these mighty regions. They swarm in Surinam, in French Guiana, in Peru, in Brazil, in the neighbourhood of the Lower Orinoco, in Nicaragua, Panama, and Cassiquiare. Twice a year they lay an immense number of eggs, and are so excessively abundant, that when the natives set fire to the brush-wood, &c. with which the country is covered, whole armies, as it were, of formidable serpents, sally forth in all directions in crowded ranks, to the number of thirty or forty thousand at a time, putting all to flight before them. But in colder climates a few individuals only are found scattered over a large extent of territory. They begin to be rare enough in Germany and Russia, still more so towards Siberia, and totally disappear as we approach the polar regions. Neither are they ever found upon high mountains, beyond an elevation of five or six thousand feet, as has been observed on the ridge of the Cordilleras, on the platforms of Santa-Fe de Bogota, on the Andes, at Antisana, and Pichincha.

But among all the known serpents, there is scarcely one-sixth, or one-fifth part of them; that may be considered of a really dangerous character. Among the forty-three species of the East Indies, described by Russel, seven alone are to be feared; and in the enumeration of the ophidians which were known in his time, by Daudin, there were eighty venomous species, and two hundred and thirty-three not venomous. In America, one race alone in five, and one in four in Europe, are redoubtable for their poison. The others are innocent animals, which creep upon the surface of the earth.

Of all the venomous reptiles of Europe, there is none whose bite is so dangerous as that of the viper (*Col. berus*). We shall describe, in the proper place, the mechanism by means of which this serpent insinuates its poison into the wounds which it has made. At present we must content ourselves with stating the following facts, relative to the poison of serpents considered in a general way.

This poison is neither acid nor alkaline. That of the viper, which has been most studied, does not redden the tincture of turnsol (*Corona Solis*), or turn green the syrup of violets. It is neither acrid, nor burning. It produces no sensation on the tongue, but one analogous to that caused by the fresh fat of animals. It has a slight odour resembling the fat of the viper itself, but much less nauseous. It produces no effervescence with acids. When put in water, it sinks in that fluid; but if mixed up with it, it muddies it, and renders it slightly whitish. It does not burn on being exposed to the flame of a candle, or thrown on burning coals. When fresh, it is a little viscous, and when dry, it sticks like pitch. It greatly participates of the nature of mucus.

This poison preserves its power after the death of the animal which has secreted it, and fixes in linen with considerable energy. That of the crotali, in particular, is said not to be destructible by lixiviation. It equally retains its properties in the fangs after the death of the reptile. A man was bitten through his boots by a rattle-snake, and very speedily died in consequence. These boots were sold successively to two other persons, who also died, because the extremity of one of the venomous fangs had remained engaged in the leather. However extraordinary such a fact may appear, it is confirmed by experiments detailed to the Philomathic Society in Paris, in the year 1827, by Dr. Cloquet. These experiments were undertaken by Dr. Emmanuel Rousseau, demonstrator of comparative anatomy at the " Jardin du Roi," who, having at his disposal a rattle-snake, which had been dead two days, discovered that the poison of this animal, even in so northern a climate, and at a very advanced time of the year, still preserved all its maleficent properties. A pigeon, in the pectoral muscles of which the doctor infixed the venomous fangs of the reptile, died within a very short period indeed.

The poison of the viper, and of some other serpents which inhabit countries remote from the torrid zone, loses its strength during winter, and in the more northern climates. Its energy, on the other hand, is augmented during summer, and in warmer regions.

The danger arising from the bite of serpents is in relation to the degree of anger with which the reptile is animated—for, in pressing with greater force, it more completely squeezes out the poison, and distils a larger quantity of it into the wound. It is also more or less great, according to the time which has elapsed since the poison vesicles were emptied by the last bite.

The bulk of the animal which is bitten, and the degree of terror which the wound has caused, also increase or diminish the quantity of danger.

The experiments of Fontana, of which he made more than six thousand, proved that the bite of a single viper was sufficient to kill a mouse, a pigeon, or other small animal. Many of them repeated, were found necessary to cause the death of an ox or a horse.

The danger of this bite besides, evidently depends on the nature of the poisonous inoculation by which it is accompanied. Notwithstanding the circumstance related by Matthiali, of a peasant who died immediately from having sucked the blood from a wound made on himself by a viper, and notwithstanding the assertion of Fontana to the same effect, it is very certain that this poison (at all events that of the viper) may be taken internally without any injury. Charas and Redi have made conclusive experiments on this subject, the results of which have been recently confirmed by Professor Mangili. This fact, however, was known as long ago as the time of Celsus, who thus expresses himself concerning it. "Neque Herculé scientiam præcipuam habent hi qui Psylli nominantur, sed audaciam usu ipso confirmatam, nam venenum serpentis non gustu sed in vulnere nocet. Ergo quis quis exemplum Psylli secutus, id vulnus exsuxerit, et ipse tutus erit, et tutum hominem præstabit. Sed ante," (adds this most sagacious and accurate observer,) "debebit attendum ne quod in gengivis palatove, aliove parte oris, ulcus habeat."

In the Pharsalia of Lucan, we find the same doctrine put into the mouth of Cato, nor shall we resist the temptation of giving to our readers the entire of the splendid passage in question.

"Jam spissior ignis,
Et plaga, quam nullam superi mortalibus ultra,
A medio fecere die calcatur, et unda
Rarior: inventus mediis fons unus arenis
Largus aquæ: sed quem serpentum turba tenebat
Vix capiente loco: stabant in margine siccæ
Aspides, in mediis sitiebant Dipsades undis.
Ductor, ut aspexit perituros fonte relicto
Alloquitur; vana specie conterrite leti
Ne dubita miles tutos harrire liquores:
Noxia serpentum est admisto sanguine pestis:
Morsu virus habent et fatum dente minantur:
Pocula morte carent: dixit, dubiumque venenum
Hausit."

"And now with fiercer heat the desert glows,
And mid-day gleamings aggravate their woes;

When, lo! a spring amid the sandy plain
Shews its clear mouth to cheer the fainting train;
But round the guarded brink in thick array
Dire aspics roll'd their congregated way.
And thirsting in the midst the torrid Dipsas lay.
Blank horror seized their viens, and at the view
Back from the fount the troops recoiling flew;
When wise above the crowd, by cares unquell'd,
Their awful leader thus their dread dispell'd:—
'Let not vain terrors thus your minds enslave,
Nor dream the serpent brood can taint the wave;
Urg'd by the fatal fang their poison kills,
But mixes harmless with the bubbling rills.'
Dauntless he spoke, and bending as he stood,
Drank with cool courage the suspected flood."

Very frequent occasions do not occur of observing the effects of the bite of serpents upon man. The terror which these reptiles inspire causes them to be too carefully avoided to allow of the multiplication of accidents of this kind. Nevertheless, there are few physicians, even in Europe, who have not been witnesses of the consequences of the bite of the viper. Sir Everard Home, in the Philosophical Transactions for 1810, relates an example of the fatal effects of the bite of a rattle-snake, a case, which he had himself the opportunity of studying and examining. Another very recent and deplorable instance of the same is too well known to render any more than a mere allusion to it necessary. We mean that which occurred at Rouen to an Englishman, in February 1827, and which terminated fatally within eight hours after the bite of the rattle-snake.

The morbid symptoms which follow the venomous inoculation, made by the tooth of the ophidians, of which we are speaking, are developed with excessive rapidity. In many animals, according to Fontana, the effects become already perceptible after fifteen or twenty seconds. In man they manifest themselves in the following manner, particularly after the bite of the viper.

An intense and stinging pain is felt in the place where the wound has been made, which soon becomes the seat of an inflammatory swelling, with a tendency to gangrene, which is indicated by livid spots. At the same time the wounded person experiences nausea, weakness, vertigo, syncopes, difficulty of breathing, confusion in the intellectual faculties, vomitings of bilious and yellowish matter, convulsive movements, and pains in the umbilical region. All these are signs of the general impression produced by the virus on the entire system; not that it coagulates the blood on the vessels, as Fontana concluded from illusory experiments, but because it exercises a special action on the principle of sensibility.

The blood which flows at first from the wound is often blackish. Soon after it is replaced by sanies, and the gangrene declares itself when the malady is about to terminate by death.

This termination is happily not the most usual, at least in the case of the bite of vipers. It is not even so common as it is universally supposed to be in the case of other species of venomous reptiles. In the sitting of the Royal Academy of Sciences, in Paris, in April 1827, Professor Bosc declared that he had seen more than thirty persons who had been bitten by rattle-snakes, not a single one of whom had died in consequence.

Fontana having ascertained that a hundreth part of a grain of the poison of a viper, introduced into a muscle, was sufficient to kill a sparrow, and that six times as much was necessary to destroy a pigeon, has calculated that it will take pretty nearly three grains to kill a man. Now as a viper has but about two grains of poison in its vesicles, which it does not exhaust even after many bites, it would seem evident that a man might receive, without dying in consequence, the bite of five or six vipers. Such, however, is not altogether the case. The experiments of the scientific Italian have had the fate of all physiological experiments founded on calculation. Ulterior facts have overturned the consequences deduced from them. Dr. Paulet, in his observations on the viper of Fontainbleau, published in 1805, says that an infant of seven years and a half old, bitten under the internal malleolus of the right foot by a reptile of this species, died at the expiration of seventeen hours; another infant, two years of age, expired in two days after having been bitten in the cheek. Still more recently Dr. Hervez de Chegoni, saw at Entrains, a small town of the department of the Nièvre, a woman sixty-four years old, of a good constitution, and in good health, expire with the most lamentable symptoms, thirty-seven hours after having been bitten in the thigh only once by a single viper.

The opinion therefore of Fontana, which is still maintained by many persons, does by no means appear to be well-founded. The medical men who hold it do not seem to take into consideration, that in the case in question, as in most pathological affections, climate, season, age, temperament of the individual, &c., are so many causes which must marvellously influence the nature and progress of the symptoms occasioned by the bite of venomous ophidians. The structure of the wounded organ and its connections are also equally deserving of attention under this point of view. Thus M. Bosc relates that during his stay in America, two horses in one enclosure were bitten on the same day by a black viper, one in the hind leg, the other in the tongue. The last died in less than an hour, while the other escaped with a swelling which lasted for a few days, and a weakness for a few weeks; the death of the first was occasioned by the violent inflammation which closed up the glottis and brought on an asphyxia.

This poison, furthermore, at least in the viper, does not

appear to be mortal if it penetrate only into the cellular tissue, and is perfectly harmless if applied only on the fleshy fibres. Injected into the veins, on the contrary, it rapidly produces death, as many experimentalists have demonstrated, and particularly Fontana.

We may add that the bite of the viper, though rarely fatal to man, gives rise to grievous and durable consequences, if not treated in proper time. One of these consequences may be an universal jaundice; it has also been known to produce a violent inflammation of the gums, dryness of the mouth, insatiable thirst, griping, dysury, shiverings, hiccoughs, and cold and colliquative sweats, and all these symptoms endured for a very considerable time.

But however terrible the accidents occasioned by the viper may appear, they are very far from equalling those produced by the serpents of the burning climates of Asia, of Africa, and of America. In a few hours, and even sometimes in a few minutes, the wounded part is seized with torpor and lividness, and the coldness of death shooting rapidly from the extremities, soon makes itself felt in the region of the heart.

Eagerly devoted to the marvellous, the ancients have admitted with perfect confidence the most absurd fables in relation to the effects of the poison of serpents. Lucan in his *Pharsalia*, and Nicander in his poem *De Theriacis*, have left us a nomenclature of serpents, and a picture of the effects of their poison, which are very admirable, considering the period in which those writers flourished. But they bear the stamp of fabling antiquity; for we read there of the transformation into formidable serpents, of the drops of blood fallen upon the Lybian sands, from the head of Medusa, cut off by Perseus. The following is Lucan's stately description of the principal venomous serpents of Africa, which are still to be found at the present day in the Ouangarah :—

"Hic, quæ prima caput movit de pulvere tabes
Aspida somniferam tumidâ cervice levavit.
Squammiferos ingens *hæmorrhois* explicat orbes;
Natus et ambiguæ coloret qui Syrtidos arva
Chersydros, tractique viâ fumante *Chelidri;*
Et semper recto lapsurus limite *cenchris*
Pluribus ille notis variatam pingitur alvum,
Quam parvis tinctus maculis Thebanus *Ophites;*
Concolor exustis atque indiscretus arenis
Ammodytes: spinâque vagi torquente *cerastæ;*
Et *Scytale* sparsis etiam nunc sola pruinis
Exuvias positura suas, et torrida *Dipsas;*
Et gravis in geminum surgens caput *Amphisbœna:*
Et *natrix* violator aquæ *Jaculique* volucres,
Et contentus iter caudâ sulcare *pareas;*
Oraque distendens avidus spumantia *Prester;*
Ossaque dissolvens cum corpore tabificus *Seps:*
——— ——— Et in vacuâ regnat *Basiliscus* arenâ."

Pausanias relates the history of a King of Arcadia, who, having been bitten by one of the venomous serpents of which we are speaking, died of a general gangrene. Ambroise Paré, the founder of French surgery, who notices this reptile after the Greek historian just cited, names it *le pourisseur*, and joins it to another serpent which he calls *coule-sang*, because, according to Avicenna, its bite, followed by a sudden gangrene and vomitings, gives rise to a flow of blood through the nostrils, mouth, eyes, anus, vulva, &c., all which is exactly referable to the *hæmorrhois* of the ancients.

According to them, the *hypnale*, or asp, caused death in producing a lethargic sleep, and Solinus, after Nicander, attributes to it the death of Cleopatra. The chelydri exhaled nausea-exciting vapours; the ammodytes concealed itself in the sand; the acontias, or jaculus, fell like an arrow on the passenger from the tops of the trees. The same was the case with the prester; the seps occasioned rottenness in the limbs and bodies of those whom it had bitten.

But among these fables, the most extraordinary and incredible is that of the basilisk, a serpent on whose head Avicenna, Pliny, Solinus, Nicander, and a number of other ancient writers, have placed a crown, affrighting all the other serpents with its terrible aspect, and appearing, says the medical poet last mentioned, to be truly their king. The sinister sound of its voice had the power of causing death, and no animal could encounter for an instant its terrific glance without ceasing to exist. Its skin, as Solinus tells us, was hung up in the temple of Pergamus, the inhabitants of which city had purchased it at a very high price, and it hindered the birds from building their nests there, and the spiders from spreading their webs. Pliny also mentions it, and Actius, who declares there is no remedy for its bite, in consequence of the instantaneous rapidity of its effect, which, according to Erasistratus, causes the muscles to fall off suddenly in shreds.

The serpents are not merely hurtful. Their flesh has been employed with some advantage by man, and certain of their organs. Without attaching any credit to stories formerly circulated on this subject, such as those relative to the extraordinary virtues attributed to the cast skin of the serpent, we may believe, on the most respectable authority, that the flesh of snakes, is by no means an inefficacious remedy in many cutaneous maladies. The anaconda, and other boas, supply the natives of the countries which they inhabit with wholesome nourishment, and the adders are used as food in many parts of the south of France.

We read in the Philosophical Magazine, December 1816, of a serpent which was found in a mine of pit-coal at Liphone, at a depth of fifty feet, and also of a living adder found enclosed in a small cavity, which had no communication with the surface of the earth, and that this adder died in ten minutes after having been taken out. Though we believe this fact to rest on respectable testimony, yet it must be confessed that it is exceedingly difficult to believe it.

We shall now proceed to consider the subdivisions of this order in detail; but according to our usual plan we shall omit all notice of those which afford nothing interesting to dwell upon.

The remarkable characteristics of the genus AMPHISBŒNA are, that the body is surrounded with rings of square scales, and that the tail is almost as bulky as the head, and as their eyes are extremely small, it is not easy at the first glance to distinguish at which end the head is situated. As these animals can move both backwards and forwards, the name of amphisbœna has been given to them, which literally signifies *double-walker*. The ancients employed this name, but it must have been for another serpent, for the amphisbœna, which are now known, are of American origin, and consequently the ancients could not have been acquainted with them. We may, indeed, aver that their amphisbœna was an imaginary animal, since they gave it, at each extremity, a head armed with venomous teeth. If, however, these fables have any foundation in nature, they must relate to the anguis, which resembles almost in all points the amphisbœnæ, excepting the form of the scales.

The amphisbœnæ are not venomous; their teeth are conical, fixed, and disposed in a single rank. Those of the upper jaw are unequal; there are sixteen in each jaw. The head is furnished with irregular scales larger than those of the rest of the body; the eyes appear there only like two little obscure points. The nostrils are two simple holes pierced at the end of the muzzle. The tongue is not extensible and forked, like that of most serpents, but flat, short, and

PUNCTATED AMPHISBÆNA.

AMPHISBÆNA PUNCTATA.—*Pr. Max.*

slightly emarginated. The upper jaw is fixed, as in the anguis, and not mobile as in other serpents. The intestinal canal is pretty long, and makes several folds ; there is a small cœcum and a large rectum. The stomach of an amphisbœna when opened, was found to be filled with insects of a moderate size.

On each side of the body a line may be remarked, which distinguishes the back from the belly ; all the scales on which this line passes are marked with two crossed wrinkles forming an X ; it extends only as far as the anus. This last has a cleft in the form of an obtuse angle, surrounded with small papillæ.

The *White Amphisbœna (A. Alba, Lin.)* inhabits Brazil, and its name in the language of that country is *Ibriaram*, which signifies Lord of the Earth. Margrave first made known this species. Its description is by no means equivocal, and it is not easy to imagine why Linnæus and his successors have applied it to a species of cœcilia to which it could never answer. It is, moreover, after what Margrave tells us of his *ibriaram*, that authors have described the manners of their *amphisbœna alba*.

He relates that this serpent wounds with its tail as well as with its head, and that the bite is extremely venomous, each of which assertions are equally erroneous. He adds that the ibriaram remains under ground, and that it attacks the ant-hills, so as to drive out the ants in large flocks as soon as they have got their wings. It is very possible that this reptile may be fond of feeding upon ants ; but it is quite superfluous to attribute the fact of their quitting their ant-hills in great numbers to the pursuit of this animal.

Captain Stedman informs us, that the common people at Surinam call the amphisbœna the king of the ants, and imagine that, when it is blind, the large ants go for the purpose of feeding it.

Many other serpents, as well as this, are partial to ants. In France, the *coluber austriaca* is sometimes found in the ant-hills, where it goes for the purpose of feeding on these insects, and, not improbably, to conceal itself during the winter season.

This amphisbœna has from two hundred and twenty to two hundred and thirty and odd rings around the body, and from sixteen to eighteen around the tail. Its length is about a foot and a half, of which the tail scarcely forms the twelfth part. The colour is an uniform white.

All the amphisbœnæ which are varied with brown and white, are known under the name of *fuliginosa* ; but there is so great a difference in the distribution of colours in different individuals, that it is not at all improbable that several species may have been thus confounded. Laurenti has endeavoured to distinguish them, and has established four; but as he had no better authority than the figures of Seba, his determinations are not to be depended on.

One of these amphisbœnæ, observed by the baron, had two hundred and twenty-two rings on the body, and eight and twenty on the tail. The head was whitish, with some brown points. The body was brown, with irregular white spots underneath. Its length about a foot and a half.

All the amphisbœnæ whose origin is distinctly known come from America. No one but Seba attributes any of them to Ceylon, an application in which he was, doubtless, as erroneous as he was in many others. The taste for ants has been attributed to fuliginosa as well as alba ; in the stomach of the one we have just mentioned, nothing was found but blattæ and crabs.

Shaw has figured in the "Naturalist's Miscellany" a rose-coloured amphisbœna, which would seem to constitute a distinct species ; but his description is rather incomplete, and does not afford a very determinate idea of the animal.

Leposternon and *Typhlops* afford no materials for discussion in this place.

The first of the serpents proper, as our author terms them, is Tortrix. Among the different species we shall notice the *anguis scytale.* The head is small, oval, depressed, and rounded in front. The eyes are very small, and, as well as the nostrils, situated on the middle of a plate. The body and tail are cylindrical, and of equal bulk throughout. The scales are smooth and reticulated, not imbricated. The teeth are small, simple, and sharp. The general colour is white, with a small tint of yellowish; and there are about sixty black bands, irregular, transversal, and forming, for the most part, interrupted rings. This reptile is about thirty inches in length at the most.

The anguis scytale is an inhabitant of South America, particularly Cayenne and Surinam. It lives on caterpillars, worms, and small insects, but especially on ants. It appears to have the habits of the amphisbœna. The negroes are very much in dread of it, but obviously without any reason.

The *tortrix,* or *anguis corallinus,* is an inhabitant of South America. Its colour is a beautiful red. According to Laborde, in Guiana, its bite is venomous and very dangerous; and if Gumilla, the author of the Natural History of Orinoco, is to be believed, it is very much to be dreaded. This does not seem very probable, though, as the Baron observes, it is by no means certain that all serpents without venomous fangs are perfectly harmless in this way.

The genus Boa, though without venomous fangs, approximates very much to the crotali. This name employed by Pliny, and subsequently by Johnston, Agricola, and Ruysch, indicated, according to those authors, the habits of this serpent, which was accustomed to follow the hinds, to fasten itself to the teats of cows, and suck their milk. Other writers think that this name comes from the Brazilian word *boa*, which is said to mean a serpent.

Linnæus was the first systematic writer who established this genus. It was subsequently adopted in part by Laurenti, Boddaert, Daubenton, and Schneider. The ancients appear to have had some acquaintance with some of its species. Aristotle speaks of African serpents as long as vessels by which a galley with three oars might have been overturned. Pliny talks of Indian serpents capable of swallowing deer. Elian mentions dragons of eighty to one hundred cubits in length; and finally, Suetonius mentions that there was exhibited at Rome, under Augustus Cæsar, a living serpent of fifty cubits in length.

The tortuous actions and enormous muscular power of the snakes result from an organization altogether their own, which we shall illustrate by figures taken from the *boa constrictor*. The vertebræ and the ribs constitute the entire skeleton of the body, as they have neither sternum, pelvis, bones, or limbs. Each vertebra has an apophysis on the upper side, which, in the boa constrictor, is separated from that of the adjoining vertebra, but in others, as the rattle-snake, these apophyses seem to touch each other; on the tail, however, of all these animals, the upper apophyses become mere tubercles. Fig. 1, will shew their shape and arrangement in the boa constrictor —the articulation of the vertebræ to each other is in an imbricated manner. The lateral or transverse apophyses are articulated to the ribs by a ball and socket joint, in the manner represented in the same figure. Thus the upper side of the mobile vertebral column is furnished with a series of laterally flattened appendages to the vertebræ, separated and at some distance from each other, evidently for the purpose of preventing a dislocation of the vertebræ ensuing from the animal inclining itself too much backward; a partial capability of such a movement is permitted by this arrangement,

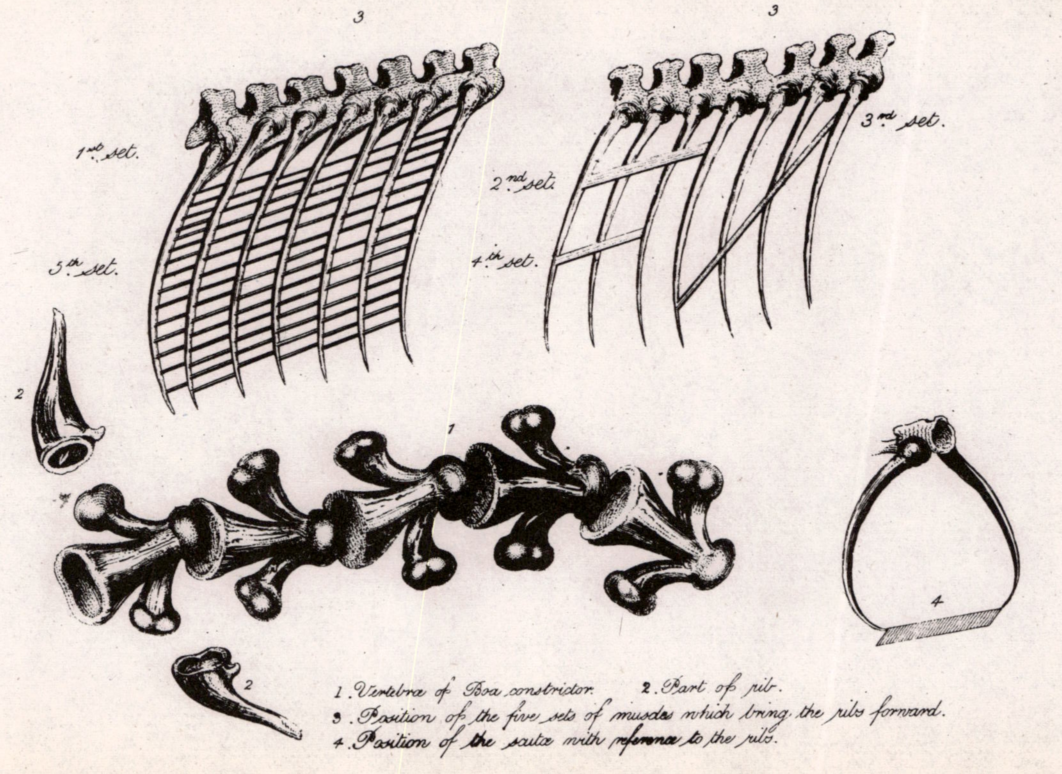

1. Vertebræ of Boa constrictor. 2. Part of rib.
3. Position of the five sets of muscles which bring the ribs forward.
4. Position of the scutæ with reference to the ribs.

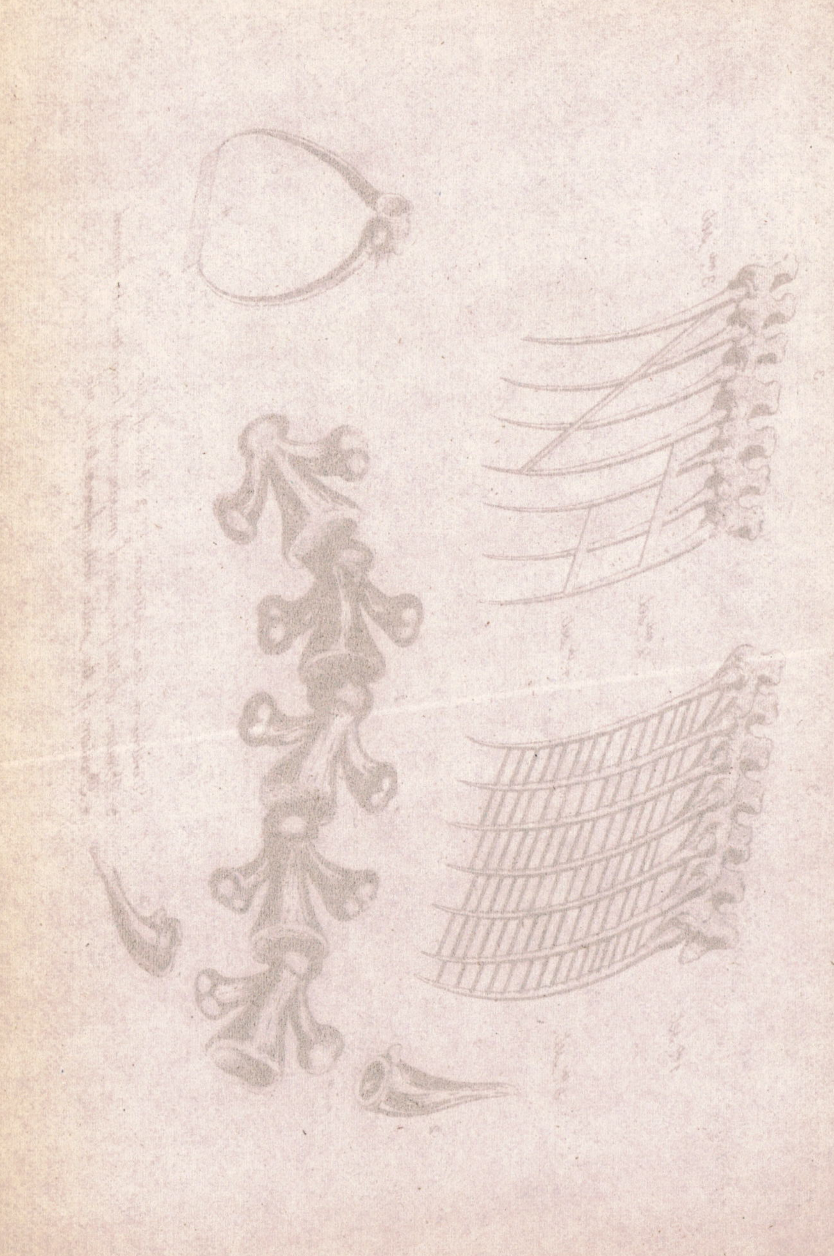

Col. Smith. delᵗ.

THE JIBOYA.

BOA. CENCHRIA. L?

but not one from which luxation might ensue, while the imbricated articulations of the several vertebræ, and the ball and socket junction of the ribs with the spine, enables the animal to execute the various tortuous gesticulations of the body of which snakes are capable.

The muscles which act on the bony frame have also a peculiar arrangement. There appears to be five different sets of muscles which bring the ribs forward—the first of these proceeds from each vertebra to the head of the rib immediately behind it—the second set are attached to the first and fourth ribs passing over two in a slightly inclined direction—the third in like manner passes over two ribs, but in a much more inclined direction, being attached to one rib near its articulation with the spine, and to the other not far from its termination beneath—the fourth passes over only one rib in a direction nearly parallel with the second, and the fifth or undermost set pass from one rib to the next for nearly their entire length.

The forward motion is moreover considerably accelerated by the scutæ, or transverse plates on the abdomen, which are attached on each side to the squamæ or scales. These scutæ are connected with the ribs by the upper edge only at each end, and when brought forward by the ribs, the scutæ have some hold on the ground by the under edge, and each becomes a sort of fulcrum to assist progression. All the five sets of muscles which bring the ribs forward, act conjointly on them, besides which, each pair of ribs, with their scutum, has a separate motion, and hence proceeds the amazing power of constriction which the boa is known to possess, as well as the capability of locomotion common to the order.

Some writers assert that the boas are found in the East Indies. But our author's opinion, as we have already seen, is different, who thinks that there are no true boas in the east. They are sometimes to be seen of thirty feet in length. They usually inhabit aquatic situations, placing themselves in ambuscade on the banks of rivers, where animals come to quench their thirst. Rolled upon themselves in spires, they form a disk of nearly seven feet in diameter, in the centre of which the head is placed. They thus await their prey in a motionless position, only raising the head occasionally some feet above this sort of spiral, to observe if any animal approaches. As soon as they imagine it within their reach, they shoot forth like a spring. They twist round its neck for the purpose of strangling it. When the animal is strangled, they break its bones, by squeezing it with the numerous folds of their body. They then extend it on the earth, cover it with their mucous saliva, and begin to swallow it, taking the head first. In this sort of deglutition the two jaws of the serpent dilate excessively, so that it seems to swallow a body larger than itself. In the mean time, digestion begins to take place in the œsophagus. Then the serpent becomes lethargic, and is very easily killed, for he neither offers resistance, nor attempts to fly.

We have seen that, from the numerous and ill-digested genus Coluber of Linnæus, many separations have been made, independently of that of those serpents which are not venomous. We shall notice some of the more remarkable.

The name Python was bestowed by Daudin on a genus very much indeed approximating to Boa, and which our author conceives to contain all the pretended boas of the ancient continent. Among the species of this genus, the one most worthy of remark is the *ular-sacra*, *Python amethystinus*, Daud; *Javan snake*; *Col. Javanicus* of Shaw. This serpent, which is as large as any boa, coming to more than thirty feet in length, inhabits the island of Java. The meaning of its Javanese name is *serpent of the rice-fields*, because it lives in them habitually. Its bite is not venomous. It usually lives on rats and birds, and sometimes on larger animals, in pursuit of which it proceeds to the mountains of the island.

Of the *Python bora*, Russel is the first who has given us any account. It is a native of Bengal, and not venomous, notwithstanding the assertion of the Indigenes, who affirm that persons bitten by it have a cutaneous eruption over their entire body in the course of ten or twelve days.

Over the rest of the separations from the COLUBER in the text, we pass, as there is positively nothing to be adduced concerning them, and proceed at once to those which might be conveniently called ADDERS.*

The head of this division is, in general, depressed. Its contour is most usually oval, though sometimes only elliptical. Some species have the faculty of widening and depressing it

* We have in the text translated the French word *couleuvre*, Latin *coluber*, by *adder*, in preference to using the term *snake*, under which the adders, as well as a number of other serpents materially differing from them, have been most improperly comprehended. The proper translation of the French word is certainly adder: the Latin word *coluber* may very properly bear the same signification, though very vaguely employed by ancient writers. *Anguis* we have preferred to translate by snake, instead of the very objectionable generic term *slow-worm*. We have had an opportunity of observing before, the difficulty, and in many cases utter impossibility of finding English equivalents for the names of Cuvier's subdivisions, and the impropriety of applying English specific names to many of them. But the word adder has the advantage of not having been generically used by our systematists, and may therefore be conveniently applied to the subdivision now under review. Indeed, the best recommendation that any name so applied can have, is that it has not been used before for a genus including very different species from those of the one for which it is employed again, or that it has not designated a species totally different from any comprehended under its new generic application. The propriety or impropriety of the translation is not of so much importance. The great objection to our translation is, that the word adder or viper is universally applied by the country people to the snake with poison fangs, which is therefore venomous.

at will. The scales which cover it, usually nine in number, are disposed, two by two, on the point of the muzzle, and on the occiput.

The tympanic bone is mobile, and almost always suspended to another bone analogous to the mastoidean, and fixed to the cranium by muscles and ligaments. The branches of the lower jaw are not united together, and those of the upper hold to the intermaxillary bones only by ligaments, so that they may be considerably separated. Accordingly they are among the number of serpents which have the power of swallowing bodies larger than themselves.

The palatine arches participate in this mobility, and are armed with sharp teeth, curved backwards, fixed, and not pierced. The branches of the two jaws are furnished with similar teeth, so that there are four ranges above, and two only below. These teeth would appear to be replaced whenever they are removed, but there are no fangs.

The tongue is forked, and very extensible. It is concealed, when in a state of repose, under a gross fleshy mass, situated at the bottom of the mouth. The œsophagus is, in general, capable of great dilatation.

These snakes are oviparous, and lay twice every year, viz. in the early days of spring, and towards the end of summer. Their eggs are oblong, and membranaceous, and are hatched by the heat of the sun.

The nature of the aliment of these reptiles varies according to the species; but they constantly seek out living animals, insects, worms, batracians, mollusca, small fish, birds, quadrupeds, &c. They never eat fruits in gardens, nor suck the milk of cows, in the fields, or stalls, as some visionary shepherds or impostors have pretended, whose tales have nevertheless spread this absurd prejudice throughout Europe.

It is probable that they live a long time; but we have nc certain data on this subject.

PERON'S PYTHON. *PYTHON PERONII.*

The colubers of cold and temperate climates bury themselves in the earth during the autumn, and remain there in a lethargic state for the whole winter.

The *common snake* of this country (*coluber natrix*, L.) is found throughout all Europe, on the banks of fresh waters, in meadows, and on the borders of woods. It is vulgarly called the water-snake. It may be handled without fear, for it never tries to bite, except when exceedingly irritated, and its bite is not at all dangerous.

If it be tormented, it hisses strongly, exhales by the mouth a fœtid vapour, slightly musked, throws fire into its looks, shoots forward in a serpentine line, and there runs out from under its scales a white humour of a most insupportable stench, and the odour of which is extremely tenacious. In moments of danger it also shoots its excrements, which are equally fœtid.

M. de Lacépède informs us that in Sardinia this serpent is brought up in a kind of domestication, and that it is not insensible to the caresses of its masters. Moreover, in that island it is regarded as an animal of favourable omen, and suffered to enter freely into all the houses.

In some countries these reptiles are eaten, and it is pretended that the flesh is exceedingly savoury. The fat is also sometimes employed in topical applications, and is said to be soothing and resolutive. Soups or broths are prepared from these animals, which are administered in scrofulous cases, in rheumatic affections, and in several cutaneous maladies.

The collared adder is easy to be reared, but it must be fed on small living animals, frogs, insects, mollusca, &c. It will refuse milk, cooked or raw meat, bread, and all kinds of vegetables.

These reptiles swim with very great facility, and traverse

ponds and streams. They also climb up trees with remarkable agility, to surprise and devour the young birds.

They lay from fifteen to forty eggs in holes on the edge of waters, in dunghills, and in hay-cocks. They are oval, as large as one's finger, and attached together in chaplets. They disclose the young about the middle of summer, and before winter the young are already six inches long. The length of the adult is three feet and a half; there are eight or nine varieties.

We insert a plate of the exuvia of this species. Fig. 1, represents the dorsal squamæ; 2, the abdominal scutor and squamæ; 3, the cellular substance; and 4, the fibrous texture of the scales—the last two greatly magnified. This cellular and fibrous construction which prevails throughout the skin, as well in the scuta as in the squamæ, seems necessary to the extraordinary power of dilatation which the ophidians possess in swallowing their food.

The *Colatrovirens*, Lac., which is called by Shaw the *French snake*, is the handsomest of the European species. Its size varies from three to four feet, and sometimes it is even five feet long. Its circumference is about three inches at most in the thickest part.

This reptile is not uncommon in the southern provinces of France, in the Bordelais, Poitou, &c.; it is sometimes found even as far as Fontainbleau. It usually fixes its habitation in woods, along hedges, or even in the midst of rocks or heaps of stones; it generally feeds on birds, mice, frogs, toads, &c.; it climbs trees and swims well. Daubenton was the first writer who has spoken of it; but the name of *common adder* (*couleuvre commune*) which he has given to it, is much more suitable to the *colubur natrix*.

M. Bosc tells us, that at the end of summer, some time before they conceal themselves, these adders are heard to hiss

THE COMMON SNAKE. COLUBER NATRIX

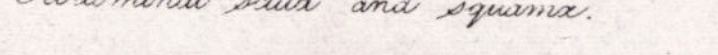

Abdominal scuta and squamæ.

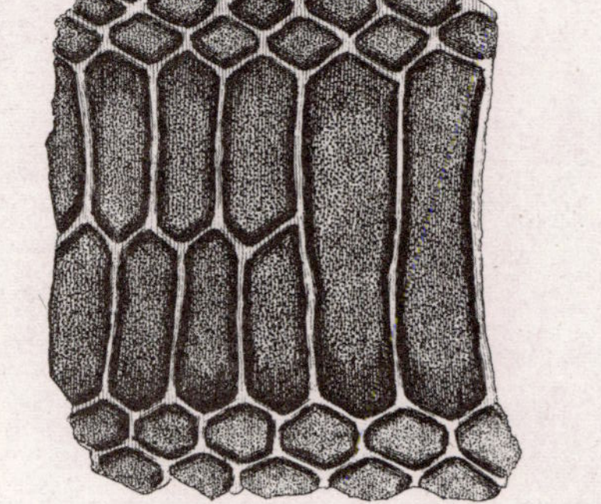

Dorsal squamæ.

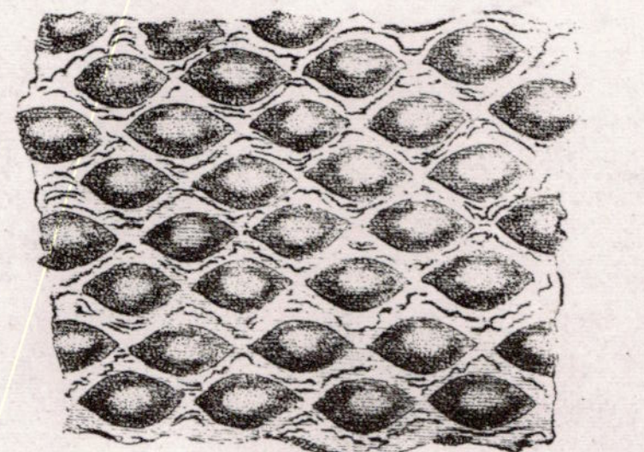

The fibres of the scale mag: 32.000.

The cellular ~~substance~~ mag: 32.000.

repeatedly during the evening; they seem to reply to each other and to be much agitated. In Burgundy, the people are persuaded that this is the season of their amours; but the truth is, that at the time of which we speak, that season is over for two or three months.

It is said that this adder is susceptible of education. Valmont de Bomare, who calls it *serpent familiar*, relates that he saw one so much attached to the mistress who fed it, that it would often glide along her arms as if to caress her, conceal itself under her garments, or repose upon her bosom. It knew her voice, would come when called, and followed her continually.

This probably is the adder which is common in Sardinia, and which Cetti has named *Colubro uccellatore.* This is the opinion of Daudin; but M. Latreille thinks that it should rather be considered as the analogue of the *anguis æsculapii niger* of Aldrovandus and Johnston.

The *Colubur Austriacus* was never described with any great exactness before the time of M. de Lacépède, and yet it is common enough in Germany and France, and even as far as Paris. According to Laurenti, it is frequently met with in the ditches and humid situations around Vienna. Daudin has observed it several times in ant-hills. It is a timid animal, and always ready to fly the moment it is perceived. It moves with considerable quickness, darting out its tongue, but it seldom hisses.

The *Coluber Constrictor* of Linnæus is very common in North America; it is found habitually from the spring in Carolina. Catesby has figured it under the name of the *black snake.* In America it may be said to hold the place of the common snake in Europe. According to the report of M. Bosc it is very strong, and defends itself obstinately when attacked; but its bite is not dangerous. It is reported in the country just mentioned, that it will fight with the rattle-snake and strangle it in its folds. Daudin tells us that it can be tamed, and that it only has recourse to flight to escape from the pursuits of man. It causes great destruction among rats and mice, and accordingly it is highly respected by the people, who are pleased to see it come into their houses. It also feeds on squirrels, opossums, frogs, lizards, and even birds of prey.

It is utterly impossible for us to follow any further the innumerable species which have been attached to this genus. We could cite little or nothing of any interest concerning them, and moreover, such are the confusion and want of authenticity in the species, that we might be delivering either that which had no truth whatever, or we might be recounting concerning one species that which properly belonged to another. The species of the figures inserted will be found described in the table.

Of Acrochordus, Shaw enumerates three species, but our author, as we have seen, allows but of one. Shaw also talks of the fruits found in the stomach of the *Acrochordus Javensis*, a fact, as the Baron well remarks, totally improbable. There is nothing concerning the habits of this animal that can be of the least interest to our readers.

We now come to the first division of the venomous serpents, and by far the most formidable.

The Crotalus is a genus of serpents celebrated from the earliest period of the discovery of America by the danger which accompanies their bite, and the peculiar appendages to their tail, which have caused them to be named *rattlesnakes*, which the name crotalus very well expresses, from κρόταλον, which signifies a bell, rattle, or cymbal. Such is the terror which they have inspired, that were we to trust to the relations of many travellers, we must conclude that America would be almost uninhabitable in consequence of their ravages. Towards the end of the last century, many

THE WANNA PAM SNAKE.

NATRIX STOLATUS. Mer. *C. FASCIATUS.* Lin.

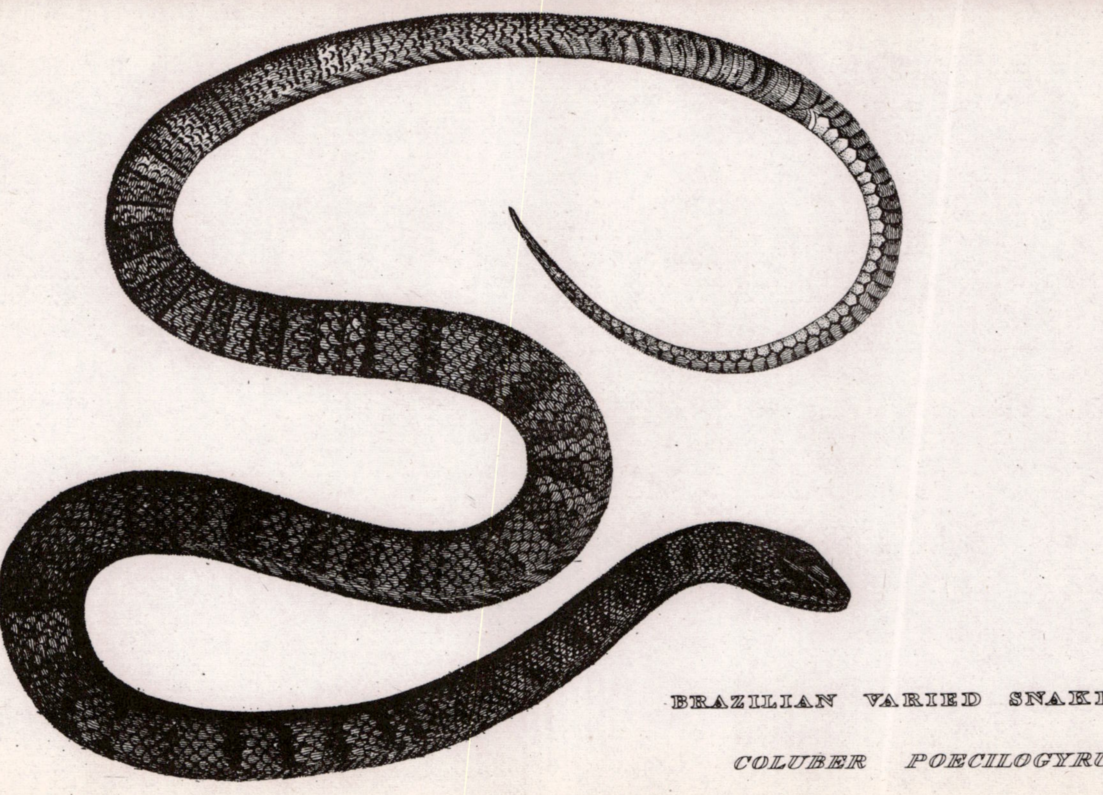

BRAZILIAN VARIED SNAKE. *adult.*

COLUBER POECILOGYRUS.

STRIPED-HEADED COLUBER.

COLUBER RABDOCEPHALUS. Pr. Max.

ORNAMENTED SNAKE.

COL. ORNATUS.

naturalists have employed themselves in making researches on the anatomy, the manners, and the poison of these reptiles, and accordingly their history is now tolerably well-known.

The head of the crotali is broad, triangular, and generally flatted in its entire extent. The scales of the cap of the cranium, and of all that part which is between the eyes and beyond them, are similar to those of the back; but those of the muzzle and those which cover the eyes, are often larger and in the form of plates. The muzzle is hollowed with a small rounded fosset behind each nostril.

Their eyes are exceedingly brilliant and provided with a nictitating membrane.

The mouth is very large; the tongue, forked at its extremity, is partly enclosed in a thin sheath, and susceptible of prompt movements in different directions. The symphysis of the lower jaw is not soldered, and the bone is armed with crooked teeth, which diminish in length in proportion as they recede from the muzzle. Enough has been said already respecting the nature of the venomous fangs, &c.

The body is robust, elongated, cylindrical, covered above with carinated scales, all of which are put in motion by a peculiar muscle; the anus is simple and transverse.

The tail is short, cylindrical, and somewhat thick. The number of little bells which terminate it, increase with age; and an additional one is formed at every moulting or casting of the skin. These bells are truncated quadrangular pyramids, broader in one direction than another, and received within each other in such a manner that only the third part of each is visible. They are hollow bones, and have each a rigid epiphysis united to them, though it is only apparent in the last or terminal bone. Thus the tip of every uppermost bone runs within two of the bones below it. Thus the bells are held without being actually bound together, and move with a rattling sound as soon as the animal agitates its tail. These various pieces receive no nourishment from the animal's body and therefore do not grow. The first which is formed is always closed and smaller than the rest; the size of the last piece of the bells or rattles, depends on the growth of the last caudal vertebræ, for those pieces are primitively moulded upon them. M. Bosc thinks that one is produced every year, "and if their number should be found to vary in the same species and at the same age, it is," adds this observer, "that they are subject to separate accidentally, and we can always, by means of calculation, find the number that is wanting, since they all grow in a regular proportion."

All these species are entirely similar to each other, not only in form, but also frequently in size. They are all of a fragile, elastic, semi-transparent matter, and of the same nature as that of the scales. Their numbers vary from one to thirty and upwards; it is for the most part between five and thirteen.

The noise which these rattles produce, when they are shaken, resembles that made by rumpled parchment, or that of two quills of a goose rubbed smartly against one another. It is reported that this noise can be heard at the distance of more than one hundred feet; but in the species which M. Bosc has observed in the living state, he says that it could not be heard farther off than twelve or fifteen paces; and when the animal was moving in its usual gait, it was necessary to be close to it, and even listen very attentively, to perceive the sound.

The crotali spread far around them a very fœtid odour. For a long time it was believed, and many naturalists still believe, that this odour has the power of stupifying, or even of fascinating the animal on which the venomous reptile is desirous of preying. When the rattle-snakes are dead, they decompose with exceeding rapidity. And such is the stench which exhales from their bodies, that it is difficult to remove them, without becoming unwell in consequence.

Excepting hogs, which feed upon them, every animal dreads the crotali; horses, and dogs especially, scent them at a distance, and most particularly avoid passing near them. "I have often," says M. Bosc, "amused myself by trying to force my horse and dog to approach one of these animals. But they would sooner have allowed themselves to be knocked down upon the spot, than come near them." Notwithstanding this, those animals often become their victims.

Although the rattle-snakes do not climb trees, yet their principal food consists of birds and squirrels. They also devour rats, hares, and small reptiles. It is imagined that by their glance alone they have the power of forcing their prey to precipitate itself into their mouths. It would seem, however, that they only seize it during the terror and confusion which the sight of them occasions in other animals.

They creep slowly, and do not bite but when provoked, or for the purpose of destroying their prey. A man can easily master them, when he perceives them at a distance, and takes proper precautions. They never attack him, and cannot possibly overtake him. It is even well known that they sound their rattles some moments before they take vengeance on their aggressors. M. Bosc was so little afraid of them, that he took all those alive which he met with, and which were not too bulky to be preserved in spirits of wine. When they are seized by the head, they cannot, like other serpents, raise their tail, and twist themselves round one's arms, nor make use of their strength to disengage themselves.

They usually rest, twisted in a spiral form, in places which are clear of grass and wood, in the customary path of wild animals, and particularly in those which conduct to water. There they remain quietly until some victim appears, and as soon as it is in their reach they dart upon it with the rapidity of an arrow. Nevertheless, it has happened more than once to travellers to pass very near a crotalus, and even to touch it with the foot without being bitten. The animal, on such occasions, immediately rolls itself up in a spiral, and waits for new provocations, before it darts out. If one removes from it, it elongates itself gently, and creeps in a right line, keeping its rattles raised up, and shaking them from time to time. If it is provoked again, it stops, and reassumes its spiral figure; it moves its bells with rapidity; its head and neck become flattened; its cheeks swell; its lips contract; its jaws, widely separated, allow the formidable fangs to appear; it darts out repeatedly its long and forked tongue; its body swells and sinks successively through rage; it threats, but it never springs forward, unless it is certain of reaching its enemy.

An animal surprised by a rattle-snake very seldom attempts to escape; it becomes petrified with terror at the aspect of the reptile, and even appears to rush upon the fate which awaits it.

These serpents are so dangerous that the slightest prick made by their venomous fangs, will kill almost the largest animals. Laurenti says, that when one has been bitten by a crotalus, the entire body is swelled, the tongue becomes prodigiously inflamed, the mouth is burning, an inextinguishable thirst takes place, a spitting of blood, the edges of the wound become gangrened; and at the end of five or six minutes, the victim dies in frightful agony. In the Philosophical Transactions we find the result of many experiments made upon the bite of this formidable animal. Captain Hall having caused a rattle-snake, about four feet long, to be attached to a stake, exposed some dogs to its bites. The first which was struck by its murderous fang died in fifteen minutes; the second perished after suffering two hours, and the third did not feel the effects of the poison until three hours had elapsed. At the end of four days the experiments were recommenced with the same rattle-snake. The first dog died in thirty seconds, and another in four minutes; three days afterwards, a frog having received the bite, died in two minutes, and a pullet in eight minutes. Some time afterwards, they presented to the same serpent a white amphisbœna, which died in eight minutes; and the rattle-snake itself, being bitten in its turn, did not survive more than twelve minutes at the most.

Kalm assures us that the crotali destroy horses and oxen almost instantaneously, but that dogs resist better. He also adds that men may be cured when a remedy is applied in time; but that if any one of the larger vessels is opened, death supervenes in two or three minutes. The fangs, says the same traveller, will pierce through leathern boots, and of course through gloves.

In the Philosophical Transactions for 1810, Sir Everard Home relates an example of the fatal effects of the bite of crotali. Among the symptoms which he enumerates, we find such a weakening of the action of the heart, that the pulse is scarcely to be felt, and so great an irritability of the stomach, that this viscous can retain nothing in its cavity. When the wound has been made in the finger, that part immediately mortifies. When death takes place, we do not find that the absorbent vessels, and the lymphatic ganglia, exhibit the alterations which poisonous substances usually produce in them. The body preserves its general aspect; the parts in the neighbourhood of the bite alone are attacked in an apparent manner. Moreover, the effect of the poison is so immediate, and the irritability of the stomach becomes so great, that remedies in general come too late: very little chance of success remains.

It is also observed in these unhappy circumstances, that

a black and fluid blood escapes through all the apertures of the body. If the wound be in the neck, death is then almost inevitable, for asphyxia is the necessary consequence of the swelling that ensues. M. Bosc has had occasion to observe a fact of this description.

It is chiefly in stormy weather, when the atmosphere is charged with electricity, and the sun shines through the clouds, that the rattle-snakes are most dangerous.

Although the wound produced by the bite of one of these animals may be more than an inch in extent, it is said to be scarcely felt in the first instance; but at the end of a few seconds, the consequences begin to be manifest.

Three kinds of remedies have been employed against the bite of the rattle-snake, suction and ligature, caustics, and internal medicines.

The first is the most certainly efficacious, when it is possible to employ it. Ligatures will, to a certain extent, contribute to retard the general absorption.

For the second the Indians employ chewed tobacco, applied to the wound; Gumilla, in his Natural History of the Orinoco, also recommends, on his own personal experience, gunpowder lighted over the part, after certain scarifications have been made.

For internal remedies many plants are used, pounded or mashed, such as certain lettuces, the root of *prehnanthes alba*, the root, stalks, and leaves of a species of *helianthus*. In desperate cases, according to M. Palisot de Beauvois, the pounded bark of the root of the tulip-tree is employed with advantage, and in the course of treatment, the root of the *spirræa trifoliata* as a purgative.

The most powerful sudorifics are also recommended. The roots of *polygala seneka* of *aristolochia serpentaria* and *anguicida*, employed in decoction and fomentation to the highest possible degree of heat, are also prescribed.

But even among those who have the good fortune to escape from death, there are few who do not retain during life some infirmity, a sorrowful remembrancer of the fatal accident which they have experienced. Swellings, periodical aches, weakness or paralysis of the part, accompany them even to the tomb.

All the species of crotali, whose country is well known, come from America, and it has been remarked that the individuals of this genus have diminished in number, in proportion to the increasing population of that mighty continent. At present in those parts, in the neighbourhood of the sea, few are found which arrive to any great size. To the religious respect in which they were held by the savages, who regarded the death of one of these serpents a public calamity, has succeeded a hatred so inveterate, that their heads have been set at a price in many settlements. They are accordingly become so rare, that M. Bosc says, that in the neighbourhood of Charlestown he saw but six or seven individuals, of the larger species, in the course of a year.

Bartram informs us that he has seen some rattle-snakes as thick as a man's thigh, and more than six feet long, and that he has heard that in the early period of the settlement of Georgia, they have been observed of seven, eight, and even ten feet in length, and eight inches in diameter.

In those parts of North America, where the cold commences to be sharp, and the winter is rigorous, the crotali pass some time in a lethargic state near the source of rivers, in covert places, where the frost cannot reach them. Many of them are often found together in the same hole, along with toads in a similar situation. They are also to be met with in the same state under masses of *sphagnum*, which grow in marshy soils. It is always before the autumnal equinox, and after they have changed their skin, that they bury themselves in their retreat, from which they do not emerge until after the vernal equinox. In winter then, of course, they do not bite; and in spring, and even to the month of July, the effects of their bite amount almost to nothing.

Rattle-snakes are still to be found, from New York as far as Savannah, and from the shores of the ocean to a very advanced distance in the west and north-west. In 1797, Messrs. Palisot de Beauvois, and Peale, of Philadelphia, took nine of them in two hours in New Jersey.

At Cayenne, and in the hot latitudes of America generally, they are in constant activity all the year, and never fall into a lethargic state.

The crotali are viviparous; at Martinique it is the general persuasion that their offspring are eaten by the vipers when they are very young, and a little after their birth. According to M. Palisot de Beauvois, this prejudice derives its origin from a fact wrongly interpreted. In the first journey made by this naturalist, in the country of the native *Tcharlokee*, he saw a *crotalus horridus* in a path, and approached it as softly as possible. At the moment when it was about to be struck, the animal agitated its rattles, opened a wide throat, and received into it five little ones, about as thick each as a goose-quill. But at the end of ten minutes, believing itself out of danger, it opened its mouth again and let the young ones out, which, however entered there again, on the appearance of a new danger. Mr. Guillemart, a countryman of our own, has verified the same fact.

The negroes eat the flesh of the crotali as well as that of other serpents. Their fat is collected and applied for the purpose of mitigating the pains of sciatica; and it is pretended that the delivery of women is facilitated by their bells or rattles.

These reptiles can live a very long time. Some have been

mentioned as having forty or fifty pieces in their rattles, and being from eight to ten feet in length. The notions, however, existing on this subject are very confused and uncertain. They have great tenacity of life.

Tyson dissected one, which lived many days after most part of the viscera had been removed, and the skin torn off. M. Bosc has tried similar experiments on those which have fallen into his hands.

The *Crotalus Horridus*, L. *Banded Rattle-snake*, Shaw, is a native of Southern America. It is named *boicininga* by Pison and Marcgrave. The Mexicans, according to Hernandez, called it *teuhtlacot zauhqui*, which means *queen of serpents*. The Portuguese of Brazil designate it under the name of *cascavela*, and the natives employ the terms *boïquira* and *boicininga*.

The boïquira is generally from four to six feet in length. It has given rise to a crowd of absurd fables. Thus Pison pretends that the point of its tail, introduced into the rectum, produces death more quickly than the poison of the fangs.

We shall copy here Dr. Shaw's abstract from Tyson's account of the anatomical peculiarities of the rattle-snake, such particulars excepted as we have already given in our general account.

" The wind-pipe, as in the viper, as soon as it enters the lungs consists of semi-annular cartilages, which, being joined at both ends to the membrane of the lungs, contitute a free or open channel, thus immediately transmitting the air to the vesicles of those organs which are of very great length, beginning near the throat, and running down three feet in length. The upper part of them for the distance of about a foot from their origin, is composed of small vesiculæ or cells, as in the lungs of a frog; and which, from the frequent branchings of the blood-vessels, appear of a florid red; this part tapers proportionally to the body: the lowest part of it near the heart being moderately blown, is about five inches and a half in circumference: a little lower, for the space of about four inches, the cells gradually disappear, so that they seem at last to form only a reticular compages of *valvulæ conniventes* in the inside of the membrane of the lungs: the greatest circumference here is about six inches; the remaining part of the organ is merely a large bladder, without any cellular subdivisions, and consists of a strong transparent membrane, the circumference of which, when inflated, is about eight inches and a half. The lungs in the water-newt, and some other animals, are divided into two large lobes, or simple bladders, without cellular subdivisions; in the frogs, crocodiles, &c. of two large lobes, with cellular divisions; while in the rattle-snake, viper, &c. both these kinds of structure are comprised, the fore-part of the organ being filled with numerous intestinal vesicular subdivisions, while the remaining part is mere lengthened bladder.

"The œsophagus, or gullet, was two feet three inches in length, and marked by two distinct swellings or enlargements of very great size, so as to represent two preparatory stomachs as it were; nor was the real or proper stomach capable of so much distension as these; the length of the true stomach, or third enlargement, was nearly similar to that of the second enlargement of the œsophagus; it was much thicker than that part, and resembled, in its fabric, that of the viper. From the pylorus the duct straitened again for about half an inch, and then formed a large intestine, the weaved rugæ of its external coat presenting a curious and pleasing spectacle. This intestine, after some small windings, terminated in the rectum, which was of much smaller diameter. In the promiscuous food which serpents take in, which they always swallow whole, and in which there are always some parts unfit for digestion, and which must therefore be returned, the œsophagus here being very long, nature has provided the above-mentioned swellings or enlargements of that part where they may be respited during the efforts made use of by the animal for that purpose, till collecting its force, it gives them, as it were, another and another lift, and at length ejects them; and if what is confidently affirmed be true, that, on occasion of danger, they receive their young into their mouths, there are fit places for receiving them.

"The heart was placed near the bottom or base of the trachea, on the right side of it; its length was an inch and a half, and its figure rather flat than round, encompassed by pericardium; the vesicle being larger than the heart itself. It had only one ventricle, the valves being small and fleshy, and the inside of the ventricle distinguished by four or five cross furrows."

"A little below the heart lay the liver, which was about an inch wide in the largest place, and seemed divided on one side by the vena cava into two lobes of an equal length; that on the left side being about ten inches, and that on the right a foot long. Its colour was a brown red, and its use, no doubt, the secreting of the gall, which was contained in a bladder, seated at some distance below it.

"The fat in this animal was very plentiful, and the membrane to which it adhered seemed to be the omentum, which encompassed all the parts contained in the lower belly, and was joined to both sides of the ribs, running from thence to the rectum, and forming a bag which enveloped the parts there, but was free, and not conjoined towards the belly. There was no diaphragm, or separation between the heart and lungs, and the abdominal viscera.

"The kidneys, which lay towards the back, on each side of the spine, were not very firmly conjoined, and were about seven inches in length, that on the right side somewhat exceeding that of the left: each was about an inch in diameter, and though forming one continued body, yet plainly distinguishable into several smaller kidneys, to the number of fifteen; all so curiously contrived, with such an elegant compages of blood-vessels and tubes, as to compose so many regularly formed bodies, which could not be viewed without admiration."

Dr. Shaw is of opinion that the individual dissected by Dr. Tyson was the *crotalus durissus*, and not the present species. This, however, makes no difference respecting the anatomical particulars which that eminently accurate observer has transmitted to us.

The *crotalus durissus, striped rattle-snake* of Shaw, inhabits the temperate countries of North America, as far as the 45th degree of latitude. An individual, four feet long, killed by Mr. Bosc, had in its stomach a hare of the species *lepus Americanus*. Indeed the habitual aliment of this reptile consists of small rodentia, such as rats, mice, squirrels, &c.

This serpent traverses with ease rivers and lakes, by swimming, swelling out its body like a bladder; and it is said that it is very dangerous to attack the durissus at such times because it shoots into the boats with great facility.

It would appear that this species and the last have been many times confounded by naturalists. At all events the names of each have been indiscriminately applied to either.

Crotalus miliarius, Lin. *Miliary rattle-snake*, Sh., is an inhabitant of Carolina. In the United States, this serpent is considered as more dangerous than the durissus. Its smallness and colour prevent its being perceived. Its rattles can hardly be heard, even when it is held in the hand, and one is thus exposed to walk, and even to sit upon it. It is fond, says M. Bosc, of remaining coiled up, on the tops of the roots of large trees, or on the fallen trunks. It lives on frogs and other small animals, such as grasshoppers, insects, and worms. It is not easily frightened, or induced to fly; but the smallest blow of a stick is sufficient to kill it. The

French traveller Lebeau, who visited the Acatapas, a tribe of Louisiana, says that the poison of this rattle-snake is more subtile than that of the other crotali, since the success of the remedy is doubtful after three hours have elapsed from the infliction of the wound; while, according to him, a person bitten by the boïquira may hope to be cured at the end of six hours. The same author tells us, that ammonia is the remedy which is employed in such cases with the greatest success.

Though a great number of the military rattle-snakes are destroyed every year, the race continues to be perpetuated with the most astonishing activity, and that even in countries which for a long period have been well populated.

The TRIGONOCEPHALUS has been separated by modern erpetologists, from the vipers of Daudin and the coluber of Linnæus. We shall here confine ourselves to the notice of the most remarkable species—*Trig. Lanceolatus*, the *Megæra* of Schneider and Shaw.

This serpent commonly attains to the length of five or six feet, and may sometimes exceed those dimensions. Colonel Moreau de Jonnès, who has given a very interesting history of this animal, tells us of one killed by an officer in 1808, which measured seven feet six inches and six lines. Pere Dutertre informs us that, in his time, individuals of this species were frequently met with of from seven to eight feet long, and as thick as a man's leg. Labat saw one that was nine feet long. It has even been advanced that some were five-and-twenty feet in length, and twelve inches in diameter; but this seems only one of many instances of the propensity to exaggeration which is inherent in the human mind.

This serpent varies much in colour as well as size; and therefore, as Count Lacépède remarks, the denomination of *yellow serpent* of the Antilles, is highly improper.

The fecundity of this reptile is terrific. The scientific

officer, whose name we have just mentioned, has always found fifty or sixty young ones in the bodies of such females as he has had occasion to open. At the moment of their birth these young ones are completely formed, very agile, ready to bite, and six or eight inches in length.

The habitat of the *trigonocephalus lanceolatus* is very much circumscribed. It does not extend throughout the entire of the Archipelago of the Antilles, nor is it found even in the majority of those islands which constitute that Archipelago. By a chance equally singular, fortunate, and inexplicable, it is confined to the islands of Martinique, St. Lucia, and Beconia alone ; and there is no proof that, as has been pretended, it is common in the American continent. Nevertheless, a tradition exists among the Indigenes that it was introduced into Martinique by the Arronages, a horde which inhabited near the mouth of Orinoco, and which, impelled by sentiments of hatred and vengeance against the Carribs of that island, made them this fatal present, and let loose in their forests this serpent, which was brought over in calabashes. But according to another popular opinion in the same country, the trigonocephalus is aboriginal of Martinico, and cannot live elsewhere, not even in Guadaloupe. Some, however, think differently, and explain the phenomenon by the existence of the dog-headed serpent, which is believed to be a boa, and which, common in Dominica and St. Vincent, has delivered those islands from the trigonocephalus.

Be all this as it may, it is very certain that this species is greatly multiplied, at the present day, in St. Lucia and Martinique, where a field of sugar canes is never cut without sixty or eighty of these serpents being found. They people the marshes, the tilled grounds, the forests, the borders of rivers, and the mountains, from the level of the sea to the region of the clouds. They may be seen creeping in the mud, struggling against the currents of rushing streams which would hurry them to the sea, and balancing themselves on the branches of forest trees, more than one hundred feet above the ground. On the edge of the crater of the naked mountain which overhangs the town of St. Pierre, in Martinique, at a height of more than five thousand feet, M. Moreau de Jonnès and his companions encountered a trigonocephalus, the more to be feared as an excessive lassitude, the consequence of their arduous exertions, had then completely overcome them. Eight days before, at the foot of this same mountain, a fisherman shooting with his canoe over the volcanic pebbles of the shore, was attacked by a similar reptile concealed among the basalts, and no effort could save his life.

The serpents of which we are speaking are seldom found in the towns, unless they have been brought there in any green fodder. They do not, however, seem to fear inhabited places; they even frequently approach them, particularly in the night, and every year a great number of them are killed in the out-works of Fort-Bourbon, in Martinique, and of Fort Luzerne, in St. Lucia.

It is not even extraordinary to find them in the very body of these fortresses. In the country they frequently penetrate into the interior of the houses, when they are surrounded by bushes and high grass, and they seem to prefer the cottages of the negroes.

But it is particularly in the plantations of sugar canes, in the close thickets formed by those large granivorous plants, that the trigonocephali find an asylum, concealing themselves under the debris of the long leaves, with which the earth is strewn, and feeding principally on lizards, small birds, and especially on rats, which the Europeans brought with them into their colonies, and which have multiplied with astonishing fecundity.

They are also attracted by poultry-yards and aviaries; they often lie in ambush in parasite plants, in the lianas which surround the trees of the forest when fallen from age, or remain covered up in the nests of birds, whose eggs or young they have devoured, also in the holes of rats, &c.

These reptiles possess an activity and vivacity of motion truly alarming. A ferocious instinct induces them to dart impetuously upon passengers, either by suddenly letting go the sort of springs which their body forms, rolled in concentric and superpoised circles, and thus shooting like an arrow from the bow of a vigorous archer, or pursuing them by a series of rapid and multiplied leaps, or climbing up trees after them, or even threatening them in a vertical position.

The effects of the bite of these serpents are in general very terrible, but vary considerably according to a multitude of circumstances. The tumefaction of the part, which soon becomes livid and gangrenous, nausea, convulsions, cardialgia, and an invincible somnolency, are the ordinary symptoms of the action of their poison, which either produces death in the course of some hours, or, at most, of some days, or causes for several years, vertigoes, paralysis, more or less extensive, phagedenic ulcers of a malignant character, and a variety of other distressing infirmities.

It is, therefore, in nowise astonishing that the trigonocephalus is an object of horror, not only to man, but also to animals. The horse trembles and prances violently in its presence; rats scud away at its approach, sending forth cries of terror; birds especially, against which it wages inveterate war, manifest their aversion to it by repeated clamours; and the *loxia indicator*, by pursuing it with its cries, often indicates to man the place of its retreat.

The African races, which form a great portion of the population of Martinique, constantly preserve certain organs of this reptile for talismans, either preservative or hurtful. These are called, in the Carrib language, *piailles*, and they are always among the materials of those magical conjurations undertaken by the negroes who are addicted to sortilege.

It is moreover an opinion among these ignorant and unfortunate people, that the trigonocephali are commissioned, as were formerly the subjects of the old Man of the Mountain, to kill the person who is doomed to destruction. They also attribute to the trigonocephalus the power of fascination, of which we have already had occasion to speak, in reference to other serpents.

The severity of the accidents produced by the bite of the trigonocephalus, varies, as in the case of other venomous serpents, according to the state of health of the bitten subject, the depth and number of the wounds, the time which has elapsed since the animal made use of its fangs, and, consequently, the quantity of poison which has penetrated into the system.

But in all possible cases, the help of art is indispensable. Unfortunately, to the present moment, the mode of treatment has been based on custom and empiricism of the blindest kind; and the prodigious number of remedies vaunted and recommended in their turn, only proves the uncertainty and insufficiency of the means resorted to.

Formerly, in the origin of the colony, recourse was had to scarifications, and the application of cupping-glasses, which are still extolled by some practitioners as efficacious against the bite of the viper. The wound was then covered with a cataplasm of theriaca, and that electuary was administered internally. In default of this, they used to pound the head of the animal, and apply it topically upon the wound. A powder made with the hearts and spleens of serpents dried, was also long in use. Embrocations of hot oil have likewise had their partisans, and applications of pounded green tobacco leaves, and several other plants.

The remedies used in Europe against the bite of the viper have also been employed, such as Eau de Luce, liquid ammoniac, opium, and arsenical preparations. But none, in fact, of the vegetable applications appear to have any certain effect; and the treatment which we have already vindicated in our observations on the crotalus, and what we shall further point out when we come to speak of the vipers and the Naia, can only be relied on with any degree of confidence.

The VIPERS were originally united by Linnæus under his genus coluber, with serpents of a totally different and perfectly innoxious character. They now very properly form a distinct genus.

The *Common Viper, Col. Berus. L.*, is an animal whose size and appearance are far from being formidable. Its forked tongue, which is very extensible, is, however, soft, and quite incapable of making dangerous wounds, and shooting a mortal poison, as has most absurdly been believed, and even been asserted in several works, otherwise entitled to respectable mention. The two sharp languets which terminate it, make it indeed resemble a double dart, which the reptile brandishes in his mouth, especially when he threatens to bite. But the resemblance is deceitful, and the comparison made in consequence of it, to Slander, totally erroneous.

Such however is not the case with the venomous fangs with which the viper is armed, and of which we have already given an ample description.

The appearance, as we have said, of this reptile, is as harmless as its poison is dangerous. Its length rarely exceeds two feet, its shape is clumsy and inelegant, its colours dull and sombre, its motions of no great agility, and in short, its exterior altogether could never attract that attention which its fatal venom has too well merited in every age of the world.

The berus is pretty generally extended in all the woody, mountainous, and stony districts of temperate and southern Europe. It is common on the borders of dry coppices, on rocks and sands exposed to the sun, and is found throughout the whole of France, the British islands, Germany, Sweden, Poland, Russia, Italy, and even as far as Siberia and Norway. But very recently it multiplied to an alarming extent in the forest of Fontainbleau, where it was known under the name of *aspic*.

It lives on small quadrupeds, mice, field-mice, lizards, frogs, toads, salamanders, young birds, and insects, such as flies, ants, cantharïdes, and even scorpions, according to Aristotle. It also feeds on mollusca and worms, and, like all the Ophidians, can support without any material suffering a fast of many months. In many shops of pharmacopolists it is said that vipers have been kept in casks for years without giving them any thing to eat.

Like the other serpents also, it casts its skin at determined periods. It passes the winter and a part of the spring in a state of lethargy, and often in society in places somewhat humid, and where the frost cannot penetrate. It is, in fact, under heaps of stones, in the clefts of rocks, in excavated roots, or in the trunks of rotten trees, that the vipers assemble close to each other, during the rigorous months of winter, and twist and interlace their bodies together as if to resist the cold with greater facility.

In the fine days of early spring, the vipers may be seen basking in the morning sun, on little hills exposed to an eastern aspect, and they speedily occupy themselves in the great work of propagating their species. The act of generation takes a very long time in its accomplishment, and its result is the vivification of from twelve to twenty-five eggs, almost as large as those of wrens or titmice. These exclude the young, in the womb of the mother, and there they remain coiled up, and come to the length of three or four inches, before they issue forth, which they generally do in the course of the fourth month after fecundation.

Having thus, by a sort of parturition, quitted their mother, the young vipers, for some time after, carry with them the remains of the egg which enclosed them, and which then have the appearance of irregularly torn membranes. But from that time, they are entirely strangers to the being which gave them birth, and do not seek refuge in her mouth, on the approach of danger, as the ancients erroneously imagined.

After the season of love, the vipers appear less frequently, and even during midsummer they are not very often to be met with. They disappear altogether when the first symptoms of cold begin.

The duration of life in these reptiles, is nearly unknown, but it may be presumed that they can live a considerable number of years; for, though they are reproductive from their third spring, they do not acquire their entire development, in less than six or seven years.

It is with considerable difficulty that the vipers are destroyed; they will resist very severe wounds, and are not even easily to be strangled. They can remain many hours in the water without perishing, and for some minutes even in brandy.

It may also be observed that the number of their enemies is but small. Excepting man, who wages continual war against them, the wild boar, whose lard secures him from their bite, and the falcons and herons which feed upon them, all other animals, wild and domestic, fear and fly them.

In certain countries of Russia and Siberia, vipers are said to be held in singular respect, in consequence of a belief among the people that if they were to kill one of these reptiles they would be immediately exposed to the vengeance of all the other individuals of the same species. In consequence of this absurd notion these animals multiply to an incredible extent, while in the more civilized countries of Europe their number progressively diminishes from day to day. About forty years ago, Professor Bosc used to kill them by dozens of a morning on the chain of mountains which runs from Langres to Dijon; whereas latterly, in the same places, it is difficult to meet with a few individuals.

Of the anatomy of the viper it is unnecessary to speak here, as we have already enlarged quite sufficiently on that subject.

We have said that the viper is an object of terror to other animals, and the danger which accompanies its bite may sufficiently explain the kind of proscription to which it is generally devoted. It is, without contradiction, the most to be dreaded of all the venomous reptiles of Europe, and the one whose bite is followed by the most grievous, and sometimes fatal consequences. These effects, however, are totally prevented by our mountebanks, by stopping the hole in each of the venomous fangs with some soft wax, without pulling them out, after the fashion of the Egyptian jugglers and the Psylli of India.

In consequence of this danger many researches have been made to appreciate the nature of their poison, to determine its effects in a precise manner, and to discover the most efficacious means of neutralizing its deleterious action. Chemists, zootomists, naturalists, physicians, and empirics, have vied with each other in endeavouring to resolve these problems, and some useful truths have escaped them in the midst of assertions more or less ridiculous, and hypotheses of greater or less absurdity.

Charas, otherwise a man of great sense and considerable knowledge, asserted that the poison of the viper, or rather the fatal consequences of its bite, resulted, not from the fluid poured into its fangs, but from what he termed the *enraged spirits* of the animal. It is not necessary to enter into any refutation of this fantasy at the present time of day; six thousand experiments performed at Florence by the celebrated Felice Fontana, have thrown this hypothesis completely into the shade.

That naturalist has established as a principle that the poison of the viper is harmless for certain animals, such as the viper itself, the anguis, the slug, the leech, &c. He has also ascertained that this poison is neither acid nor alkaline to any marked degree.

Its taste is not very easy to determine, and very opposite statements have been made on that subject. The correct one, however, appears to be that it leaves in the mouth a sensation of an intermediate kind between that produced by a narcotic substance and that of an astringent salt taken simultaneously.

Its consistence is a medium between that of olive oil, and the aqueous solution of gum Arabic.

By dessication it grows yellow, and appears to crystallize, or rather to concrete, in the manner of mucus or albumen.

It remains a long time in the cavity of the tooth, whether separated or not from the bone which supports it, and from the membranes which envelope its base, just as we have already mentioned to be the case with the fangs of the crotali.

It is only with animals of a small volume that this poison is constantly mortal. It appears to be more dangerous for larger species in proportion as the serpent at the moment of biting has a greater portion of venom in reserve, as the bites have been more multiplied, as they have been made at a greater distance from each other, and as the temperature of the climate or season is hotter. *Cæteris paribus*, we find a sparrow succumb under this bite in five or eight munites, a pigeon in eight or twelve minutes, while a cat will sometimes resist it, and a sheep escape from its effects very frequently. These effects are always less severe in England and France than in Italy.

A hundredth part of a grain of this poison introduced into a muscle, kills instantaneously a warbler or a canary, while Fontana tells us that it requires six times as much to destroy a pigeon. Nearly two grains was found to have no effect upon a raven; according to this calculation three grains at least would be necessary to destroy a man, and twelve to produce the death of an ox under ordinary circumstances.

The organ which is wounded has a great influence on the nature and severity of the symptoms. Wounds in the neck, for example, are far more perilous than wounds in the limbs, in consequence of the neighbourhood of the larynx, the pharynx, the pneumo-gastric nerves, the multiplicity of absorbent vessels and lymphatic ganglia in that part, and its connections with the head and the principal centres of sensation, and with the respiratory and digestive viæ.

A considerable number of experiments and observations have, besides, demonstrated that the poison of the viper may be swallowed with impunity, if the mouth be not the seat of any excoriation or ulceration. Dr. Cloquet had occasion to verify this fact in his own person: being bitten by a reptile of this species in one one of his zoological excursions, on the forefinger, he instantly sucked the wound, which healed as quickly as would the prick made by a sharp pin, and no unpleasant consequences ensued.

The symptoms which, in man, succeed the bite of a viper, follow each other in this order: a sharp pain is first felt in the bitten part, which swells, becomes shining, red, hot, violet, then livid, cold, and almost insensible. The pain and inflammation seem to follow the course of the large nervous trunks and those of the lymphatic vessels. They acquire greater and greater intensity; fierce shooting pains are felt at some distance from the wound, and a sort of fire seems to insinuate itself into the intermuscular spaces. At the end of a few minutes the eyes, which are red and burning, water abundantly. This is speedily followed by swoonings, nausea, bilious vomitings, pains in the stomach, difficult breathing, heart-burn, a cold and colliquative sweat, tympany, sharp pains in the loins, a relaxation of the sphincter of the anus, a paralysis of the neck of the bladder, and, consequently, involuntary stools and evacuations of urine. The pulse is small, hard, concentrated, intermittent, and even convulsive, and the skin becomes as pale as virgin wax, or as yellow as lemon-peel, while a black, watery, and sanious blood flows from the wound, which presents a gangrenous appearance.

If these fearful symptoms be not speedily assuaged by the force of nature, or the assistance of art, the scene changes and assumes a still more terrific character. A yellowish serosity, and without consistence, takes place of the blood which was escaping through the wounds, the neighbourhood of which is then invaded by a soft œdema, and is covered with small pellucid vesicles announcing the approaching development of a sphacelus which is the precursor of death. A period, however, is not put to the sufferings of the unfortunate victim, until he has been left some time longer a prey to a violent cephalalgia, fatiguing vertigoes, excessive prostration of strength, an overwhelming and involuntary terror, a devouring thirst, repeated bleedings at the nose, and passive and disgusting hæmorrhages from the gums and the mucous membrane of the rectum.

To these symptoms are often united the most insupportable fetidity of breath, convulsive hiccoughs, and inexpressible anguish, and death does not terminate them until the most extreme degree of prostration has taken place.

As these symptoms vary in intensity, so are they developed with a greater or less degree of rapidity. Dr. Hervez de Chegoin saw a woman die in thirty-seven hours after having been bitten by a viper, and Dr. Pruina mentions the case of a man who gave way in eight hours under the influence of the venom of this reptile. This difference is by no means surprising to those who give themselves the trouble of a little reflection. It may be explained, on the one hand, by circumstances relative to the aggressing animal, such as its strength, size, the degree of anger by which it is animated, the quantity of poison which it has poured into the wound, the number of bites which it has made, the country more or less southern which it inhabits, and the temperature of the season. On the other side, the difference is referrible to causes belonging to the wounded animal, as constitution, age, nervous susceptibility, degree of fright, the fulness or vacuity of the primæ viæ at the moment of the accident, the nature of the injured part, and its more or less vascular structure. Thus we may ascertain why some infants, women, and even adult men, have perished rapidly in consequence of the bite of a viper, while others have been scarcely incommoded by it.

The experiments of Fontana, confirmed by the more recent ones of M. Mangili, have clearly demonstrated that the poison of the viper may not only be swallowed without inconvenience, when there is no excoriation in the cavity of the mouth, but even be placed with impunity in contact with other mucous membranes. Some experiments mentioned by M. Cloquet, and performed in Paris on the 28th of September 1828, with the poison of a dead rattle-snake, by Dr. Rousseau, anatomical preparer to the King's garden, are confirmatory of the assertion of the learned Italians on this point. A drop of the poisonous liquor was let fall on the conjunctiva of the eye of a frog; another drop of the same fluid was spread on the pituitary membrane of a second frog, and these two reptiles were not sensibly incommoded, while a third individual of the same species, having had the edge of the nostrils just scratched, gave way in a very little time, as did also some pigeons, underneath whose integuments the poison was introduced with a needle.

The treatment for the bite of the viper, has been very well indicated by Celsus in his treatise of medicine, and it merits some attention, especially as naturalists are frequently exposed to accidents from this reptile.

The first precaution to be observed in a case of this kind, is, when the disposition of the parts will permit, to fix a ligature above the wounded place, and not to tighten it too much, for fear of giving rise to mortification.

Immediately after, a cupping-glass is applied on the wound, the parts adjacent being first scarified, and this mode, highly praised by Celsus, has very recently been attended with happy results in the hands of Messrs. Mangili, Barry, and Bouillaud. This method, from analogy, affords an additional recommendation to employ the plan of suction, which has received the further confirmation of professional experiments tried by a number of physiologists and physicians.

When the cupping-glass has performed its office, the lips of the wound, already scarified, should be cauterized deeply and extensively. This should be done with a red-hot iron, chlorine of antimony, or concreted potassum.

According to Dr. Barry, the cupping will produce a good effect, in considerably mitigating the symptoms, even though they have declared themselves previously to its application. Dr. Barry's experiments were repeated in the presence of a commission composed of the members of the Royal Academy of Medicine, in Paris. One fact, indeed, was opposed to invalidate the results of those experiments. It was the case of a man mentioned by M. Richard, in which the cupping did not prevent the manifestation of the symptoms consequent on the wound. But the details of this case were very insufficient to found any opinion on, and we are left in ignorance whether the cupping-glass was properly applied or not, or suffered to remain for a sufficient time. The following case, on the other side of the question, was stated by Doctor Piovey at a sitting of the Royal Academy of Medicine.

"A man of forty-five years of age was bitten on the right hand by a viper. In an hour and an half after, pain, enormous swelling, and numbness of the wounded part, and the corresponding member supervened. Then followed a reduction of temperature, and a diminution of the action of the heart, and the radial pulse and that of the carotids were imperceptible. Then nausea, vomitings, spontaneous defecation, and enormous tumefaction of the face. The cerebral symptoms scarcely amounted to anything. Incision was made in the wounds on the hand, and a cupping-glass applied immediately, for half an hour. There flowed at first some drops of serous fluid, with which a cat was inoculated without inconvenience, and then several table-spoonsful of a fluid analogous to the serum of the blood. The internal affections were instantaneously suspended, and the local accidents diminished. It is true that a phelpnonous erisypelas seemed inclined to manifest itself on the following day, but it was removed by the application of forty leeches, and the patient was cured."

A variety of different substances, taken internally, has been lauded from time to time as efficacious against the bite of the viper. To mention them all would be out of the question; we shall, however, notice a few.

Sudorifics have been especially recommended, and among them the flesh of the lizard, of the coluber, and the viper itself, have been preferred, in consequence of the great quantity of ammonia which it has been ascertained to contain.

An alexipharmic virtue of the same kind has been also attributed to theriaca and other analogous electuaries.

It is very certain that ammoniac admixture, internally and externally, is a most powerful auxiliary in the treatment of the viper's bite. Jussieu tried it with signal success on a pupil, who was bitten by a viper in one of his herborizing excursions. Dr. Piorrey, above cited, relates that an individual bitten by the very same viper that bit the man whose case has been stated, applied to the wound a mixture of oil and volatile alkali, and experienced no ill consequences. What is said here of pure ammoniac applies entirely to such preparations of it as *Eau de Luce*, &c.

It has also been often stated that olive-oil possessed undoubted alexipharmic virtues. To prove the efficacy of this remedy, Mortimer exposed himself to the consequences of the bite of a viper. This courageous experimentalist was visibly benefitted by unctions made with this oil on the abdomen and on the wound, and by drinking two glasses of the same liquid when the symptoms exhibited the greatest severity. Experiments on this were instituted by the French Royal Academy, and their execution entrusted to two of its members, Messrs. Geoffroy and Hunauld. They appear to prove that the olive-oil cannot save small animals which have been bitten by the viper, from death, and is of little benefit to the larger in mitigating the symptoms. This, however, has been manifestly contradicted by Mr. Miller, in the case of the crotalus, so much more dangerous than our viper; and this ought to stimulate observers to new researches on the subject.

When all the means which we have enumerated prove inefficacious, recourse should be had to stimulants administered internally, to increase the action of the heart and vessels, and the general energy of the vital forces, to bring on a copious diaphoresis, and direct the action of the system from the centre to the periphery. Generous wine, mingled with theriaca, is very useful for this purpose.

The ancient chemists having supposed in vipers the exist ence of an *active* and *penetrating salt*, and an *exciting oil*, recommended these animals, and various preparations of them, against leprosy, elephantiasis, itch, tetters, scrofula, and malignant and pestilential fevers, because they accelerated the circulation of the blood, and were a solvent for lymphatic concretions. But these notions are now almost universally exploded.

The *Chersæa* (*Swedish viper*, Shaw) is common in the environs of Upsal, in Sweden : also in Smaland, Scania, and Pomerania, where it retires into thickets, under hedges, and to the foot of tufted trees. It is sometimes seen in Prussia, Poland, Denmark, and in the Pyrenees.

In Sweden it is known under the name of *œsping*, which seems evidently a corruption of *aspic*, and it has been sometimes called the *red viper*. It is erroneously located by Linnæus, Wolf, and Laurenti, in the genus coluber. Its resemblance to the common viper is very striking.

The chersæa of Sweden is a small reptile, about six inches long, and as thick as one's little finger. That of Switzerland and France differs from it much both in size and number of plates. Herpetologists, however, admit the identity of these animals, and make but one species of them.

Be this as it may, the Swedish viper is a reptile of the most dangerous kind. Its bite is often mortal, and its deleterious effects are manifested with greater rapidity than those which follow from the bite of the common viper. Acrell tells us that Linnæus beheld a woman perish in consequence of it, and this in spite of all the assistance which he could afford her. We are also told by the latter, that many of the inhabitants of Smaland are destroyed in this kind of way.

The accidents which follow the bite of the œsping are,

with the exception of their greater intensity, pretty much the same as those caused by the common viper. But the bitten part becomes the seat of a more considerable swelling, the wound acquires a tint of a more lively red, and its environs are covered with spots and bladders. A horrible anguish suddenly seizes on the wounded person: vomitings of greenish matter take place: the tongue swells and stiffens: the body becomes full of pains: a death-like coldness spreading more and more from the extremities, seizes at length upon the region of the heart. A peasant, according to the report of the Swedish physician Lars Montin, was bitten by an œsping in the little toe of the left foot. At the end of about six hours, the foot, the leg, and the thigh were very red, and considerably swelled. The pulse was small and intermittent. The patient complained of head-ache, of violent pains in the abdomen, of lassitude, and oppression. Tears flowed abundantly from his eyes, and his appetite was absolutely gone.

It is generally known that the poison of the viper does not preserve all its force during winter, and in northern countries, and that its energy, on the contrary, is augmented during summer, and in hot countries. The violence of the accidents observed in Sweden, towards the north of Europe, would seem to prove that the œsping is a species altogether different, even though that point was not proved by its zoological characters.*

* Such is the observation of M. Cloquet, but we are not informed respecting the time of year in which these accidents have occurred. It is presumable that they happened in summer, and the summer we know to be very hot in the north of Europe. Quere, also, whether after their long winter slumber, revivified by an almost tropical temperature, these reptiles may not possess more energy than in climates not subject to such extremes?

E. P.

In Smaland, where, as we have already said, this serpent is common, they have a custom of burying the bitten part, also of putting the crushed animal upon the wound, which they scarify, to make the blood flow. But these remedies, and many others, rarely succeed, and most of the inhabitants when they are bitten in the toe, prefer having it immediately amputated. Olive oil, in the case of the bite of this reptile, has been used without success.

The patient of Lars Montin, above mentioned, appears to have received some relief by the application of a cataplasm of ash-leaves on the bitten part, and by the administration every half hour of a mixture of the juice of these leaves, and wine, a wine-glass full to each dose. In the same case theriaca and hot oil were also employed.

Bergius has recommended internally a warm infusion of the stems of the *aristolochia trilobata* of Jamaica, taken internally, and unctions externally with camphorated linseed oil.

The *ammodytes* is a native ot all the south of Europe. It is found in Dauphiné, and in the neighbourhood of Lyons in France, and in the east of Europe, in the mountains of Illyria. It habitually frequents the rocks which border on the Danube, the neighbourhood of the city of Gorice and the Japidian mountains.

This reptile passes the winter concealed in clefts and crevices of rocks, from which it issues forth when the warmer rays of the sun announce the return of spring. Then it casts its skin, and the time of reproduction begins.

Its habitual food differs in nothing from that of the common viper, and its bite is not less dangerous. It instantaneously produces dazzlings of the eyes, giddiness, and a sort of insensibility, which lasts a long time, and is followed by intense pain. The wound is inflamed, the parts adjacent swell, and become livid and black in succession. Paleness follows. The pulse appears small, frequent, weak, and interrupted. The extremities of the limbs grow cold, the lips swell, and the patient is soon affected by all the symptoms of an adynamic bilious fever. Matthiolus informs us that he has seen persons succumb under the effects of a bite of this kind in less than three hours; but this, as in the case of the common viper's bite, must depend on many individual circumstances, on the present state of the system, or on a special idiosyncrasy.

The poison of this viper may also be swallowed with impunity, and the best method, in case of a bite, is to suck the wound as soon as possible, and with a certain degree of force. This method was tried with success in his own case, by Charas, in 1692, and it was on this reptile that most of his researches and experiments were made.

In the fields in the neighbourhood of Vienna, when any one is bitten by an ammodytes, a ligature is immediately applied both above and below the wound, which is scarified with a thorn of *paliurus acculeatus*, then rubbed with garlic, and fomented with a vinous decoction of rue and rosemary. Internal medicaments are also recommended, but as they are of the same nature as those already indicated, it is needless to repeat them here.

This viper was well known to the ancients, but its synonimy is very much confused, for Belon speaks of it under the name of *dryinus*, Aetius under that of κεγχρίας; Laurenti, sometimes agreeing with Aldrovandi, calls it *vipera Illyrica*, and sometimes makes a different species of it, under the denomination of *vipera Mosis Charas*. Gmelin, in his edition of the *Systema Naturæ*, describes separately, under the appellation of *coluber aspis*, the very same reptile which Linnæus and himself had already described under *coluber ammodytes*, which is the one under our present consideration.

The *Cerastes* has received its name from the Greek word κέρας, in consequence of the eminences which surround its eyes, and which, from the most ancient times, have been erroneously compared to the horns of mammiferous animals;

> "Cornua prætendens immania fronte cerastes
> Dum torquet spinam sibilat ecce vagus."

It attains to the length of about two feet, and the resemblance of those eminences just mentioned to a grain of barley, must probably give rise to the fable recounted by Pliny and Solinus, who tell us that the cerastes, concealing in the earth, or under leaves, the rest of their body, put these little horns in motion to attract the birds which they want to devour. This absurd assertion has received further amplification from Bishop Isidore, who, in a collection of all the popular stories in circulation in his time, has written that the horns of these serpents were curved like those of rams.

The cerastes partakes with the Naja the dominion of the deserts in the hottest regions of Northern Africa. Shunning humid and marshy situations, it is found only in the burning and arid sands of Egypt, Arabia, and Syria—sands in which it remains concealed during the entire day, and notwithstanding its great agility, it waits patiently until some victim presents itself to its insatiable voracity. It sometimes in this way gets possession of the jerboa, whose hole, according to Bruce, is very often contiguous to its own.

The singularity of the horned head of this serpent, and the danger which accompanies its bite, have caused it to be remarked even from the heroic times, by the inhabitants of those countries which are watered by the Nile. Accordingly, the Egyptians, friends equally zealous and blind of the marvellous, have often figured it among the hieroglyphics of their sacred monuments, and pointed it out to strangers as one of the most redoubtable of beings. They also pretended that at a period of antiquity the most remote, an invasion of serpents of this kind depopulated a part of the country, which probably gave occasion to Lucan to say, in the ninth book of his Pharsalia,

> "Pro Cæsare pugnant
> Dypsades, et peragunt civilia bella cerastes."

At the present day, when this reptile has been examined by observers of a sound and unprejudiced judgment, the bite of the cerastes is regarded as very dangerous; and yet we are not in possession of facts sufficiently precise to determine its effects, nor of positive information respecting the remedies opposed to them. We must consider as nothing, all that is told us on this subject, by Dioscorides, Aetius, Nicander of Colophon, Pliny, Paulus of Egina, and Celsus. Sautés de Ardoynis merely informs us that this serpent is of an excessively hot temperament, and that its poison is of the most subtile kind. The first part of this description as applied to any serpent is most absurd, and particularly false as regards the cerastes. Actuarius says, that its bite occasions delirium. Avicenna recommends to give to the wounded person a grain of horse-radish in wine, or to cover the wound with an onion, pounded in vinegar. It is not likely that such a remedy can be of the least use. Bruce alone has afforded us any data on this subject; and as his statements on this, as well as many other points, have excited severe criticism, more particularly among French writers, we shall present them to our readers as much abridged as possible, and then offer a few remarks on that author himself and his commentators.

After telling us that the cerastes is a great lover of heat ("for though the sun was burning hot all day, when we made a fire at night, by digging a hole, and burning wood to charcoal in it, for dressing our victuals, it was seldom we had fewer than half a dozen of these vipers who burnt themselves to death by approaching the embers"), Mr. Bruce proceeds to give so accurate a description of the animal as evidently proves his great familiarity with it. He then goes on:—

"The poison is very copious for so small a creature: it is fully as large as a drop of laudanum, dropt from a vial by a careful hand. Viewed through a glass, it appears not perfectly transparent or pellucid. I should imagine it hath other reservoirs than the bag under the tooth, for I compelled it to scratch eighteen pigeons upon the thigh as quickly as possible, and all died nearly in the same interval of time; *but I confess the danger attending the dissection of the head of this creature made me so cautious that any observation I should make upon these parts would be less to be depended upon.*"

"I kept two of these last mentioned creatures (the cerastes) in a glass jar, such as is used for keeping sweetmeats in, for two years, without having given them any food—they did not sleep, that I observed, in winter, but cast their skins the last days of April. The cerastes moves with great rapidity, and in all directions, forward, backward, and sideways. When he inclines to surprise any one who is too far from him, he creeps with his side towards the person, and his head averted, till judging his distance, he turns round, springs upon him, and fastens upon the part next to him; for it is not true what is said that the cerastes does not leap or spring. I saw one of them at Cairo, in the house of Julian and Rosa, crawl up the side of a box, in which there were many, and there lie still, as if hiding himself, till one of the people who brought them to us came near him, and though in a very disadvantageous posture, sticking as it were perpendicular to the side of the box, he leaped near the distance of three feet, and fastened between the man's forefinger and thumb, so as to bring the blood. The fellow shewed no signs either of pain or fear, and we kept him with us full four hours, without his applying any sort of remedy, or his seeming inclined to do so. To make myself assured that the animal was in its perfect state, I made the man hold him by the neck, so as to force him to open his mouth, and lacerate the thigh of a pelican, a bird I had tamed, as big as a swan. The bird died in about thirteen minutes, though it was apparently affected in about fifty seconds; and we cannot think this a fair trial, because a very few minutes before it had bit the man, and so discharged a part of its virus, and it was made to scratch the pelican by force, without any irritation or action of its own."

"A long dissertation," adds Mr. Bruce, "would remain on the incantation of serpents. There is no doubt of its reality, the scriptures are full of it; all that have been in Egypt have seen as many different instances as they chose. Some have doubted that it was a trick, and that the animals so handled, have been first trained, and then disarmed of their power of hurting; and fond of the discovery, they have rested themselves upon it, without experiment, in the face of all antiquity. But I will not hesitate to aver, that I have seen at Cairo (and this may be seen daily without trouble or expense), a man who came from above the catacombs, where the pits of the mummy-birds are kept, who has taken a cerastes with his naked hand, from a number of others, lying at the bottom of the tub, has put it upon his bare head, covered it with the common red cap he wears, then taken it out, put it in his breast, and tied it about his neck like a necklace; after which it has been applied to a hen and bit it, which has died in a few minutes, and to complete the experiment, the man has taken it by the neck, and beginning at the tail, has ate it as one would do a carrot, or a stock of celery, without any seeming repugnance."

"We know from history, that when any country has been remarkably infested with serpents, the people have been screened by this secret. The Psylli and Marmarides of old undoubtedly were defended in this manner,

'Ad quorum cantus mites jacuere cerastes.'

"To leave ancient history, I can myself avouch that all the black people in the kingdom of Sennaar, whether Funge or Nuba, are perfectly armed against the bite of either scorpion or viper. They take the cerastes in their hands at all times, put them in their bosoms, and throw them at one another, as children do apples or balls, without having irritated them by this usage, so much as to bite. The Arabs have not this secret naturally, but from their infancy they acquire an exemption from the mortal consequences of the bite of these animals, by chewing a certain root, and washing themselves (it is not anointing) with an infusion of certain plants in water. One day, when I was sitting with the brother of Shekh Adelan, prime minister of Sennaar, a slave of his brought a cerastes, which he had just taken out of a hole, and was using with every sort of familiarity. I told him my suspicion that the teeth had been drawn, but he affirmed that they were not, as did his master Kitton, who took it from him, wound it round his arm, and, at my desire, ordered the servant to carry it home with me. I took a chicken by the neck, and made it flutter before him; his seeming indifference left him, and he bit it with great signs of anger—the chicken died immediately. I say his seeming indifference; for I constantly observed, that however lively the viper was before, yet upon being seized by any of these barbarians, he seemed as if taken with sickness and feebleness, frequently shut his eyes, and never turned his mouth towards the arm of the person who held him. I asked Kitton how they came to be exempted from this mischief? He said they were born so, and so said the grave and respectable men among them. Many of the lighter and lower sort talked of enchantments, by words and by writing, but they all knew how to prepare any person by medicines, which were decoctions of herbs and roots. I have seen many thus armed, for a season, do pretty much the same feats as those who possessed the exemption naturally; the drugs were given me, and I several times armed myself, as I thought, resolved to try the experiment, but my heart always failed me when I came to the trial, because among these wretched people it was a pretence they might very probably have sheltered themselves under that I was a Christian, and that therefore it had no effect upon me. I have still remaining by me a small quantity of this root, but never had an opportunity of trying the experiment."

On this account the following remarks are made by M. Cloquet.

"All the absurdities which have been formerly put forth, on the subject of the psylli of Africa, and the marsi of Italy, who possessed the secret of handling venomous serpents, and escaping the accidents produced by their bite, have been repeated in modern times respecting many venomous ophidians, and by Bruce in particular, in the case of the cerastes. At the present day, we positively know what to believe in relation to this magic faculty, and no one is ignorant that it is founded on nothing but the dexterity with which mountebanks extract the inoculating fangs, or empty the poison bags, by making the animal repeatedly bite a spongy body. It is therefore impossible to attach any credit to the different facts reported by the traveller whom we have just mentioned, who says, among other matters, that he saw one at Cairo, which had bitten, without any evil consequence, the hand of a man so as to draw blood, and afterwards destroyed a pelican," &c. &c.

Now, we are very ready to give up Bruce's statement

respecting the *natural* exemption possessed by the inhabitants of Sennaar from the fatal effects of the bite of the cerastes, but the fact which he states so very clearly respecting the pelican, of which he was an eye-witness, and in which he was an actor, does not, we think, deserve to be dismissed so unceremoniously as M. Cloquet has dismissed it. In fact, when we consider how the bite of the most venomous serpents is known to vary according to circumstances, we shall not find any thing so very marvellous in this story of Bruce, nor need we have recourse for its explanation to any natural exemption, or medicated preservative on the part of the wounded man. As to the usage of herbs, we shall have occasion to see that that notion is not altogether so very unfounded as M. Cloquet would have us to believe. But the truth is, that poor Bruce seems to be regarded as an authority, by our continental friends, much in the same way as he was considered in this country immediately after the publication of his travels. As a philosopher, and as a systematic naturalist, his claims to attention are perhaps not very high; but the gross injustice of the imputation of wilful falsehood, so ignorantly and maliciously brought against him, has been clearly proved by the testimony of subsequent travellers—a fact with which the French *savans* in general appear to be totally unacquainted.

If the story of the pelican be not a downright lie, it is clear that, on that occasion, the fangs of the cerastes were not drawn, nor its poison exhausted. On the other occasion, Bruce merely relied as to this point on the assertions of Kitton and his slave, and Dr. Shaw very judiciously queries whether the chicken might not have died in consequence of the spinal marrow having been pierced, a result that would have as certainly followed from the insertion of a pin. We beg to call the attention of our readers to the passage which we have marked in Italics, as a proof of the caution and love of truth which distinguish this much abused writer.

The usual symptoms following the bite of the cerastes are a tumefaction more or less considerable of the part, a general icterus, a discharge of blackish sanies from the wound, swelling of the face, priapism, delirium, violent convulsions, and, in most instances, death: respecting the therapeutic treatment we know next to nothing.

Laurenti was the first who established the genus Naja, comprising two very celebrated species of serpents, the *Cobra di Capello Naja* of the Indians (*Spectacled Snake*, Shaw), and the *Haje* of the Egyptians, which is without doubt the true *aspic* of the ancients.

The *naja* or *cobra di capello* is equally remarkable for the elegance of its forms, the strength of its body, and the danger which accompanies its bite. It has received the name of spectacled snake, in consequence of a black mark which more or less exactly represents spectacles on the extensible portion of its neck. When the animal is in a state of repose, the neck has no greater a diameter than the head, but under the influence of passion, the skin of this part extends in the form of a coif or hood.

This serpent inhabits Coromandel, and is not found in Peru or Mexico, as many modern naturalists have erroneously stated, after Seba. It is very formidable from its envenomed bite, which is as dangerous as that of any other species of reptiles. When surprised by some imprudent traveller, it slowly raises its head, swells its neck, and advances against its aggressor in undulating movements executed solely by the tail.

The male does not appear to differ from the female. When Seba advanced that the latter had not the spectacled mark, he committed an evident mistake, confounding a difference of variety in the species with a difference of sex. The varieties, in fact, of this species are very numerous in India, and are distinguished by the inhabitants under many different names. Most of them are figured, and characterized in the admirable work of Russell.

It rarely happens that any one who has been bitten by a naja escapes death. In the moment of anger, this serpent unrolls himself, erects his body, moves with velocity his sparkling eyes, swells his neck, opens his mouth, and darts upon his enemy. Russell has made numerous experiments on this species, from which it appears that the bite of this singular reptile will kill dogs sometimes in seven-and-twenty minutes, and pullets in a minute and a half. The symptoms manifested by the dogs are plaintive cries, difficulty of sustaining themselves on their legs, impossibility of walking, agitation, tremblings, hard breathing, convulsive motions, paralysis of the hinder limbs, a stupor, and death following in general about twenty minutes, though sometimes this does not happen for some hours.

The same observer caused several pullets to be bitten by the cobra di capello. Concentrated sulphuric acid having been applied on the wound, the subjects of the experiment perished much sooner than those which had been bitten at the same time, and on whose wounds this caustic had not been applied.

A pig was bitten by the same reptile in the internal part of the thigh. During the first ten minutes it experienced no sensible effect; but after this it lay down, its respiration became laborious, convulsions followed, and it died in an hour after being bitten.

The effects produced on man by the bite of the cobra di capello, appear to be almost in all respects the same as those which we have already detailed in our accounts of the crotalus and the vipers. It is, therefore, unnecessary to repeat them, but we shall say a few words on the mode of treatment generally adopted in India for wounds of this kind.

Having dressed the wound, many practitioners are in the habit of administering a purgative *lavement*, and a potion consisting of two drams of Fowler's arsenical solution, of ten drops of Thebaic tincture, an ounce and a half of pepper-mint water, and half an ounce of lemon juice, put in at the moment in which the potion is given, which the patient should swallow during the slight effervesence then produced in the mixture.

The arsenical solution of which we speak, is an arsenite of liquid potash, two drams of which contain a grain of arsenic and the same of potash.

This potion is to be repeated each half hour for many hours in succession, and in the meantime the suffering parts are frequently to be rubbed and fomented with a liniment, composed of half an ounce of oil of terebinthine and liquid ammoniac, and an ounce and a half of olive-oil. The treatment is concluded by keeping the bowels free for a few days, and dressing the wound in a suitable manner.

The pills of Tanjore are also an Indian preparation greatly in vogue, for the cure of the bite of the naja and other venomous reptiles in general. Russell does not inform us of their composition, but he tells us that the arsenious acid constitutes their basis, and that a six-grain pill contains somewhat less than three-fourths of a grain of it. These pills are, perhaps, the same which the Indian physicians employ in the treatment of elephantiasis, the receipt for which has been given us by Dr. R. Thomas, of Salisbury. In them arsenic is combined with pepper in proportions of one part of arsenic to six parts of black picked pepper. These two substances, to which are added, in a receipt given by Daudin, quicksilver, almonds of Newalan, and roots from Velli-navi and Neri-visham, are pounded for a considerable time in an iron mortar, and then reduced to an impalpable powder in a stone mortar. Being thus completely pulverized, a little water is added to them, and pills are made about the thickness of a pea, which are preserved in a dry and dark place. One of these pills must be taken night and morning, in a betel-leaf, or if the latter cannot be procured, with cold water.

In the sixth volume of the "Asiatic Researches," Mr. Boaz, after having examined with care the ancient curative processes for the bite of venomous serpents, recommends in the last place as a specific in the dreadful malady caused by the poison of the cobra di capello, nitrate of silver, which was a remedy long since proposed by Fontana in the case of the viper. In the second volume of the same collection, Mr. J. Williams has inserted a paper on the caustic volatile alkali against the deleterious effects of the bite of different serpents, and particularly that of the cobra di capello. It seems that this medicament should be applied both externally on the wound, and at the same time administered internally.

A remedy much boasted of by some of the ancient missionaries, who had travelled in the East Indies, and who pretended to cure the bite of the naja, as well as that of other venomous animals, has been proved useless by the experiments of the learned Redi. This is a calculous concretion, which according to them is formed near the head, or in the body of the serpent in question, and which is named serpent or cobra-stone. This account of the origin of this pretended stone is assuredly false, and it is nothing but a factitious medicament composed by some charlatan. It appears to be nothing but a blackish or greenish argillaceous earth, which has the property of absorbing with great facility the humours which are formed at the surface of any wound whatever. But in India it is believed to imbibe quickly the poison discharged recently into the body of an animal bitten by the naia.

Kæmpfer has highly praised the virtues of the *ophiorriza mungo*, which grows in the warm countries of Asia, in cases of this kind. The remarks of Russell are in direct contradiction to this assertion, and confirm on the other hand the singular virtues of the Tanjore pills.

Notwithstanding the dangers to be apprehended from the cobra di capello, the Indian jugglers contrive to tame in some measure these redoubtable reptiles so as to shew them in public, and cause them to execute certain movements in cadence. These pretended enchanters carry these serpents from house to house, making them dance to the sound of the flute.

The fellows, who in Hindoostan are called by the Dutch name *snake-mans*, pretend to have the power of charming serpents by the effect of music, and carry about with them remedies which they aver to be of marvellous efficacy against the bite of these reptiles.

They even voluntarily suffer themselves to be bitten by the cobra di capello, whose fangs are in all probability first extracted. There are to be seen daily in the streets and public places, seated on the ground, squatting upon their knees with a small reed flute in the left hand. They open a round basket in which the reptile is kept, and at the sound of the instrument he issues forth, raises himself by degrees, rears upright, shouts out, moves in cadence, keeps his eye fixed upon his master, imitates his gestures in some sort, and follows all the motions of his right hand, in which he folds the cover of the basket.

In an anonymous work published at the commencement of the eighteenth century on the *conformity of the customs of the East Indians and Jews*, the author, who gives a figure of the kind of dance which we have just described, says that when in a house or garden, people are disturbed by the presence of an animal of this kind, they address themselves to an enchanter to expel him. This operation the conjurer performs by making the serpent come creeping to his feet, which appears charmed by the sounds of his flute or by his magic incantations, and he takes the reptile in his hands without receiving any hurt, taking care, however, not to kill him. "On a certain day," says the writer, "a soldier having knocked down a viper, which was thus charmed out of a guard-house by an enchanter, the latter appeared in a strange consternation, took it up, and proceeded to inter it with great ceremony in a hole, where he placed a little rice and milk, as an expiation of the injury committed against the viper race."

We know generally that the Phœnicians, the Egyptians, the Greeks, and the Romans, were accustomed to worship serpents. Some modern Egyptian and Indian enchanters pretend to tame, handle, and even eat with impunity reptiles of this species. In Malabar, in particular, the naja, is a kind of object of reverence and adoration. Its figure is one of the most usual ornaments of the pagodas; prayers and offerings are addressed to it. The Bramins conjure and exorcise it, and the pious souls fetch it milk, and various aliments in the forests which it infests and the roads which it frequents. The traveller, Dillon, relates that during his stay at Cananor, a secretary of the prince was bitten by a cobra di capello. The prince, terrified, convoked an assembly of the priests, who, after having explained to the reptile the reasons why a great value was attached to the life of the individual in question, concluded by threatening to burn it alive unless the man recovered; prayers and menances, however, produced no effect. The victim expired, and the animal, instead of being punished, was set at liberty, with a profession of excuses for having detained him, and a multitude of profound reverences

The name of *aspic* has been given amongst all civilized nations to a serpent rendered ever memorable by the death of Cleopatra, whose beauty, glory, honours, and deplorable end, have occupied the historians and poets of all times and of all nations.

It is universally known that this illustrious princess, abandoned by Fortune, who had so long smiled upon her, commanded that a reptile of this species should be brought to her, concealed in fruits and flowers, and caused it to bite her, to put a period to her misfortunes. But after the fall of the Roman empire, though Egypt still preserved some traces of the high renown of Cleopatra, and though the name of the aspic was not pronounced without some degree of horror by all the people of Europe, still for a long series of ages the true species of the serpent was unknown, and the cerastes, the Egyptian viper, the ammodytes, and the lebetina, were taken for it. Bruce declared for the first of these opinions, Forskal for the last, and Laurenti, Hasselquist, Daudin, and Count Lacépède, for the second, which undoubtedly has some plausibility, for it is well proved that under the name of ασπις, the ancients were acquainted with many venomous serpents aboriginal of Egypt.

It has been only since the expedition of the French to Egypt that the true species of the aspic has been ascertained. During the period of that expedition, the French philosophers attached to the army observed a species of ophidian, regarded as harmless by Linnæus and most herpetologists, but considered as extremely venomous by the traveller, Forskal. This ophidian is called *hajé* by the inhabitants, and recent travels have incontestibly proved that it is the true aspic of the ancients which never inhabited Europe; for the reptile which some years since infested the forest of Fontainbleau, and was called by this name, was nothing but a variety of the common viper, and the *æsping* of the

Swedes, as we have already seen, is quite another species from the one in question.

Forskal informs us that when the hajé is provoked, it swells and extends its neck greatly, and then springs with a single bound upon its enemy. This habit of rearing up when it is approached, caused the ancient inhabitants of the countries watered by the Nile, to believe that this serpent guarded the fields which it inhabited. They made it in consequence the emblem of the protecting divinity of the world. They sculptured it on the two sides of a globe, on the portico of all their temples. It is often exhibited by the jugglers at Cairo, apparently metamorphosed into a rod or wand, which is done by pressing its nape with the finger, and thus causing a sort of catalepsy. They take care, however, to remove the fangs, which might cause very serious accidents.

The ancients were zealous friends of the marvellous, and at least as much inclined as ourselves to adopt as true every thing which deviates from the ordinary course of things. They have accordingly informed us that the poison of the aspic, though inevitably mortal, produced no pain, and merely occasioned the progressive loss of strength, which was followed by a quiet and lethargic sleep that "knew no waking." Galen tells us, that in Alexandria, to shorten the punishment of criminals condemned to death, they were bitten in the breast by an aspic, of which he declares himself to have been an ocular witness, and that it was with much difficulty the traces of the wound could be discovered. Dioscorides, in mentioning that the bite of the aspic produced immediate dimness of sight and epigastraglia, remarks that it is unaccompanied by any local tumefaction, and they are so fine, that they appear to be made with a very slender needle.

All that we know on this subject is, that the poison of the hajé is excessively violent, and far more deleterious, for example, than that of the European viper, which it resembles in its yellowish tint, and its transparence. Forskal relates, that having taken a very small drop of it, and introduced it into a slight incision made in the thigh of a pigeon, he saw this unfortunate bird perish in a quarter of an hour in convulsions and vomitings.

The modes adopted against the bite of the hajé are the same, for the most part, as those used in the case of that of the viper, particularly cauterization by fire, alcholized potassum, &c., and the administration of sudorifics internally.

But simple processes of this kind have not always been sufficient to satisfy practitioners. In the case before us, Pliny has recommended the *clematis* of Egypt, the anis, lupine, and henbane bruised in wine. Atheneus has spoken of the good effects of lemon-juice, and has related on this subject an anecdote which is either fabulous, or founded on some ill-observed fact, and which we may leave in the too copious annals of credulity along with the assertions of Pliny concerning the efficacy of vinegar; of Galen in that of theriaca; of Aetius on the virtues of sea-water, and the tepical application of dock, &c. It would be better to believe, with Aristotle, that all means are useless, than to have recourse to such as those.

Elaps is a name of Greek origin, and signifies, as Nicander tells us, a non-venomous serpent. Aetius uses it in the same sense, which will not at all apply to the species of this genus.

The *elaps lemniscatus* is not an inhabitant of Asia, as many naturalists have pretended. It is a native of Guiana and Surinam, where it is much dreaded, and has occasioned the tortrix scytale, and the black-banded adder to be also objects of terror, in consequence of the similarity of their forms, &c., although these latter reptiles are perfectly harmless. It is probably the serpent called *oroucoucou* by the

BANDED BUNGARUM.

BOA. (Sohn. & Ruſs.) PSEUDOBOA. (Oppel.) FASCIATA.

negroes of Surinam, whose poison is very active. Stedman relates that a slave having been bitten in the foot by one of these animals, had the leg swelled in less than a minute, experienced the most acute pain, and soon fell into convulsions, which preceded his death. The same traveller relates that the gall of this serpent is regarded as a specific against its bite; but he never witnessed the success of this remedy. He also remarks that, in general, the smaller the serpent is, the more is its bite to be feared. He thinks that this elaps is the same animal as the small *labarra* mentioned by Dr. Bancrost in his History of Guiana, and which, according to him, is *fourteen feet* in length, which appears, however, as Daudin well observes, to be a typographical error, for which we should read fourteen inches. He assures us that the violence of its poison is so great, that it causes death in less than five minutes, in the midst of convulsions, accompanied by a discharge of blood through the natural apertures of the body.

The subdivisions immediately following Elaps, presents no materials for any observations here—and of Bungarus or Pseudoboa we can add nothing of any interest to the text.

Of the genus Hydrus, the species of Hydrophis, which are known, are tolerably numerous. Their name sufficiently indicates their singular habit of living in the water, and by Linnæus they are classed with *anguis*. They seem to have been partly known by the ancients. Elian informs us, that hydri with flat tails were found in the Indian seas, and that they also existed in the marshes. He also tells us that those reptiles had very sharp teeth, and appeared to be venomous. According to Ctesias, the serpents of the river Argada, in the province of Sittacene, remain concealed at the bottom of the waters during the day, and by night that they attack persons who go to bathe, or wash linen. In the *Periplus*, or circumnavigation of the Erythrean sea, Arrian mentions hydrophides, or pelamides in three different places. We shall slightly notice the most interesting species.

The *hydrophis obscurus*, Daud., has been found in the saline waters of a river near Calcutta, which divides into two parts the country of Bengal, and is called *Sunderbunds*. This serpent swims with great facility, but moves with difficulty on the ground, and soon dies there, which it also does if plunged into fresh water. It has been figured by Russel. The Hindoos call it *kalo-shoutur-sun*.

The *hydrophis cloris* has the same habitat and manners as the foregoing reptile. It appears to be ovo-viviparous, for Russel, who has figured it in the seventh plate of his supplement, found in the belly of a female two well-formed young ones, and an egg not yet enclosed. The Indians call it *shoutur-sun*.

The *hydrophis nigrocinctus* inhabits the same places as the last two. The Indians call it *keril-patee*, and it appears very venomous. A bird bitten by it in the thigh died in convulsions at the end of seven minutes.

The *hydrophis melanurus* of Wagler, a figure of which is inserted, is an inhabitant of India. It is an extremely elegant snake, whose specific characters we shall not insert here. It is nearly allied to, if not a variety of, the *hydrus spiralis* of Shaw, Zool. iii. 564.

For the rest of the division hydrus we have nothing to add, excepting that some of the species appear to be harmless.

The last and the most singular genus of the ophidians is the Cœcilia, which has been sometimes placed among the batracians, and in this, as in many other cases, it is exceedingly difficult to draw the line of distinction.

The first description of the cœcilia was given by Linnæus at Upsal, in 1748, from a preserved specimen, but it is but

SPIRAL HYDRUS._*Shaw?*

HYDROPHIS MELANURUS _*Wagler.*_ *HYDRUS SPIRALIS.*_*Sh.*

lately that any precise notions have been attained respecting this genus of animals. Schneider has given us some particulars respecting their skeleton, and he approximates them to the fish. M. Dumeril considers them as very near the tailless batracians, resting upon some characters of their organization, that is on the viscosity and nudity of the skin, on the absence of ribs, on the presence of two occipital condyles, and on the rounded form and position of the anus. M. Appel has made of them his family of apode batracians, and our author places them last of the ophidians. But in fact, until the mode of generation in these animals is ascertained, and until it is known whether they undergo metamorphoses or not, there will be much embarrassment in their classification. It is certain, however, that they cannot be confounded with the fish, since no gills have been observed in them.

To the anatomical description in the text, it is quite unnecessary to add anything, and respecting the manners of the cœcilia, little is known. According to the observations of Peron, they seem to approximate considerably to those of the tritons. It is supposed that the food of these animals is small insects and worms. The position of the anus leads to the supposition that there cannot be any real copulation.

THE FOURTH ORDER OF REPTILES.

The Batrachia,

Have but a single auricle and a single ventricle to the heart. They all have two equal lungs, to which are united in their early age, gills which have some relation with those of fishes, and which have, on the two sides of the neck, cartilaginous arches, which are attached to the hyoid bone. The majority of them lose these gills, and the apparatus which supports them, when they arrive at the perfect state. Three genera only, the *siren*, the *proteus*, and *menobranchus*, preserve them all their lives.

As long as the gills subsist, the aorta proceeding from the heart is divided into as many branches on each side as there are gills. The blood of the gills returns by the veins, which unite towards the back in a single external trunk, as in the fish. It is from this trunck or immediately from the veins which form it, that the greatest part of the arteries which nourish the body spring, and even those which conduct the blood to be respired in the lungs.

But in the species which lose their gills, the branches which repaired thither become obliterated, excepting two, which are united in a dorsal artery, and each of which sends a small branch to the lungs. This is the circulation of a fish, metamorphosed into the circulation of a reptile.

The batracians have neither scales nor carapace, a naked skin covers their bodies.* With the exception of a single genus, they are without claws to the toes.

The envelope of their eggs is simply membranous. The male disposes the female to lay them, by very long embracings, and in many species they are not fecundated but at the moment of their coming forth.

These eggs swell greatly in the water after having been laid. The young does not differ from the adult merely by having gills. Its feet are developed only by degrees, and in many species there are a beak and tail which are to be lost, and intestines of a different form from those of the adult; some species are viviparous.

* M. Schneider has proved that the scaly frog of Walbaum appeared such only by accident, some scales of lizards, which were kept in the same vessel, having attached themselves to its back.

The Frogs, Rana, Lin.,

Have four legs and no tail in their perfect state. Their head is flat, their muzzle rounded, their mouth greatly cleft. In the majority the soft tongue is not attached to the bottom of the mouth, but to the edge of the jaw and folds back within. The front feet have but four toes, the hinder sometimes exhibit the rudiments of a sixth.

Their skeleton is entirely destitute of ribs. A cartilaginous plate on a level with the head, is in lieu of a tympanum, and enables us to recognize the ear without. The eye has two fleshy lids and a third concealed under the lower one, transparent and horizontal.

The inspiration of air is performed only by the movements of the muscles of the throat, which, dilating, receives the air through the nostrils, and contracting while the nostrils are cleared by means of the tongue, obliges this air to penetrate into the lungs. Expiration on the contrary is performed by the muscles of the abdomen; thus when the belly of one of these animals alive is opened, the lungs dilate without being able to sink down, and if one of them be forced to keep its mouth open, asphyxy takes place, because the air of the lungs cannot be renewed.

The act of coition continues for a long time; there is a spungy sort of swelling on the thumbs of the male, which increases during the spawning time, and assists in grasping the female. The eggs are fecundated at the moment in which they are laid: the little being which springs from them is called a tadpole; it is at first provided with a long fleshy tail, a small horny beak, and has no other apparent members but little fringes at the side of the neck; these disappear at the end of some days, and Swammerdam assures us, that they then sink under the skin to form gills there. There are small and very numerous tufts, attached to four cartilaginous arches, placed on each side of the neck, adherent to the hyoïd bone and enveloped in a membranous tunic covered by the general skin. The water which comes through the mouth, passing into the intervals of the cartilaginous arches, issues out sometimes by two apertures, sometimes by one, either pierced in the middle or at the left side of the external skin according to the species. The hinder feet of the tadpole are developed by little and little, and visibly. The fore-feet are also developed, but under the skin, which they afterwards pierce. The tail is re-absorbed by degrees. The beak falls and lets the true jaws appear, which at first were soft and concealed under the skin. The gills become obliterated and leave the lungs alone to exercise the function of respiration which they had shared with them. The eye, which was only observable through a transparent place in the skin of the tadpole, is now discovered with its three lids. The intestines, at first very long, slender, and turned spirally, grow short and acquire the necessary inflations for the stomach and the colon. The tadpole accordingly lives only on aquatic plants, and the adult animal on insects and other animal matters. The limbs of the tadpoles are regenerated, almost in a similar manner with those of salamanders.

The time of these particular changes varies according to the species.

In temperate and cold climates, the perfect animal remains during winter under ground, or under the water in the mud, and lives there without eating and without respiration; but during the fine season, if it is hindered from respiring for some minutes, by preventing it from closing the mouth, it perishes.

The FROGS, properly so-called, (RANA, Laur.), have the body slender and the hind feet very long, very strong, and more or less well palmated. Their skin is smooth, their upper jaw is furnished all around with a row of small fine teeth, and there is a transverse interrupted range of them in the middle of the palate. The males have on each side, under the ear, a slender membrane which swells with air when they cry. These animals swim and leap extremely well.

The Green Frog, (*Rana Esculenta*, L. Rœsel. Ran. pl. xiii. xiv.)

Of a fine green, spotted with black. Three yellow stripes on the back, the belly yellowish.

This is the species so common in all stagnant waters, and so troublesome in summer from the continuity of its nocturnal clamours. It furnishes a wholesome and agreeable aliment. It spreads its eggs in packets in the marshes.

The Common Frog, (*Rana Temporaria*, Lin. Rœsel. Ran. pl. i. ii. iii.)

Reddish-brown, spotted with black, a black band proceeding from the eye, and passing over the ear.

This is the species which appears the earliest in the spring. It keeps more on land than the preceding, and croaks much less. Its tadpole also grows somewhat less before the metamorphosis.

The South of France produces a frog, (*R. Cultripes*, Nob.) altogether sown with blackish spots, the feet amply palmated, and especially remarkable for having the vestige of the sixth toe covered with a corneous and trenchant lamina.

Among foreign frogs we may distinguish,

The Paradoxical Frog, (*Rana Paradoxa*, L. Seb. I., lxxviii. Merian. Surin., lxxi. Daud. Gren. xxii. xxiii.)

Of all the species of this genus is that whose tadpole grows the most. The loss of an enormous tail, and of the envelopes of the body even makes the adult animal have less volume than the tadpole, which made the earliest observers believe that it was the frog which was metamorphosed into a tadpole, or, as they said, into a fish. This error is now completely refuted.

This frog is greenish, spotted with brown, and particularly recognized by irregular brown lines along its thighs and legs. It inhabits Guiana.

There are many other foreign frogs, some of which are very large, and as yet, but badly determined.*

We may remark in the number,

The Bull-Frog of the Anglo-Americans, (*Rana Pipiens*, Lin. Catesby. II. lxxii.)

Green above, yellow underneath, spotted and marbled with black.†

Certain species have the hinder toes, almost without palmation, but always greatly elongated.‡

* N. B. A deeper examination, and the inspection of numerous batracians brought to the Museum within a few years, have caused me to change the favourable opinion which I had pronounced on the labours of Daudin. What he has done is neither complete nor critical, and one half of his figures, made after mutilated individuals, cannot serve for any precise determination of the species. We may, however, except his *hylæ*, which are much better made out than his frogs and toads.

† I am convinced that in the United States many species have been confounded under this name, similar in size and colours, but differing among other characters, in the relative magnitude of the tympanum. That in which the tympanum is the largest has been designated by Merrem under the name of *Mugiens ;* but its synonyms are not certain. The fig. of Daudin, xviii., with a yellow stripe along the back, is a species of the Indies. Add, *Rana Palmipes*, Spix, V. 1; *R. Tigrina*, Daud. xx.; *R. Virginica*, Gm. Seb. I. lxxv. 4; or *Halecina*, Daud, or *Pipians*, Merrem; Catesby, lxx.; *R. Clamitaus*, Daud. xvi.

‡ *Rana Ocellata*, L. Seb. I. lxxv. 1, Lacep. I. xxxviii., Daud. xix;

WHITE-FACED HORNED FROG.

CERATOPHRYS BOIEI. Pr. Max.

AGUA TOAD.

BUFO AGUA.—Daud.

CERATOPHRIS, Boié,

Are frogs with a large head, skin grained altogether, or in part, and each eyelid has a membranous prominence like a horn.*

There are some whose tympanum is concealed under the skin.†

They all come from South America. The South of Africa produces batracians, similar to the frogs in their teeth, their smooth skin, pointed toes, the hinder ones widely palmated, and the three internal ones having their extremity enveloped in a conical claw of a black and corneous substance. Their head is small and mouth moderate. Their tongue, attached to the bottom of the throat, is oblong, fleshy, and very large. Their tympanum is not visible. These numerous characters have determined us to

R. Gigas, Spix, I; *R. Pachypus*, id. II.; *R. Coriacea*, id. V. 2; *R. Silulatrix*, Pr. Max. liv.; *R. Maculata*, Daud. xvii. 2; *R. Rubella*, ib. 1; *R. Typhonia*, ib. 4, which is not, as Merrem believed, the *Virginica* of Gm.; *R. Punctata*, ib. xvi. 1; *R. Mystacea*, Spix, III. 2, 3; *R. Miliaris*, et *R. Pygmæa*, id. VI.; *R. Labyrinthica*, id. VII.

* *Ceratrophis varius*, B. or *Rana Cornuta*, Seb. I. lxxii. 1, 2; Tiles, Mag. de Berl, 1809, deuxieme-trim. pl. iii.; and Voy. de Kreusentz, pl. vi., or *Ceratophris Dorsata*, Pr. Max. deuxieme liv.; *Cerat Spixii*, Nob.; or *R. Megastoma*, Spix, IV. 1; *Ran. Scutata*, ib. 2; *Cerat Daudini*, Nob. Daud. xxxviii.; *Cerat Elipeata*, Nob.

† *Ceratophris Granosa*, Nob. It is of the horned frogs, with concealed tympannm, that Gravenhorst has made his genus STOMBUS, but they have teeth like the others, and should not be approximated to the toads, as Fitzinger has done.

form a genus of them under the name of DACTYLETHRA.*

HYLA, Laurenti, CALAMITA, Schn. et Merrem,

Do not differ from the frogs, but because the extremity of their toes is widened and rounded into a sort of viscous ball, which allows them to fix to bodies and climb on trees. They remain there in fact all the summer, and pursue insects; but they lay their eggs in the water, and bury themselves in the mud in winter like the other frogs. The male has a pouch under the throat, which swells when he cries.

The Tree Frog, (*Rana Arborea*, L. Rœs. Ran. pl. ix. x. xi.)

Green above, pale underneath, a yellow and black line along each side of the body. It does not reproduce until four years of age and couples at the end of April. Its tadpole completes its metamorphosis in the month of August.

The foreign hylæ are tolerably numerous. There are many of them rather pretty. One of the largest and handsomest is,

* From δακτυλίθρα (thimble). Their claws have this form. The *Crapaud Lisse* of Daudin, pl. xxx. f. 1, is a bad figure of it, in which the hind feet are altogether wanting. Merrem has made of it his *Pipa Lævis*. The *Pipa Bufonia* of Merr., or pretended male *Pipa*, pl. enl. No. 21, f. 2, is again the same species represented without claws. M. Fitzinger has made of these species of Merrem, ENGYSTOMA, but the true engystoma, or breviceps, Merr., have no teeth or claws.

The Blue and Yellow Frog, (Sh. *H. Bicolor*, Daud. viii. and Spix xiii.)

Celestial blue above, rose-coloured underneath; from South America.

One still larger is,

The Zebra Frog, (Sh. *R. maxina*, Linn. *Hylæ palmata*, Daud. xx.)

Striped irregularly across with red and fawn-colour; is of North America.*

We may also remark in consequence of the singular property which is attributed to it,

The Tinging Frog, (*Rana Tinctoria*, L.)

Whose blood, inserted into the skin of parrots, in the places where some feathers have been plucked from them, causes red or yellow feathers to come instead, producing on the bird a parti-coloured plumage. We are told that this is a brown species, with two whitish bands, united crosswise in two places (Daud. pl. viii.); its hind feet have the toes almost free.†

* Add in palmated species, *Hyl Venulosa*, Daud., xix., or *Cul. Boans*, Merr. Seb. I. lxxii.; *Hyl. Tibicen*, Seb. ib. 1, 2, 3; *H. Marmorata*, Seb. I. lxxi. 4, 5, Daud. xviii.; *H. Lateralis*, Catesb., II. lxxi., Daud. II.; *H. Bilineata*, Daud. III.; *H. Verrucosa; H. Oculata; H. Frontalis*, id. and in Spix, *Hyl. Bufonia*, xii.; *H. Geographica*, xi. 1; *H. Albomarginata*, viii. 2; *H. Papillaris*, 2; *H. Pardalis*, 3; *H. Cincrascens*, 4; *H. Affinis*, vii, 3.

† Add in species, with the hind feet but little palmated, *H. Femoralis*,

The Toads, Bufo. Laur.,

Have a corpulent body, covered with warts or papillæ, a thick pad behind the ears, from which is expressed a milky and fœtid humour. No teeth; the hinder feet but little elongated. They leap badly and remain in general remote from the water. They are animals of a hideous disgusting form, which have been erroneously considered venomous from their saliva, their bite, their urine, and even the humour which they transpire.

The Common Toad, Rana Bufo, L. Rœs. Ran. xx.,

Reddish or brown gray; sometimes rather olive or blackish, the back covered with many rounded tubercles as large as lentiles. The belly furnished with tubercles smaller and more rounded. The hind feet are semi-palmate. It remains in obscure and sheltered places, and passes the winter in holes which it excavates. Coupling takes place in the water, in March and April; when it occurs on land, the female trails herself to the water, carrying the male along with her. She produces small and innu-

Daud. IV. *H. Squirella*, id. V.; *H. Trivittata, &c* Spix, IX.; *H. Abbreviata*, id. IX. v. 4.

The *Rainette Blue* of New Holland, *Hyla Cyanea*, Daud. should have, according to White, p. 248, but four hinder toes, and M. Fitzinger, who appears to have seen it, has, in consequence, made of it his genus Calamita. We have got one, and from the same country, altogether similar, but which certainly has five toes.

merable eggs, united by a transparent jelly in two cordons, often from twenty to thirty feet in length, which the male draws along with the hind feet. The tadpole is blackish, and of all those of our country is the smallest when its limbs appear and it loses its tail. The common toad lives more than fifteen years, and produces at four years of age. Its cry has some resemblance to the barking of a dog.

Reed Toad, (*Rana Bufo Calamita*, Gm. Rœs. xxiv. Daud. xxvii. 1.)

Somewhat olive; tubercles like the preceding; but not such large pads behind the ears. A yellow longitudinal line over the spine, and a reddish denticulated one on the side. The hind feet without membrane. It exhales a disagreeable odour of gunpowder, lives on the land, does not leap at all, but runs tolerably fast; climbs walls for the purpose of withdrawing into their clefts, and is therefore provided with two small osseous tubercles on the palms of the hands. It goes to the water only for the purpose of coupling in the month of June. It lays two strings of eggs like the common toad. The male cries like that of the tree frog, and has likewise a pouch under the throat.

The Brown Toad, (*Rana Bombina*, Gm. 7. *Bufo Fuscus*, Laurenti. Rœs. xvii. xviii.)

Clear brown, marbled with deep or blackish brown.

The tubercles of the back are not numerous, and are about as big as lentiles. The belly smooth; the hind feet with elongated toes and entirely palmated. It leaps pretty well, remains, by preference, in the neighbourhood of waters, exhales an odour of garlick when disturbed. Its eggs issue from the body in a single cord, but thicker than those of the common toad. Its tadpole is later than the others of this country in passing to the perfect state, and is still very large before it parts with its tail and the front feet have appeared. It even appears to be smaller when it loses altogether its envelope of tadpole. In some places it is eaten as if it were a fish.

The Variable Toad, (*Rana Variabilis*, Gm. Ball. Spic. VII. vi. 34. Daud. xxviii. 2.)

Almost smooth; whitish, with strong contrasting spots of deep green; remarkable for the changes of shade in its skin, according to whether it is asleep or awake, in the shade or in the sun. It is more common in the South of France than in the neighbourhood of Paris.

Obstetric Toad, (*Bufo Obstetricans*, Laur. Daud. pl. xxxii. f. 1.)

Small, grey above, whitish underneath, blackish points on the back, whitish ones on the sides. The male assists the female in getting rid of her eggs, which are pretty large, and attaches them in packets

THE VARIABLE OR GREEN TOAD. *R. VARIABILIS.*

Mus. Brit.

on the two thighs, by means of some threads of a glutinous matter. He continues to carry them until the eyes of the tadpole become apparent through the envelope which contains it. When they are about to exclude the young, the toad seeks some dormant water in which to deposit them. They immediately open, and the tadpole issues forth and swims. It is very small and lives on flesh. This species is common in the stony places in the neighbourhood of Paris.*

A toad is found in Sicily two or three times as large as ours, brown, and with flat and irregular tubercles. It remains, by preference, in the tufts of palm-trees. We name it, *Bufo Palmarum.*

The foreign toads have been hitherto but badly determined; there are many of them remarkable for size.

The Marine Toad, (*Rana Marina,* Gm. Daud. xxxvii. Spix. xv.)

Brown, varied with deeper brown, unequal tubercles, not much projecting. The triangular parotids are more than an inch broad, in some individuals of from ten to twelve inches long, not including the feet. It lives in the marshy countries of South America.†

* We cannot tell why Merrem has placed this toad in his genus Bombinator. The tympanum in the present species is perfectly visible.

† Add *Bufo Maculiventris,* Spix, xv., if, indeed, it differ from Marina; *B. Ictericus,* id. xvi. 1; *B. Lazarus,* id. xvii. 1; *B. Stellatus,* id. xviii. 1; *B. Scaber,* Daud. xxxiv, 1., which is not the same as the *B. Scaber,* of

Some subgenera have been recently separated from that of the toads; thus,

The Bombinator, Mer.,

Do not differ from the others, but because their tympanum is concealed under the skin. Such is in our country,

The Yellow-bellied Toad, (*Rana Bombina*, Gm. Rœs. xxii. Daud. xxvi.)

The smallest and most aquatic of our toads, greyish or brown above, black blue with orange spots below. The hind feet completely palmate, and almost as elongated as those of the frogs; accordingly it leaps nearly as well as they do. It continues in marshes and couples in the month of June. Its eggs are in small pellets, and are larger than those of the preceding species.*

Oxyrhyncus, Spix,

Have the muzzle pointed in front.†

We should approximate to them,

Spix, x. 1; *B. Bengalensis*, id. xxxv. 1; *B. Musicus*, id. xxxiii. 2.; *B. Cinctus*, Pr. Max. trois liv.; the *B. Agua* of the Prince of Wied. Sept. liv.; does not appear to be the same as that of Spix.

* Add *Bufo Ventricosus*. Daud. xxx. 2. A species represented with an exaggerated swelling.

† *Bufo Proboscideus*, Spix, xxi. 4, the neighbouring species represented in the same plate, *B. Semi-lineatus*, *B. Granulosus*, *B. Acutirostris*, and those of pl. xiv. *Nasicus* and *Nasutus*, connect, too intimately, this subgenus with the common toads, to be easily preserved.

The Otilophes, Cuv.,

In which the muzzle is also angular, and in which the head has on each side, a crest which extends over the parotid. *The Pearled Toad, (Rana Margaritifera,* Gm.) Daud. xxxiii. is the type.

Breviceps, Merr., Engystoma, Fitz. in part,

Are toads without visible tympanum or parotid, with an oval body, head and mouth very small, feet but little palmated.*

A more essential difference is that which has caused the separation from the whole great genus rana, of

The Pipa, Laur.,

Which are distinguished by their body flatted horizontally; by their large and triangular head; by the absence of all tongue; by a tympanum concealed under the skin; by small eyes placed towards the edge of the upper jaw; by the front toes being cleft each at the end into four small points, and by the enormous larynx of the male, made like an osseous

* *Engystoma Dorsatum*, Nob. or *Bufo Gibbosus.* Auct. Seba, II. xxxvii. No. 3. Daud. xxix. 2; *Eng. Marmoratum*; *Eng. Granosum.* Nob. New Species, one from India, the other from the Cape. The *Eng. Surinamense.* Daud. xxxiii. 2; has the mouth more ample, as well as the *Bufo Globulosus* and *Albifrons*, Spix, xix.—N.B. The *Engystoma Ovale*, Fitz. is a Dactylethra; his *Eng. Ventricosa* is a Bombinator. N.B. The *Bufo Ephippium*, Spix, xx. 2, of which M. Fitz makes his genus Brachycephalus, because but three toes are visible on all its feet, may be nothing but a young individual badly preserved, or badly represented.

triangular box, within which are two mobile bones, which can close the entrance of the bronchiæ.*

The species anciently known, (*Rana Pipa*, L.) Seb. I. lxxvii, Daud. xxxi. xxxii., inhabits Cayenne and Surinam, in the obscure places of houses, and has the back grained, with three longitudinal ranges of thicker grains. When the eggs are laid, the male places them on the back of the female and fecundates them with his milt, then the female betakes herself to the water, the skin of her back swells and forms cells in which the eggs exclude the young. There the little ones pass their tadpole state, and they do not emerge until they have lost their tail, and their feet are developed. Then the mother returns to land.

M. Spix represents one of a species more or less approximating to this, (*Pipa Curururu*, Spix,) from the lakes of Brazil, and assures us that the female does not carry her young.†

The SALAMANDERS, SALAMANDRA, Brongn.,

Have the body elongated, four feet, and a long tail, which gives them the general form of the lizards; accordingly Linnæus has left them in that genus, but they have all the characters of the batracians.

* This is what M. Schneider has described under the name of *Cista Sternalis*.

† In the cabinet of the King, is a true pipa of the Rio Negro, entirely smooth, with the head more narrow than usual. This is my *Pipa Lævis*, very different from that of Merrem, which is a Dactylethra.

Their head is flatted; the ear concealed entirely under the flesh, without any tympanum, but only with a small cartilaginous plate on the fenestra ovalis; the two jaws furnished with numerous and small teeth, two longitudinal ranges of similar teeth in the palate, but attached to the bones which represent the vomer; the tongue as in the frogs; no third eyelid; a skeleton with very small rudiments of ribs, but without an osseous sternum; a pelvis suspended to the spine by ligaments; four toes in front, and almost always five behind. In the adult state, they respire like the frogs and tortoises; their tadpoles respire at first by gills, of a tufted form, three in number on each side of the neck, which are afterwards obliterated; they are suspended to cartilaginous arches, some parts of which remain in the hyoïd bone of the adult. These apertures are covered by a membranous opercle, but the tufts are never enclosed in a tunic, and float externally. The front feet are developed before the hinder. The toes push out one another in succession.

The LAND SALAMANDERS, SALAMANDRA, Laur.,

Have, in the perfect state, the tail round; they remain in the water only during their tadpole state, which does not last long, or when they bring forth. The eggs exclude in the oviductus.

Our terrestrial species have on each side, on the occiput, a gland analogous to that of the toads.

The Common Salamander, (*Lacerta, Salamandra*, Lin., *Salam Maculosa*, Laur. Lacep. II. pl. xxx.)

Black, with great spots of a lively yellow; on its sides are ranges of tubercles, from which, in time of danger, oozes a bitter milky fluid, of a powerful odour, and poisonous to weak animals. This probably has given rise to the fable that the salamander can resist the flames. It remains in humid places, and retires into subterraneous holes; feeds on worms, insects and humus; receives the seed of the male internally; produces its young alive, and deposits them in marshes. In their early age, these animals have the tail compressed vertically, and gills.*

In the Alps is found a salamander, like the common, but entirely black, and without spots. (*Sal. Atra*, Laur. pl. 1. Laurenti pl. 1. f. 2.)

The Spectacled Salamander, (*Sal. Perspicillata*, Savi.) Has but four toes on the hind feet, as well as on the fore. It is black above, yellow spotted with black underneath, and has a yellow line across over the eyes. It is a small animal of the Appennines.†

* See, Ad. Fred. Funck. *De Salem terresti. vita, evolutione, formatione*, Berlin, 1827.

† *We* have ascertained that the *Sal à trois doigts*, (Lacep. II. pl. 36.) is but a dried individual, a little mutilated, of the *Sal. Perspicillata*. Add *Sal. Savi*. Gosse.

North America, which possesses many more Salamanders than Europe, has several land ones, with round tails, but without glands on the occiput.*

The AQUATIC SALAMANDERS, TRITON, Laur.,

Always preserve the tail vertically compressed, and pass almost their entire life in the water.

The experiments of Spallanzani on their astonishing force of reproduction, have rendered them celebrated. They re-shoot several times in succession the same limb, when it is cut off, and that with all its bones, muscles, vessels, &c. Another faculty, not less singular, is that which Dufay has recognized in them, of remaining a long time encompassed by ice, without perishing.

Their eggs are fecundated by the milt spread in the water, and which penetrates along with the water into the oviductus; they come forth in long chaplets; the young ones are excluded in fifteen days after the laying, and preserve their gills a longer or shorter time according to the species. Modern observers have recognized many of them in our country. But some doubt remains as to their determinations, seeing that these animals change

* *Sal. Venenosa*, Daud.; or *Subviolacea*, Barton; *Sal. Fasciata*, Harl. *Sal. Tigrina*, id.; *Sal. Erythronota*, id.; *S. Bilineata*, id.; *Sal. Rubra*. Daud. viii. pl. 191. f. 2; *Sal. Variolata*, Gilliams, Sc. Nat. Philad. I.; pl. xviii., fig. 1; and many new species. The *Sal. Japonica* of Hourtouin Bechstein, trad. de Lacep. II., pl. 18., f. 1, is very much approximating to Erythronota.

colour according to age, sex, and season, and that the crests and other ornaments of the males, are not well developed until spring. When winter surprises them with gills, they preserve them until the following year, always increasing in size.*

The best characterized are,

The Marbled Salamander, (*S. Marmorata*, Latreille. *Triton Gesneri*, Laurenti,)

With chagreened skin, pale green above, with large irregular brown spots; brown, pointed with white underneath; a red line along the back, which in the male forms somewhat of a crest, and has black spots. Not very aquatic.

The Salamander with spotted sides, (*S. Alpestris*, Bechst. trad. de Lacep. pl. xx.)

With chagreened skin; slate-coloured, and brown above; belly orange, or red; a band of small black crowded spots along each side.

The Crested Salamander, (*Sal. Cristata*, Lat.,)

With chagreened skin; brown above, with blackish spots; orange underneath, spotted with the same; the sides painted with white. The crest of the male is high, cut into sharp indentations, and bordered with violet in the season of love.

* It is of an individual which had thus preserved its gills, that Linnæus has made his *Proteus Tritonius*.

The Punctated Salamander, (*S. Punctata*. Lat.)

Skin smooth ; clear brown above ; underneath, pale or red ; some black and round spots throughout Some black stripes on the head. The crest of the male festooned. Its toes a little enlarged, but not palmated.

The Palmipede Salamander, (*Sal. Palmata*, Latr.)

Back brown ; upper part of the head verniculated with brown and blackish ; sides more clear, with round blackish spots ; belly without spots. The male has three small crests on the back ; the toes dilated and united by membranes ; the tail terminates by a small thread.*

North America also possesses many aquatic salamanders.†

Among the schists of Œningen, are found skeletons of a salamander three feet in length. One of them is the pretended fossil man of Scheuzer.

At the end of the salamanders come to be ranged many similar animals, some of which are considered

* This characterization of European species, is that which has appeared to me the most conformable to nature ; but it would be very difficult for me to refer to it exactly the synonimy of authors, so little do I find their descriptions and figures in accord with the objects which I have under my inspection.

† *Sal. Symmetrica*, Harl. which appears to me already represented in the Lacep. of Bechstein, II. pl. 18, f. 2, under the name of *Sal. Punctata;* and many species whose descriptions I have not been able to recognize, and which well deserve a monograph, accompanied by good figures.

as never having gills, but it is probable that they have them, as soon as our terrestrial salamander; others, on the contrary, preserve them all their lives, which does not prevent them from having lungs like the batracians, so that they may be regarded as the only vertebrated animals, which are truly amphibious.*

Among the first (those in which no gills are visible) we arrange two genera:

The MENOPOMA, Harlan.†

Which have altogether the form of the salamander: apparent eyes, feet well developed, and an orifice on each side of the neck. Beside the range of teeth around the jaws, they have a parallel range on the front of the palate.

Such is the reptile a long time named,

The Great Salamander of North America, (*Salamandra Gigantia*, Barton. *Hellbender* of the United States. An. of the Lyce. of New York, I. pl. 17.)

Fifteen to eighteen inches long, of a blackish blue.

* The simultaneous existence and action of the branchial tufts and the lungs in these animals, can no more be contested than the most certain facts in natural history. I have under my eyes the lungs of a siren three feet long, in which the vascular apparatus is as developed and as complicated, as in any reptile; and, nevertheless, this siren had its gills as complete as the others.

† Dr. Harlan had first named these ABRANCHUS; Leukard and Fitzinger, CRYPTOBRANCHUS; others, PROTONOPSIS.

It inhabits the rivers of the interior and great lakes.

AMPHIUMA, Garden,

Have also an orifice on each side of the neck, but their body is excessively elongated; their legs and feet, on the contrary, are very little developed, and their palatine teeth form two longitudinal ranges.

There is one species with three toes on all the feet, (*Amphium, Tridactylum,* Cuvier) and one with two toes only, (*Amphium Means.* Gard. and Harlan, Mem. du Mus. XIV. pl. 1.)*

Among those which always preserve their gills,

The AXOLOTS

Resemble in every point the larvæ of the aquatic salamander, having four toes before, and five behind, three long gills in the form of tufts, &c. Their teeth are smooth at the jaws, and in two bands on the vomer. Such is,

The Axolotl of the Mexicans. (*Siren Pisciformis,* Shaw. Gen. Zool. vol. III. part ii. pl. 140. Humboldt Obs. Zool. I. pl. 12.)

* Linnæus was acquainted with the Amphiuma, but too late to insert it in one of the editions of his system, which appeared in his life-time. It has been described since, by Dr. Mitchell, under the name of *Chrysodonta Larvæformis;* and by Dr. Harlan, under that of *Amphiuma.* I have published the species *Amphiuma Tridactylum,* which is of Louisiana, and attains to the length of three feet. See the Mem. of Mus. tom. XIV. 1. I suspect that it is of this species that Barton speaks, as of a siren of three feet, in his letter on the siren.

From eight to ten inches long; grey, spotted with black. It inhabits the lake which surrounds Mexico.*

The MENOBRANCHUS of Harlan, or NECTURUS of Rafinesque,

Have but four toes on all the feet. There is one range of teeth in their intermaxillaries, and another parallel but more extended, in their maxillaries.

The species most known (*Menobranchus Lateralis* Harl., *Triton Lateralis*, Say, Ann. of the Lyc, New York I. pl. 16,) inhabits the great lakes of North America, and grows very large, attaining, as is said, to the size of three feet. It was first brought from Lake Champlain.

The PROTEUS, PROTEUS, Laur., HYPOCHTON, Merr.,

Have only three toes before and two behind.

Hitherto but a single species has been known, (*Proteus Anguinus*, Laur. pl. iv. f. 4. Daud. VIII. xcix. 1. *Siren Anguina*, Schn.) an animal more than a foot long, as thick as one's finger, with the tail compressed vertically, and four small limbs. Its muzzle is elongated and depressed. Its two jaws furnished with teeth. Its tongue but little mobile, and free in front. Its eye is excessively small, and concealed by the skin, as in the rat-mole. Its ear

* It is with some doubt that I place the Axolotl among the genera with permanent gills; but so many eye-witnesses assure us that it does not lose them, that I am obliged to do so.

is covered by the flesh, as in the salamander. Its skin is smooth and whitish. It is only found in the subterraneous waters, through which certain lakes of Carniola communicate together.

Its skeleton resembles that of the salamanders, except that it has many more vertebræ, and less rudiments of ribs. But its osseous head is altogether different from theirs in its general conformation.

Finally, there are some which have the front feet, and entirely want the hinder. These are,

The SIRENS, SIREN, L.,

Elongated animals, almost of the form of eels, with three branchial tufts, without hind feet, or even any vestige of pelvis. Their head is depressed, their mouth little cleft, their muzzle obtuse, their eye very small, and the ear concealed. Their lower jaw is armed with teeth all round, but the upper one has none, and there are several ranges of them which adhere to two plates pasted under each side of the palate.*

* It is in vain that some recent authors have desired to renew the ancient supposition, that the siren is a tadpole of the salamander. There are individuals of the siren, larger than any known salamanders, and whose bones have acquired a perfect hardness, without exhibiting the least vestige of hind feet. Their osteology, besides, is altogether different from that of the salamanders. The vertebræ are more numerous (90), and differently formed, and the ribs are much fewer (eight pairs). The conformation of the head, and the connection of the bones which compose it, are altogether different. See my Researches on the Fossil Bones, Vol. V., Part 2.

The Lacertine Siren, (Siren Lacertina, Lin.)

Attains to three feet in length, and is blackish. Its feet have four toes, its tail is compressed into an obtuse fin. It inhabits the marshes of Carolina, and particularly those where the culture of rice is established. It remains there in the mud, from which it sometimes goes to land, and sometimes to the water. It lives on earth-worms, insects, &c.*

Two smaller species are known,

The Intermediate Siren, (S. Intermedia, Leconte. Lyceum of New York, II. Dec. 1826. pl. 1.)

Blackish; and with four toes like the large species, but the branchial tufts are less fringed. It is not more than a foot in length.

The Striped Siren, (S. Striata id. ib. I. pl. iv.)

Blackish; two longitudinal yellow stripes on each side; three toes only on the feet. The branchial tufts but little fringed. Its length is only nine inches.†

* Dr. Barton disputes the habit of feeding on serpents, and the cry like that of a young duck, which Gordon attributes to the siren.

† The gills of these two species are considered as performing no part in their respiration, and, in consequence of this, Mr. Gray has formed of them his genus Pseudobranchus. It is not, however, difficult to see, at their lower face, certain folds, and a vascular apparatus, whose use does not appear to us to be doubtful. It is now clearly demonstrated, by the observations of M. Leconte, that these sirens, like the Lacertina, are perfect animals.

SUPPLEMENT ON THE BATRACHIA.

The fourth order of the class reptilia has received its denomination from the Greek word Βάτραχος, which signifies a frog. To call the animals which compose it frogs would be by no means justifiable, but they all possess a sufficient analogy to the genus rana, to warrant the kindred epithet of *batracians*.

In this order are ranged all reptiles with naked bodies, without carapace or scales; the head without any distinct neck, or division; the toes always distinct, without claws. Finally, all which in general have no real copulation, and most usually undergo metamorphoses.

The batracians proceed from eggs which have a membranous envelope, and which must remain in the water before the young can be excluded. The animal which proceeds from this egg has the form and structure of a fish. It has no feet, and its body is terminated by a very long and compressed tail formed like a fin; it is then named a *tadpole*. On this subject it is indispensable that we should enlarge a little on the observations of the text.

The *tadpole* then is a young batracian, from the moment in which it issues from the egg, until, after various metamorphoses, it passes to the adult state, without preserving either its form, structure, or even its mode of living.

When we examine the different periods of its evolution in the eggs of frogs, (which of all the eggs of reptiles have been the most carefully studied, as to the development of germs,) we find that during the three or four days which follow the fecundation, the tadpole is nothing but a kidney-formed mass of small granulations. Towards the middle of the fourth day, these little grains are confounded one with the other: the embryo becomes distinct. It is divided by a contraction into two parts, one of which comprehends the head and thorax, the other the abdomen and tail. It is immersed in a fluid, which Swammerdam has compared to that of the *amnios*.

Moreover, according to the same observer, we then perceive in the eggs in question, an allantoïs, a chorion, an amnios, and umbilical vessels.

During the fifth day, the embryo increases a little, and towards the evening of the sixth, we see besides the head, thorax, abdomen and tail, a gill appear on each side of the neck, and answer the purpose of respiration, for the little animal, of swimming, and of reposing itself in the glairy fluid.

In the course of the seventh, and at the commencement of the eighth day, the fœtus successively leave the albuminous fluid of the milt; and from thence until the thirteenth day, they exhibit no change of form, and merely augment in volume.

On coming out of the egg, the little batracian is blind and without feet. It has a tail even in the anourous species of frogs and toads; it respires by gills; it has a large and globulous belly; its intestines are excessively long. It lives solely on vegetable substances, with the exception, according to M. Dumeril, of the obstetric toad.

This is the state in which it is named by u. *tadpole*, a word which literally signifies the young of a toad. The French call it *tetard*, from *tete*, (head,) in consequence of the volume of the anterior part of the body. At this time it inhabits the water as a matter of necessity.

But it soon changes its skin; its eyes begin to shew themselves. First its two hinder feet, then the fore-feet appear on the sides of the trunk, and finally, the fall of the tail is speedily followed by the loss of the gills, while at the same time the digestive canal loses much of its dimensions.

Then the animal respires the atmospheric air, and acquires the form which it is destined to preserve for the rest of its existence.

In the egg of the salamander, the tadpole, curved on itself, and enveloped by a vitelline membrane, is free, and unprovided with an umbilical cord.

The spinal marrow of this embryo is divided and composed of two nervous cords, in front of which is a very small vessel, dilated at one of its extremities, and which has been supposed to represent both the heart and the aorta. Here then we find, according to M. Roth, some approximation between this arrangement and that of the nervous system in the annelides.

The limbs make their first appearance in the shape of papillæ.

Under the head of the hyoïd apparatus, the examination of the tadpole of the frog is of great importance in comparative anatomy, since, as the Baron has demonstrated, it leads us to the knowledge of what is truly the hyoïd bone in fishes; and the more so because the frog, having in its first state respired like fishes, undergoes such modifications of organization that its branchial apparatus becomes by degrees, and visibly, a true hyoïd bone.

Such a metamorphosis, well ascertained, involves the highest results for osteological theories. Accordingly we find, that it has peculiarly arrested the attention of Messrs. Cuvier, Geoffroy St. Hilaire, and other anatomists of celebrity. It has been singularly overlooked by M. Steinheim in a work written *ex professo*, on the development of frogs, in 1820, and by M. Van Hasselt, in a Latin dissertation printed at Groningen in the same year, and on the same subject.

The Baron tells us in his "Ossemens Fossiles," that when the tadpole is taken at the moment in which its gills are in full activity, and its lungs are no more than a blackish tissue not yet recipient of the air, the ranges of teeth attached to its lips, and the corneous laminæ which invest its jaws, serve alone for mastication. Its jaws, scarcely cartilaginous, are very little developed. Its tympanic bones on the contrary are very much so. To them is suspended the branchial apparatus on each side by a pretty thick and angular branch, which represents that which in the fish, composed of three bones, suspends the whole branchial apparatus to the bone, which our illustrious anatomist considers as analogous to the temporal, and which supports the branchiostegous rays.

Between these two branches is an odd piece, which corresponds with the chain of odd osselets placed in the majority of fishes between the first two branchial arches. At its lower point are attached laterally two small rhomboidal pieces, at the external edge of which are suspended the arches which support the gills. These two pieces represent the even osselets which terminate the chain just mentioned, and which in many fishes support the two last branchial arches.

If we afterwards examine tadpoles of more advanced ages,

we find the branches which support this apparatus grow longer and longer and narrower and narrower, until they finish by changing into those two long cartilaginous threads, which attach the hyoïd bone to the cranium, a little below the *fenestra ovalis*. The angle which their anterior extremity forms in front, becomes a little crotchet or hook of this thread.

At the same time the odd piece and the two rhomboidal pieces become united by synostosis, extend, grow slender, lose by little and little the branchial arches which were attached to them, and which are re-absorbed; and finally become a large disk or buckler, the anterior angles of which are widened like a hatchet, the posterior often retain in an emargination which divides them, the trace of the branchial arches which proceeded from them, and the posterior edge has two osseous horns, which are found in the posterior angles of the even pieces, and which may well correspond to the lower pharyngian bones of fishes.

The eggs of the batracians are not fecundated by the males until the moment in which they are deposited by the female, or in a little time after. In the deposition the male often assists the female; and then the eggs are placed in succession like a chaplet, and retained either by a glutinous matter which pastes and unites them into a packet, or by a substance which dries up, becomes elastic, and thus retains the embryones twisted on the thighs of the males; or otherwise these eggs are placed by the male on the back of the female, whose skin swells, and forms around each of them a sort of alveolus, within which the little tadpole undergoes all its metamorphoses as in a matrix. Sometimes the eggs are laid separately, and the male fecundates them in succession.

At the period of generation, which usually takes place but once a year, and in the spring, the males undergo a change of form. Some exhibit membranous crests upon the back, others experience in the skin of the anterior feet very notable changes, which seem to be made for the purpose of enabling them to retain the female more closely.

All the batracians in their first state can swim ; some even appear to remain all their lives under the tadpole form, only that they then possess both lungs and gills. The majority walk upon the earth, climb, and even leap with facility. The form of their body varies considerably and indicates as it were beforehand the nature of their movements. Thus, for example, all the batracians which preserve the tail in their latest state, walk slowly, can merely drag their bodies along the ground, and live usually in the water : such are the salamanders, the protei, and the sirens. Those, on the contrary, which lose this member, as the frogs, hylæ, and toads, walk on the ground, climb trees, and leap perfectly well.

We may conceive that the osseous frame-work of these animals must present considerable differences. Their spine consists of a series of vertebræ not very distinguishable from each other ; the bones of the head, or rather their forms, present differences still more marked. The head, in general, has little mobility, and is articulated in a single point by the assistance of a tubercle, or condyle with three facets. The bones of the upper jaw are always united together by synostosis, and are not dilatable. Some species have no ribs whatever, and very short ones are observable in some genera. The number of feet varies ; sometimes only the front feet are visible, but most generally there are four. Their respective length differs according to the genera, as also does the number of toes. The batracians have very strong and very irritable muscles ; a notorious example of this is to be found in the thighs of frogs. These animals exhibit many other peculiarities in their locomotive organs, which we shall have occasion to notice in treating of the genera.

Although the nerves are very distinct and very robust in the batracians, in proportion to the other organs, the cavity of the cranium which contains their origin is in general very small. The eye is contained in a very large orbit; it is protected by three lids in some species, and moistened by a fluid analogous to tears. The pupil is very dilatable, usually of a rhomboïdal form, elongated, and in a vertical direction. The ear of these reptiles does not appear externally; a tympanic box is found, however, under the skin, and sometimes two osselets of the ear. The nostrils are very simple, carried in front of the muzzle, which they generally cross, and prolonged into a small membranous tube, in the interior of which is observed a valvule which is intended for respiration. The tongue is mucous, adherent to the mouth in the salamanders, and attached to the front of the lower jaw in the tailless batracians. The sensation of touch appears to be perfect in these animals. All have a naked skin, with a mucous epidermis, and often furnished with an assemblage of glands and follicles in the form of warts. Their toes are often more or less cleft, especially on the front feet.

All the reptiles of this order, when arrived at the perfect state, subsist on living animals, and never on carcases. Their mouth is very wide and without moveable lips; their teeth, very short, are implanted in the jaws, which appear firmly crenulated. In this same perfect state their intestinal canal is short, and in that same cavity of the belly, are observable a liver with its vesicle, a spleen, an epiploon, kidneys, and a bladder.

The circulation of the batracians may be regarded as simple; the heart has but one ventricle and one auricle. A part of the blood passes through the lungs or gills, and then returns into the general torrent; the lungs float in the cavity of the belly: they are formed of very large cells. Sometimes there is but a single one, as is observed in the salamanders, in which this organ resembles a bladder; the trachea is always very simple; there is no epiglottis nor inferior larynx. Respiration is carried on by the muscles of the throat, which perform the office of the diaphragm, an organ not existing in these animals. For the inspiration of air, it is necessary that the mouth should be closed, so that a frog or toad, placed in the water, and the mouth kept forcibly open, would perish immediately. The majority of the batracians, however, have a voice which is termed croaking, which is produced by means of certain air-sacs, or extended membranes, on which the air expelled from the lungs impinges and vibrates.

Linnæus comprehended the major part of the animals of this order in his genus rana. The salamanders he placed among the lizards; of the siren, that great naturalist formed his order *Meantes*, which subsequent naturalists rejected; and in Dr. Turton's Linnæus, the siren forms a separate genus after lacerta, and before the serpents.

We shall commence our more peculiar investigation of the batracians with the genus RANA, taking first the FROGS, properly so called.

There is so great a relation between the first three subdivisions of the batracians, that so far from blaming Linnæus for having united them together at the time in which he wrote, we cannot think that he could reasonably have done otherwise. Many species are, indeed, sufficiently well distinguished by their characters; but many others are so little so, that it is a matter of some difficulty to determine where they should be placed. It is principally in the species of toads and frogs that this uncertainty exists, for the subdivision of the hylæ is much more neatly characterized.

"This great resemblance to ignoble beings," (says Count Lacépède, in the true spirit of a Frenchman,) "is a great misfortune. The frogs conform in external appearance so much to the toads, that we cannot easily consider them without thinking on the others. We are disposed to confound them all in the disgrace to which the toads are doomed, and to refer to the first the revolting habits, disgusting qualities, and dangerous properties of the second. We shall have considerable difficulty in assigning to the frogs the place which they should occupy in the minds of our readers, such as it really is in nature; but it is not less true, that if toads had never existed, if we had not before our eyes this horrid object of comparison, which caricatures by its resemblance, as much as it defiles by its approach, the frog would appear to us as agreeable from its conformation as distinguished by its qualities, and interesting from the phenomena which it exhibits at the different periods of its existence. We would behold it as a useful animal, from which we have nothing to fear, whose instinct is harmless, which unites an elegant form with supple and slender limbs, and which is adorned with pleasing colours, rendered more vivid from a kind of natural varnish, with which this animal is constitutionally provided."

"And who," continues this amiable, and somewhat enthusiastic writer, "can regard with pain, a being whose form is light, whose movements are nimble, and whose attitudes are graceful? Let us not deprive ourselves of an additional source of pleasure; and in our peregrinations through the smiling fields, let us not regret to see the banks of rivulets adorned by the colours of these harmless animals, and animated by their light and lively gambols. Let us contemplate their little manœuvres, observe them in the midst of stilly lakes, the solitude of which they diminish without troubling the repose; see them exhibit beneath the sheets of water the most agreeable tints, cleave the bosom of the tranquil stream, and vary its silvery surface with many a circling furrow."

Such are certainly the reflections of a man of refined feelings and a lover of nature. But, philosophically speaking, we cannot consent to this marked sort of preference of one genus or species at the expence of another. The toad fills its proper place in nature as well as the frog. And however our peculiar prejudices may operate, we cannot doubt for a moment that they are looked down on with an equal eye by "the Lord of all," whose wisdom and power are equally displayed in their conformation.

Our countryman Bradley was the first who started the idea of the separation of the toads and frogs into two distinct genera. After him Laurenti supported the same theory, but without much success. Messrs. de Lacépède and Dumeril have better established the characters of this genus. M. Schneider adopted, with little change, the principles of those writers, merely adding, that in the toads, the thumb of the front extremities is separated from the other toes, and that the index is very short. All the naturalists of our own days have admitted this division of the genus rana.

The term *rana* is one of old usage in the Latin language, as is proved by this line from the Georgics:

"Et veterem in limo ranæ cecinere querelam."

It corresponds to the Greek expression Βατραχος, from which the whole order has been named, as the frog is justly considered the prototype of this grand division of the reptile class.

The etymology of both these words appears to be somewhat uncertain. Isidore derives *rana* from *garrulitas*, in consequence of the noise which the frogs make in the neighbourhood of waters. This, we confess, appears to us somewhat farfetched, and not very unlike M. Menage's derivation of *chez* from *apud*. Others derive it from the Hebrew רנן, *vociferavit*. Βατραχος, according to Aldrovandi, is a sort of onomatopeia, to represent the rough croaking of these animals,

(Βοήν τραχεῖαν εχων,) The French word, *grenouille*, does appear to be an onomatopeia. Our word *frog* is of Teutonic growth, and also sounds like onomatopeia.

The muzzle of the frogs terminates more in a point than that of the toads. The nostrils are visible at its summit. The teeth are very small, and the eyes are large and brilliant, and surrounded with a circle of gold. The ears are placed behind them, and covered by a membrane, &c.

The frogs, when in a state of repose on the ground, carry the head very high, and their hind legs are then doubly folded on themselves, forming an angle of forty-five degrees, with the length of their body. Their muscles are considerable in relation to their bulk, and most peculiarly elastic. The leaps which they take to avoid real or supposed danger, for they are timid beyond measure, sometimes extend several feet. Their ordinary mode of progression is likewise by leaps, but not so much elongated. When they are taken by the hind feet, their body curves rapidly, and they give themselves such strong and repeated shocks, that one is frequently obliged to let them go. The glutinous matter with which their skin is varnished, favours considerably, in such cases, the action of their muscles, and causes their feet to slip through one's fingers.

As the organization of these reptiles is exceedingly curious, we shall make a few observations respecting it, in addition to the text of Cuvier.

The cranium of the frog is nearly prismatic, flatted above, and widened behind. It is less rounded than in the toad. The frontal bones are elongated rectangularly, and fill the interval of the orbits.

With the exception of the symphysis of the chin, and of the intermaxillary bones, which are free in all parts, all the bones of the cranium and face are totally united by synostosis in the adult individuals.

The head is articulated by two condyles, on an atlas of no great capacity of motion. The vertebræ are altogether ten in number. The eight which are extended from the nape to the pelvis, are furnished with pretty long transverse processes, which, in the last vertebra, extend as far as the bones of the ilia.

The sacrum is long, pointed, and compressed, and there is no coccyx.

The coxal bones are united into a single piece in the adult subject. Their iliac portion is very much elongated. The ossa pubis and the ischia being short, and synostosized into a single solid piece, form a crest more or less rounded at the place of their symphysis. There is no foramen formed by these bones.

No traces of ribs are at all visible.

The sternum forms in front a cartilaginous appendage, terminated by a disk situated under the larynx. It afterwards receives the clavicles. Then it widens, and finally terminates by another disk placed below the abdomen, and serving for the attachment of the muscles.

The bones of the furca, and the clavicles are on each side intimately united on one part to the sternum, and on the other to the omoplate. There is nothing very peculiar in bones of the arm.

The carpus is formed of eight bones, in three ranks; two bones in the first, and three in the second and third ranks. The largest bone in the second range bears the rudiment of a thumb with two articulations. The two toes which follow, have each two phalanges, and the two others three each.

There are no trochanters in the femur. The osseous piece which follows the femur is a peculiarity in the skeleton of tailless bactracians, and is longer in the frog than in the toad. The majority of anatomists have been wrong in considering this as a representative of the two bones of the leg. These last, as in the toad, seem separated in their entire length. Some anatomists, considering that they form the third articulation of the pelvian member, regard them as being the astragalus and calcaneum.

The tarsus is composed of four bones, the last of which is crooked. There are five bones in the metatarsus.

The toes of the hind foot are five. The thumb, placed inwardly, has but one phalanx. In the two following toes there are two in one, and three in the other. The next has four, and the last three.

The muscles of the frog are very strong, very irritable, and very sensible to the action of galvanism. It is not necessary to enter into a detailed account of them here. We shall simply notice those of the abdominal members, which are better characterized, and have very important functions to fulfil in this genus.

There is but one gluteus muscle, the *medius*. It descends from that elongated part which is the iliac portion of the coxal bones, and is fixed below the head of the femur. The pyramidalis comes directly from the point of the sacrum, and is attached towards the upper third of the femur.

The gemelli, the obturator internus, and the psoas magnus and parvus do not exist.

The quadratus femoris is elongated. It comes from the posterior symphysis of the ischium, and is attached to the internal side of the femur, towards its upper third.

The iliacus is very much elongated.

The pectinalis descends towards one half of the femur.

The three adductores have attachments analogous to those which they have in man.

An obturator externus is observable, notwithstanding the absence of the foramen ischii et pubis. It comes from the symphysis of the pubis, and its fibres are inserted on the articular capsula.

The thigh of the frog is rounded like that of man. The muscles of the leg are strongly marked. They exhibit some peculiarities. Among others, the triceps femoralis is formed only of two very distinct portions. The anterior rectus is wanting. The triceps cruvis has but one belly. It comes from the interior under part of the ilium, and descends in front and externally.

The semi-tendinosus is formed of two bellies, one of which is attached to the symphysis of the pubis, and the other to that of the ischium.

The semi-aponeurosis is the same as in man. The sartorius is couched directly in front of the thigh, and is not contorted. There is no popliteus.

The gastrocnemius has but one belly; but it is inserted by a little isolated tendon to the external edge of the crest of the tibia, or, more properly speaking, of the bone which follows the femur. Its terminating tendon spreads under the foot to form the aponeurosis plantæ.

The general sensibility of the frogs appears to be tolerably obtuse. They are killed with difficulty, even by the severest wounds. The heart and entrails may be plucked out without destroying them immediately; but their power of apparent organic contractility is very great. The heart contracts and dilates a long time after the death of the animal, and even when it has been extracted from the body, and separated from the other organs.

The bones of the nose and the intermaxillaries are very short, and, being broader than they are long, give a roundness to the face of the frog. The cavity of the cranium is very narrow, and the encephalon of very small volume. The cerebral hemispheres are elongated and narrow; the optic beds are large and hollowed with a very marked ventricle. Their volume exceeds that of the cerebral hemispheres. There is no pons Varolii.

The olfactory nerves come from the anterior extremity of the cerebral hemispheres. The foramen through which they issue from the cranium is double. The tube of the nasal fossæ is represented by a simple hole, and there is no cavity in the parietes of the cranium, or in the thickness of the facial bones, that can be compared to the sinuses in man and the mammifera. In the interior of these fossæ, nothing is found but some tubercles instead of the projecting laminæ which furnish those of other animals. The pituitary membrane is coloured by a net-work of blackish vessels. The nostrils are tubulous.

The orbits are not separated from the temporal fossæ but by an incomplete osseous branch. Their base is directed upwards. The optic foramina are very much separated. A large tunnel-formed auricle embraces the optic nerve, and is divided towards the globe of the eye into three portions only. Two other muscles, one depressor, and the other oblique, serve also for the motions of the eye. The lids are three in number, and are all horizontal. The upper one is a mere projection of the skin; the lower one is more mobile; the third, which moves from bottom to top, is more in action than the others. It is very transparent. The obliquus magnus does not exist. Two small blackish glands lodged in the orbit, seem to occupy the place of the lachrymal gland. The ciliary processes are few in number. The pupil is rhomboidal. The axis of the crystalline is to its diameter as 7 : 8.

The ear of the frogs is conformed like that of toads, of which we shall speak in its proper place. Gautier has observed that the cavity of the tympanum is transversed by a kind of cord, which separates it into two equal parts, and which can stretch at the will of the animal, and to different degrees, the membrane which closes this cavity, and which is apparent externally, smooth, and oval. The tympanic box, as in the toads, communicates immediately with the hinder mouth by a large hole, which can be perceived only in opening simply the mouth.

The epidermis is a kind of mucous epithelium like the cuticle of the lips, or internal parts in man, which drops off by shreds at several seasons of the year.

In examining with the microscope the bed of the skin covered by the epidermis, it appears to be composed of globules, which may be separated from each other, and which seem to be small glands for the preparation of the bitter and viscous humour which continually lubricates the surface of the body in the animals in question.

To this mucous-cutaneous tissue is owing the colours which adorn the surface of the body in frogs.

The chorion is a tissue of a very compact and dense character, and, as is also the case with the toads, it is united to the cellular tissue only in certain determined points; for example, at the circumference of the mouth, the middle line of the body, the arm-pits, and the groins.

From the observations of Mery we learn some curious details concerning the skin of the frog. This skin appears to cover four cavities, separated from each other by a very thin membranes, united on one side to the teguments, and on the other to the muscles of the body. These four cavities correspond to the back, the abdomen, and the sides. The skin of the thigh is not attached to its muscles, except in the folds of the articulations, and it forms two kinds of sacs, one before and the other behind. The same thing takes place with the skin of the legs.

There is no general cuticular muscle: we merely observe some fleshy fibres under the throat. These fibres descend from the compass of the lower jaw, and are lost in the cellular tissue which unites the skin to the origin of the breast.

The tongue is entirely fleshy, and differs from that of the toad only by being bifurcated in the point. It is composed for the most part of a thick glandulous mass, formed by a crowd of little tubes united by their base, and separated, like hairy papillæ, at the surface of the organ.

The lower jaw is a very open arch, composed of six pieces, the two middle ones of which are less thick than the others. This jaw alone is moveable. There is no eminence observable for the attachment of the digastric muscle, as in some other reptiles; neither is there any coronoid process.

The upper jaw alone is armed with teeth: these are about forty in number on each side, the eight intermaxillary ones being slender, pointed, fine, and crowded.

There is no epiglottis. It is clear that the hyoid bone, by means of the muscles which raise it, contributes to deglutition. It would be perfectly beside our purpose to enter into the anatomical minutiæ of this muscular and osseous apparatus. It equally contributes to the inspiration of air, which is performed by a sort of deglutition, while its expiration is the consequence of the action of the abdominal muscles.

In many species of frogs, the rectum is more or less conical and pyriform. The aperture of the anus is placed at the extremity of the back, and consequently above the animal—a singular arrangement, which is referrible to that of the pelvis. There is but one sphincter in this part, which is the only muscle.

On the organs of circulation and respiration, it is not necessary to enlarge here.

The eggs of frogs, when fresh laid, are globular, black on one side, and whitish on the other. They are placed in the centre of a glary and transparent mass, which must serve for the nutriment of the embryo. This matter is contained in two membranous envelopes, which represent the shell of the eggs of birds. These eggs swell greatly in the water after being laid. The experiments of Spallanzani have proved that they could support thirty-five degrees of heat without undergoing any alteration, and without ceasing to be productive.

The frogs feed on the larvæ of aquatic insects, on worms, small mollusca, flies, &c., and always choose a prey which is living and in motion. Every dead or motionless animal is rejected by them. To obtain this prey, they remain fixed in one situation with wonderful patience, watching it until they believe it to be sufficiently near: then they dart upwards, with the rapidity of lightning, putting out their tongue to catch it by means of the viscous fluid which invests that organ. This fluid retains it, while the two points of the bifurcation of the tongue seem to twist around it. When this prey is once thus seized, it is speedily swallowed. According to Daudin, the frogs sink it into the œsophagus with the thumbs of their anterior extremities. Sometimes, however, this gluttony is punished. Rœsel presented a wasp to a frog which he had reared. The reptile immediately swallowed it; but presently he began to struggle, and succeeded, by great efforts, in re-gorging it, but, doubtless, not without having been severely stung.

From the peculiar nature of the aliment of frogs, some naturalists have reasoned, not without justice, that they ought not to be persecuted in gardens. In fact, they are useful in such places, by destroying an immense quantity of those small slugs which are so detrimental to young plants of every species.

Daudin tells us that they also devour the spawn of fresh water fishes when they get near them.

Our countryman Townson, has made some curious experiments on the faculty possessed by frogs of absorbing water by the surface of their bodies. He is convinced that these animals instead of drinking water with the mouth, absorb it by the sole means of their skin, and instead of returning it by the urinary organs, return it by transpiration. If living frogs be placed on wet paper, at the end of an hour and a half their weight will be doubled. Such, at least, is the result of the observations of Townson and Daudin.

These reptiles are usually found on the ground in humid places, in the grass of meadows, and on the banks of streamlets, into which they continually leap and dive. They swim well and without difficulty by means of their hinder feet, the toes being united by a membrane. But they seldom sustain themselves between two waters. They are almost always seen either at the bottom, or at the surface, and constantly when the weather is fine upon the banks.

It is only during the summer that the frogs are to be seen springing about our fields and meadows in the neighbourhood of waters, or swimming in our ponds and rivulets. Frequently, at the close of warm rains in the fine season, they spread themselves through the country, and are so numerous as to be pressed and crowded against each other in places where they were not observed before. To this phenomenon is owing the existence of the popular belief in the *rains of frogs*, a very ancient prejudice, and still fully accredited in the provincial parts of many countries in Europe. Elian informs us that going from Naples to Puzzuoli, he observed a rain of this description. Aristotle had noted this fact before, and would even seem to consider that those frogs which appeared in this sudden manner were a peculiar species, under the name of διοπετής, that is, *sent by Jupiter*.

These frog-rains, of which a great number of other writers make mention, have occasioned much embarrassment to those who, considering the phenomenon as real, have been desirous of explaining the cause. Cardan, in his book *de Subtilitate*, asserted that it was the violence of the winds which carried the frogs from the tops of the mountains, and caused them to fall in the plains, or " stoop to th' vale." The wind also, according to the same author, could carry off the eggs of frogs, which opened in the air. But Scaliger demonstrated the utter impossibility of this, by shewing that the first production of the egg is a tadpole, and not a perfect frog. Besides, had these frogs been engendered in the clouds with the rain which brings them, if even, by virtue of this rain, they had been instantaneously evolved from the dust which it moistened, how shall we account for the aliment found in their stomachs, and the excrements which fill their intestines? We must believe then, with Redi, that their birth was anterior to their appearance, an observation developed with much talent by that learned Italian, but the honour of which is originally due to Theophrastus, the disciple of Aristotle, who lived under the reign of the first Ptolemy, King of Egypt, and who has written a treatise on animals which suddenly appear, entitled περὶ τῶν αθρὸον φαινομενῶν ζώων. It is therefore clearly proved, that the rain only draws them from the retreats in which they had lain concealed.

The frogs are distinguished by a very peculiar and sonorous cry, which we term *croaking*, and the French *croassement* or *co-assement*. Aristophanes has tried to imitate it by the barbarous and discordant combinations, *Brekekekex, coax, coax*. It is particularly during the time of rain, and in hot days, in the evening and morning, that the frogs indulge in this harmonious concert. The noise which they then make becomes sometimes insupportable. During the feudal regime, in France, when all the castles were surrounded with water, it was the occupation of the slaves or villains, to strike the water of those dykes, morning and evening, to prevent the frogs from disturbing the repose of my lord and lady. Even up to the period of the revolution, this custom existed in many places, and might be truly termed, without the affectation of a pun, to be a ***villanous*** employment. Yet, perhaps, after all, it was one of the most harmless (except to the frogs) of the "*droits de Seigneur*."

It is principally the males which croak. Their voice is much stronger in consequence of the two sacs which they have on the sides of the neck, and which dilate when the animal cries out. As for the female, she can only swell the throat, and produce a feeble sort of grunting.

Love among frogs, as well as among men, has its peculiar accents. It is an indistinct and plaintive sort of sound, named *ololo* or *ololygo* by the Latins, after the Greeks, because the pronunciation of this word imitates the cry in question. As this is peculiar to the males, the ancients have named them *ololyzontes.* In spring it is that this cry is uttered, as the tocsin of copulation. These animals when grasped, or held by the foot, send forth a short and sharp hissing sound.

Aristotle tells us that at Cyrene, a town built on the coast of Africa, there were formerly no croaking frogs. Pliny, after recounting the same story, with this addition, that croaking frogs were transported thither from the continent, and perpetuated their race, tells us that in his own times, those of the island of Serpho, one of the Cyclades, remained mute, but if carried elsewhere, incontinently began to croak lustily. But Tournefort assures us that the frogs of Seriphos, the ancient Serpho, possess the usage of their voice in as high perfection as those of any other country. Peradventure this may be a step in the Batracian "march of intellect."

Linnæus and some other naturalists have pretended that

the red frog of Europe had no voice. This, however, is only true when the animal is out of the water. Daudin has ascertained that in spring it sends forth some smothered cries from the bottom of the marshes.

As soon as the summer season is over and the weather begins to be a little cold, the frogs lose their natural voracity and give over eating. When the cold becomes more considerable, they protect themselves from its rigour by sinking into the mud of deep waters, in the holes of fountains, and sometimes even in the earth. The quantity of frogs which sometimes thus collect in one place, is so considerable that they cover the soil to a foot in depth, and thousands of them may be taken in a few moments. They interlace together more closely in proportion to the intensity of the cold. This seems to indicate that they find in their approximation an augmentation of heat.

In his voyage to the Icy Sea of America, Hearne informs us that he has many times, under the moss, found frozen frogs, whose legs might be broken without their exhibiting any sign of life, but which resumed their motions when exposed to a gentle heat.

The reptiles of which we are speaking, thus pass the winter in a state of most profound lethargy. Malpighi thinks that during their time of retreat they are nourished by a fatty matter contained in the vena-porta. This opinion is erroneous. The fat destined for alimentation in this case is contained in peculiar kinds of epiploa, which we shall notice when we come to speak of toads.

But this state of torpor, which is comparable to death, is soon dissipated in the early days of genial spring. On the first returning warmth and radiance of the sun, the frogs begin to bestir themselves, and obey the grand call of nature in the reproduction of the species. Even before the conclusion of the frosts, they are sometimes thus employed in the recesses of the waters. The season of love is announced in the males by the presence of a black papillose wart, which grows on the fore-feet. At the same time their belly swells; on opening it there is found in that of the male a mass of white jelly, and some black grains enveloped in a mucosity in that of the female.

If love be early in its diagnostics among these reptiles, it is slow in its effects. The intercourse lasts for many days, sometimes even fifteen or twenty. Bartholin has observed it not to finish until the fortieth day. The feet of the male become considerably enlarged, growing stiff and curved, so that he cannot separate himself from the female. Under such circumstances the head of the male has been cut off without his ceasing to fulfil his destination, namely, that of fecundating the eggs. But if the tubercles of the thumbs be removed, he can no longer maintain his position.

This intercourse, which takes place but once a year, is followed by the exclusion of the eggs from the body of the female. At the moment of this exclusion, they are bedewed with the fecundating fluid of the male. Some hours after the operation is terminated, the male separates from the female, and at the end of one or two days his feet recover their usual flexibility.

The eggs of frogs are always abandoned in the water, where they float at the surface, while those of the majority of the toads are deposited in the mud. They are connected with each other in the form of chaplets.

Repeated observations on the mode of generation in the frogs, clearly prove that it is not performed in the same way as that of the preceding classes of animals, but in a mode analogous to that of fishes. There is also not the least truth in the assertion of Mentzius, a professor of Leipsic, that the fecundating fluid proceeds from the caruncles on the feet of the male, a notion of the most absurd kind. Intromission of any kind is impossible, as the male is provided with no organ for the purpose.

But even these mistakes are pardonable compared with the ridiculous fables repeated by Pliny and Cardan on the reproduction of frogs. Those authors pretend that every six months these animals melt into a sort of slime, and that in spring they regenerate of themselves in the bosom of the waters.

The frogs are extremely multiplied. Their intercoure is very rarely indeed unproductive of a fruitful result. Daudin observed this failure but once in eleven times.

Each female lays annually from six to twelve hundred eggs. Swammerdam once reckoned eleven hundred, the production of a single female, and Montbeillard thirteen hundred. Be it also remarked, that the frogs can live a great number of years, if fortunate enough to escape the tooth or beak of their enemies.

These enemies are very numerous. A number of quadrupeds, birds, reptiles, and fishes, live habitually at the expence of the frogs. Serpents, pikes, vultures, and storks, destroy an immensity of them. Without the intervention of the last-mentioned birds, Egypt in particular would be covered with frogs. In some countries they are even sought after by man as a wholesome and agreeable aliment. Their only means of defence is the fluid which they eject from the anus, and which has but small effect in deterring those animals which approach them with hostile intentions.

According to some writers, Rœsel in particular, the frogs in summer moult every eight days; but at each moulting they only lose their epidermis. Daudin has frequently observed that they change colour, and grow brown, from terror, when they are in the presence of a snake. The frogs do not reproduce beyond the third or fourth year of their existence. It seems probable that they live a long time; but we have no positive data on this subject.

Living frogs have been found in thermal waters beyond the thirty-fifth degree of the thermometer of Reaumur. Spallanzani mentions an example of this kind on the authority of one of his friends, who has seen them alive in the baths of Pisa, though exposed to a temperature of 37° ⋈ 0. R.

We have already mentioned man among the enemies of the frogs, which in certain countries furnish food for his table. In England we hold this kind of aliment in detestation, which appears to be the result of nothing but a most unfounded prejudice. In France a very great consumption of frogs takes place; and they are caught in various ways, either with lines, or little nets, or by means of a rake, which brings them to the shore along with the mud. Sometimes they are pursued by night, and with torches, the light of which attracts them.

It is in autumn, at the moment in which they plunge into the water, where they are about to pass the winter, that their flesh is most in estimation, for at this time it is fatter, and of a more delicate flavour. Nevertheless, a greater quantity of it is eaten in spring than in any other season, for the frogs are then more easily caught.

There are places in which depôts of frogs are kept in reserve in gardens furnished with pieces of water, and closed round by walls, to be sold at all times to amateurs. About a century ago or more, they were in great request in Paris. A native of Auvergne, named Simon, made a considerable fortune by fattening up, in a suburb of that city, the frogs which he had collected in his own country. At present they are much less generally eaten in France; but they are constantly to be found in the markets of that country ; and those of Italy, in like manner, superabound with them for a certain period of the year.

The Romans do not appear to have made much use of them as food. Galen says nothing about them in his works. The physicians of the middle ages were in general opposed to their introduction as an aliment, and attributed to them deleterious properties. Aëtius, and Juan Rodriguez de Castellobranco, have particularly declared themselves to this effect. Others would fain have established a distinction between frogs, of such as are poisonous and such as are harmless. Matthioli, and the celebrated Ulysses Aldrovandi, were of this opinion. The latter even mentions a great number of delicate culinary preparations of which frogs constitute the basis. As, however, we are not writing a book on cookery, and, even if we were, as we should despair of overcoming the gastronomic prejudices of Englishmen in this particular, who consider their abstinence from frogs as an honourable anti-Gallican distinction, we shall resist our inclination to give any additional illustrations of the excellent definition of man, "a cooking animal," though we have no doubt there is as much *reason* in the fricasseeing of frogs as there is in the roasting of eggs.

In the sixteenth century, on the continent at least, frogs were served up at the best tables. Champier complains of this taste, which he considers fantastic. It does not, however, appear that this custom was a very ancient one, for, in 1550, the author of a book, entitled "*Devis sur la Vigne*," tells us that he laughed "de Perdix quand on lui apporta des grenouilles *en façon de poulletz fricassez*." Thirty years afterwards, Palissy, in his treatise "Des Pierres," thus expresses himself: "*Et de mon temps j'ai veu qu'il se fust trouvé bien peu d'hommes qui eussents voulu manger ni tortues ni grenouilles.*"

In Germany, all parts of these animals are eaten, the skin and intestines excepted. In France, epicures confine themselves to the hinder quarters, which are dressed with wine, like fish, or with white sauce. Sometimes they are fried, or even spitted.

Cooks are not the only persons who have studied the art of appropriating frogs to the benefit of man. For a considerable period, the continental physicians have employed the flesh of these reptiles, variously prepared, in the treatment of different diseases. Broths are made of it, which are considered restorative, diluent, analeptic, and anti-scorbutic, and are prescribed in affections of the chest, pulmonary consumption, cutaneous disorders, and many other maladies.

Even supposing the utility of such preparations in the cases we have mentioned, it is certain that a great number of physicians have adopted the most absurd notions and practices on this subject. Timotheus, for instance, would apply frogs cut in two on the kidneys of hydropic patients, to attract externally the superabundant serosity in the abdomen. We have heard of the application of a brick-bat, in certain cases, to a peculiar part of the human body, and we presume that its curative efficacy is fully equal to that of the cataplasm here recommended.

According to Dioscorides, the flesh of the frog, cooked with salt and oil, is an antidote for the poison of serpents; and Arnold informs us that the heart of this animal, taken every morning in the form of a pill, cured a fistula in the epigastric region, which had resisted many other remedies.

With a like degree of νους, some doctors have recommended, in epilepsy, the liver of a frog calcined in an oven on a cabbage-leaf between two plates, and swallowed in peony-water.

We may observe, notwithstanding these absurdities, that

frog-spawn may be usefully employed in external inflammations, as soothing and emollient.

In the old pharmacopeia, oil of frogs is mentioned, which is now generally abandoned, even by the empirics of the continent. A plaster composed of these animals was the invention of a French surgeon, Jean de Vrigo. But while we smile at the errors committed by our predecessors in the art of healing, we should not forget that the discovery of the most efficacious remedies has been owing to accidental experiment, and not to any speculative reasoning on the economy of the human system. We should also remember that no satisfactory account can even yet be given of the *modus operandi* of such remedies; and those who have reasoned from the fact of their being general or partial stimuli, and, in consequence, made experiment of corresponding stimuli, have failed in producing the same results. No one can yet tell why sulphur should cure the itch, and why other agents on the skin should not succeed equally well in removing that troublesome disease; and the same is true of the exhibition of mercury in syphilis.

Klein, who separated the frogs from the toads, has described some foreign species, as have also Seba and Catesby. Linnæus but little extended the catalogue of this genus comparatively with the moderns. Latreille, in his Natural History of Reptiles, has enumerated a dozen genuine ranæ, and Daudin has at least doubled the number.

We shall now cast a rapid glance over the more remarkable species of this genus.

The green frog, (*rana esculenta*,) is called by our Gallic neighbours the common frog. It is about two or three inches long, without reckoning the hinder feet. It is found abundantly in stagnant waters both in Europe and Asia, though it is much less common in England than the *rana*

GREEN FROG.

RANA ESCULENTA. Lin.

temporaria. It seldom comes to land, and never removes from the banks of streams, ponds or lakes. Motionless, either on the surface of the water, or fixed on some aquatic plant, it pours forth during the summer season the most intolerable croakings.

Its eggs are spread in packets on the marshes. Its thighs are in great request among the amateurs of good cheer. A very considerable consumption of these frogs is made in Vienna, where they are fattened up in *froggeries*, (*grenouillières*) constructed for the express purpose.

They are frequently taken during the heats of summer, with a line baited by a small bit of scarlet cloth, which is kept in motion, so as to give it the appearance of a living being.

There are many varieties in this species, which have been pointed out with great care by Daudin. One of these was employed by Spallanzani in his experiments on generation. The back of this variety is of an uniform green, and it inhabits the river and dykes of Lombardy. Another has the edge of the lips black, rounded black spots on the sides, and the belly entirely white. It has been observed by Van Ernest in Holland. A third is of a sombre green, with transverse brownish spots on the limbs; it was found by Daudin in the neighbourhood of Beauvois. A fourth inhabits Provence, and is remarkable for its reddish belly.

The common frog of this country, *rana temporaria*, is called the red frog, by the French. It is tolerably abundant throughout all Europe, and is different from the last species, both in colour and habits, though it has not always remained unconfounded with it. It prefers wooded and mountainous situations, and frequents meadows and gardens during fine weather. It is most generally found on land in the summer season, and while the green frog rarely abandons the bosoms of sleeping waters, this species must be sought for among bushes and plants with long stems, and in places remote from the banks of streams.

Some authors have given the epithet *muta* to this species, because it does not croak. It is not, however, destitute of the power of uttering sounds; at the time of reproduction, or when it is tormented, it sends forth a sort of grunting noise. It also croaks, but only at the bottom of the water, which is contrary to the habits of the other species.

It also possesses the faculty of shooting from the anus an acrid fluid, and in much greater abundance than the green frog.

At the approach of winter, it retires into fountains and ponds of pure water. It never goes but through absolute necessity into marshes and miry waters. It does not bury itself in the mud like the preceding species, and numbers of these frogs may be taken during the winter by making holes in the ice.

It lays its eggs at a later season than the green frog, and the developement of its tadpole is more slow.

This species is most commonly eaten in the central parts of France. The hind quarters are as good *en fricassée*, as those of the green frog.

Like the preceding species, it presents many varieties not necessary to be noticed here.

The *rana punctata* is said to be susceptible of a change of colour under the influence of terror. It is also believed that it croaks at the bottom of the water like the common frog.

Rana clamitans, was found by M. Bosc in the fresh waters in Carolina in the neighbourhood of Charlestown. Its obscure tints give it some resemblance to a toad, but it is soon distinguished by the extreme vivacity of its movements, being by far the most lively of all known frogs, and it is extremely difficult to catch it again after it has once made its escape. It croaks continually in an insupportable manner, and never removes far from the shore; when one attempts to seize it, it shoots into the water, uttering a sharp cry.

The *rana pipiens* is one of the largest species of the genus, being three or four inches broad, and six or eight in length, without taking in the feet; when it is measured with the limbs extended, its entire length may be about eighteen inches.

It inhabits North America, and more especially Carolina. It is not quite so frequent in Virginia. There it often abides at the entrance of its hole, which is placed near the water of some fountain, into which it precipitates itself as soon as it hears any one approaching

The inhabitants of Virginia give the name of *bull-frog* to this species, and will not destroy it because they pretend that it purifies the waters in which it lives. In Pennsylvania it is called *shad*-frog, because it makes its appearance at the same time that the shads do in spring.

Catesby affirms that it utters sounds very much resembling the bellowing of a bull, and with greater force when it is at the bottom of the water. During the summer evenings and in dry weather, it makes a terrible noise. It is exceedingly partial to young ducks and goslings, which it swallows whole. According to Bartram, it proceds in pursuit of prey to some distance from its retreat, and abounds in the rivers, marshes and lakes of the southern regions. As the voracity of this species is proportioned to its bulk, it is rare to find more than a couple of them in each marsh.

It is extremely difficult to catch this frog: it is only during the night and when it removes a little from its retreat, that it is possible to procure an individual. When it is on a level ground, it makes leaps of from six to eight feet in length.

It would seem that this species has been confounded with the *ocellata*, *clamitans*, and *grunniens*, in consequence of the Anglo-American name, bull-frog, which is common to them all. There is another frog mentioned by Bartram, called the *bell*-frog, and which is provisionally referred by Daudin to this species; its voice exactly resembles the sound of one of those little bells hung to the neck of cows. They usually croak in bands, one commencing and another replying; the sound is then repeated from troop to troop, to a very considerable distance, during some minutes; this sound increases or diminishes according to the intensity of the wind; it afterwards ceases almost altogether, or is prolonged in the distance by other troops which reply to the first. It is momently renewed, and when the ear is accustomed to it, it is found not to be altogether devoid of harmony, though at first it appears troublesome and disagreeable to strangers.

The *rana grunniens*, if it be distinct from the preceding species, is at least fully at large; Daudin thinks that it has been seen by Bartram in Florida and in Carolina, in humid marshes, and on the banks of lakes and great rivers, where it sends forth a strong and unpleasant sound, very much resembling the grunting of a pig, but not so sonorous as that of the pipiens.

This enormous frog is also called in English, bull-frog, and is improperly named *crapaud* by the French colonists in the West Indies. It is found in most of the western isles, where it has been very accurately observed by M. Moreau de Jonnès, and it has been called a toad, because it inhabits shady and humid places, as toads do in France and elsewhere, and not stagnant waters like the frogs.

It does not quit its retreat until night. Its strength is so great that it can clear in a leap a wall five feet high. In the dry season it is very torpid, but resumes its vivacity when the rains set in.

These frogs are domesticated in the Antilles for the use of the table, and become tolerably familiar. The flesh is white and delicate. It is fricasseed like fowl, and two frogs constitute a good dish.

The *ocellata* is found in Florida, and some countries of South America. It has been confounded by several naturalists with the bull-frog, to which it is not inferior in size. It is six or eight inches in length, without comprising the feet.

The *rana halecina* of Kalm, very much resembles the common frog, but is smaller, its body being seldom more than two inches long: its muzzle is also much more pointed.

It is very common in Carolina, where it is tiresome from its continual croaking, and is said to announce the approach of rain by its cries during the nights of spring. It seldom comes to land; but its leaps there are exceedingly rapid, and are from fifteen to eighteen feet in length. In its general manners it resembles the *rana esculenta*.

The *rana paradoxa* is found at Surinam, and in other countries of South America. Its tadpole grows to the largest size before the metamorphosis takes place. From the ambiguous aspect which it exhibits in the latter part of its progress towards its ultimate form, it was long considered a paradox by European naturalists. As the adult animal, in consequence of the loss of an enormous tail, and of some envelopments of the body, is less in size than the tadpole, Mademoiselle Merian, Seba, and some other ancient observers, were led into the error of supposing that the animal passed from the frog to the tadpole state, and was subsequently transformed into a fish. This error, though long consecrated by time, is at the present day completely refuted.

Laurenti was the first who separated the HYLÆ from the frogs and toads. This genus is now universally adopted by naturalists, and the viscous disks or cushions by which the toes are terminated in the animals which compose it, constitute their most obvious and leading characteristic.

By means of these the hylæ are enabled to attach themselves to smooth bodies, to climb, to leap from branch to branch, and to traverse the moving leaves of trees, agitated by the wind. More agile than the frogs, and endowed with extraordinary suppleness, they make their way with dexterity and nimbleness on the most flexible branches, where they are, besides, steadily retained by the conformation of their toes. Notwithstanding this, they are more tranquil than the frogs, and are observed to await for entire days in the same place in expectation of their prey.

The hylæ feed on all kinds of worms and small insects, and during the fine weather they proceed on the leaves of trees in the woods in search of aliment. Later in the season, they retire to the bottom of waters, and, like the frogs, pass the winter there in a state of lethargy. They also continue there during a part of the spring, to couple, and to lay their eggs.

In the day-time, and particularly when the heat of the sun is greatest, they shelter themselves in shady places, where the trees afford a thick foliage. They put themselves in motion when twilight comes on, and sport in security.

The cushions with which their toes are provided, are simply fleshy, and of the form of a lentile. Examined with the microscope, they appear porous sieves, from which an unctuous humour very slowly exudes. They are generally a little concave, and sometimes furnished with a very distinct fold.

The croaking of these animals has much analogy with that of the frogs. It is only less sharp, and occasionally stronger, especially in the males, which have under the throat a pouch, which then swells. It is heard under the same circumstances as that of the frogs, and especially during rain, and in the middle of the fine nights in summer. Often, during that season, the hylæ are to be found assembled on the tops of trees, sending forth in chorus their hoarse and discordant music.

In Europe, there is but a single species belonging to this genus. It is the *tree-frog* (*rana arborea*, Linn.). This batracian is very common in the southern countries of Europe ; but more rare as we proceed northwards. M. Bosc thinks that he has met with it near Charleston, in North America. It seems to avoid dry situations and mountainous forests. It is only to be found in humid woods, in hedges bordering on marshes, in parks and gardens ornamented with pieces of water.

Its internal organization, its mode of reproduction, the phenomena relative to its fecundation and eggs, the growth of the tadpole, the metamorphosis, exhibit no difference from the frogs, except that the coupling takes place a little later in the season.

It appears that it is not until the third or fourth year of its existence that this species is fit for propagation. Until this epoch, the male is nearly mute. It also seems that two months' time and more is required for the tadpole to undergo its metamorphosis, and come to the perfect state in which it is enabled to leave the water. It has been observed by M. de France that these hylæ swallow their skin at each moulting.

The *hyla lateralis* has been observed by Catesby and M. Bosc in Carolina. It is said also to have been seen at Surinam. It is usually found attached underneath the green leaves of trees, to conceal itself, and lie secure from the birds and serpents, which are among the most dangerous of its enemies.

These hylæ are sometimes found in very numerous troops; the bushes and weeds are completely covered with them, and their croaking may be heard at the distance of entire leagues.

The hyla lateralis makes prodigious leaps. When young, it is called in the United States the cricket of the Savannahs, its cry there resembling the noise made by that insect.

It is seldom seen in the day-time, but it proceeds to some distance, and makes itself heard during the night. It leaps from branch to branch, ascending to the very top of the highest trees, to catch flies and other insects.

Linnæus, and many other herpetologists, have made of this species only a simple variety of the preceding.

The *hyla tinctoria* is found in various parts of South America, especially in Surinam, and the interior of Guiana. It inhabits woods, living on the trees almost the entire year, conceals itself under their bark during the cool nights, and seldom visits the water but for the purpose of coupling, and leaving its eggs.

Lacépède and Buffon state that this is the species employed by the Indians to change the plumage of parrots from green to red, in some parts. For this purpose it is said, that they pluck out the green feathers from these birds when young, and rub the wounded skin with the blood of the hyla. The feathers which grow after this, are of a fine red or yellow.

A species called by Laurenti the *hyla tibiatrix*, is said by Seba to be American, and to croak in a most melodious manner during the very hot weather after the setting of the sun, which usually betokens fine weather. During cold and rainy periods the animal is silent, and conceals itself at the bottom of the water.

Daudin has enumerated some others of this division, but there is nothing in their history which could interest our readers.

The TOADS are easily distinguished from the two preceding subgenera, as the parotids are wanting in the latter, and their hinder feet are longer than the body. They are also easily distinguished from the pipas which have the toes free, and want the tongue altogether. It must however be granted that the greatest relations subsist between them and the frogs, to which they were united by Linnæus, whose example has been followed by the majority of naturalists. Many frogs have the hinder feet considerably shortened. Many have also the body covered with tubercles. The presence of *parotids* appears, in fact, to be the only distinctive character, that can be relied on with any degree of certainty.

The word *bufo*, is one of considerable antiquity in the Latin language. *Inuntusque cavis bufo*, says Virgil, in the first book of the Georgics. Hermolaus thinks that this name was bestowed on the toad from the faculty possessed by this animal of swelling itself when angry, and uttering a hissing sound resembling a sigh. The Greeks seem to have designated the toad by the word μυοξὸν, although Scaliger thinks this an error, and by the words φρυνος, and βατραχος ελειον (Marsh-frog). Of the French word *crapaud*, the origin is not known, and our word *toad* is Saxon.

The toads have, in all ages, and in all nations, been regarded as revolting and disgusting animals. They are universally considered as repulsive, and even sometimes are objects of horror. They are usually believed to be venomous, and in consequence of this prejudice are subjected to general proscription and extermination. We shall soon see, however, that these animals are comparatively harmless, that the study of their organization involves much interest, and that their history presents a crowd of facts, equally curious and important.

We shall first speak briefly concerning their organization, as far as it differs from what we have already described as peculiar to frogs, and generally, with respect to those points in which the two genera agree, and which we have thought proper to omit in our notice of rana, and reserve for the present head.

The bones of the upper region of the head in toads are, for the most part, rugous at their surface. The intermaxillary, jugal, and tympanic bones alone are beneath. The bones of the lower region do not present the inequalities observed in the others.

With the exception of the symphysis of the chin, and of the intermaxillaries which are free in all parts, all the bones of the cranium and face are completely united by synostosis, in adult individuals. The osselets of the ear are but two, the malleus, and the stapes. They are very large and cartilaginous.

The toads in general are without teeth. In some species, however, teeth tolerably large and curved have been observed. The head is articulated by two condyles with the atlas.

In the toads of Europe, the vertebræ are eight in number; there are but seven in some foreign species; their processes are generally strong and pretty long; the transverse processes are broad and securiform.

The sacrum has some transverse processes, prismatic, triangular, and very robust; it is long, pointed, and compressed, without coccyx.

The coxal lines are united in a single piece, in adults, as is the case with frogs in general; there is no appearance of ribs.

The sternum is broad, it is united in front with the furca and clavicles; it is emarginated behind and provided with two cartilaginous pieces in the horned toad; in the other species it is terminated by a disc, which serves for the insertion of the muscles.

The bones of the furca and clavicles are entirely joined, on one side with the sternum, and on the other with the scapula; at their point of junction, those three pieces have a wide oval aperture between them, which communicates by a short canal into the scapulo-humeral articulation; this peculiarity is very evident in the horned toad. It is not met in most of the others.

The omoplate is formed of two articulated pieces, the upper one of which looks towards the spine.

The bones of the fore-arm are synostosized, so as to form but a single one, hollowed on each side and underneath, with a furrow of no great depth.

The carpus is generally composed of eight bones; the metacarpus of four; the front toes are four in number; there is only a vestige of a thumb supported by a metacarpal bone.

The femur is straight and without trochanters; after it, comes the piece erroneously supposed to represent the two bones of the leg, which we have noticed before. The real tibia and fibula which most authors have represented as the astragalus and calcaneum, are separated in their whole length.

The tarsus has four, the metatarsus five bones.

The myology of the toad exhibits in its examination the same irritability, strength, and sensibility to galvanic action in the muscles, as we find in the frog. It is unnecessary here to describe these muscles in detail, as enough has been said on this subject in treating of the frog.

Though the nerves in the toad are very distinct and large, the cavity of the cranium where they originate is small. The brain itself is of very small volume.

The hemispheres are smooth, without circumvolution,

elongated, and narrow. The cerebellum is flatted, triangular, and thrown back on the medulla oblongata.

The olfactory nerves come from the anterior extremity of the cerebral hemispheres; the hole by which they are transmitted from the cranium is double. The nasal fossæ are of no great extent; they have neither cornets, nor any sinus communicating with them; they merely exhibit a few tubercles. The nostrils are tubulous and furnished with a small valve, intended to oppose the exit of air during the movements of respiration.

The orbits are separated from the temporal fossæ, only by an incomplete osseous branch, &c. (See Frogs.)

Toads enjoy the sense of hearing. Aëtius tells us that they were distinguished by the ancients into deaf toads, and such as could hear; the former were regarded as venomous; the skin which covers the tympanum is finer than that on the rest of the body.

The tympanic box is entirely membranous in its lower part. The three semi-circular canals are situated above the membranous labyrinth. The labyrinthian sac contains a stone of an amylaceous consistence, as in chondropterygian fishes.

Of the epidermis and dermis, we have already spoken under the frogs, as well of everything relative to the cuticular system.

It would appear that the toads can, at will, increase the excretion of the viscous humour, with which their bodies are lubricated, and cause it to run out like a dew from all their pores. The use of the mucus which thus invests their body is manifest: it serves to defend them against the dryness of the air and the heat of the sun; it is the same with frogs. One of the latter animals was destroyed by Bartholin, by first rubbing the head and back with oil, and then exposing it to the sun. M. Schneider has also observed the ill effect of the sun on toads, and we are informed by Adanson, that the evaporation from the skin of these animals is so great, that the negroes in traversing the burning sands of Senegal, are in the habit of applying one of them alive to the forehead, for the purpose of cooling it.

The sense of touch must be delicate in these animals, from the nakedness of the toes, which are without claws.

The point of the tongue is not bifurcated, as in most of the frogs.

The length of the intestines to that of the body is in the proportion of two to one. The rectum is cylindrical.

The structure of the heart is the most simple imaginable. There is but one rounded auricle, wider than the basis of the heart, and strengthened by fleshy columns ; and there is but a single conical ventricle, whose cavity has fleshy columns adherent to it, and opens into the common trunk of the arteries by a single orifice below the auriculo-ventricular aperture.

The aorta is soon divided into two branches, each of which produces a pulmonary, a common carotid, an axillary, and a vertebral artery. Then they approach one another and unite into a trunk, which furnishes the cœliac and all the other arteries of the abdominal aorta. In this manner a part of the blood only passes into the lungs. The veins have a similar distribution.

The toads do not possess the sonorous vesicles which are in the mouth of the male frogs, and which give to their croaking its resounding noise.

The toads, like the other batracians, have no generative organs of intromission.

The young, on quitting the egg, have the head and belly united in a spherical mass, and terminated by a fish-like tail. Their metamorphoses are similar to those of frogs. The eggs acquire a marked development in the ovaries, and sometimes singularly enlarge the belly of the animal. Their colour is blackish while in the ovary, as Camper has observed, except the smallest, which are yellow and white.

The toads feed on small mollusca, worms, and insects, and never touch dead or motionless animals. When Linnæus tell us, "delectantur *cotula*, *actœa*, *stachide*," we must not suppose that the learned Swede meant that they lived on vegetables, but merely that they were pleased with the odour of these fetid plants.

It is during night that the toads issue forth from their sombre retreats. They also abandon them after the hot rains of summer, and at such times they are often seen almost to cover the surface of the earth in places where they were not observed before. This has given rise to the same absurd opinion concerning them which we have already noticed to prevail concerning the frogs, namely, that they fell in rain from the sky.

Toads live a long time without eating. They have been known to remain shut up for years in walls, hollow trees, or in the earth, without being able to get out, and without losing life. In 1777, Herissant undertook some experiments to ascertain the truth of facts of this kind, which might appear fabulous. He shut up three toads in sealed boxes in plaster, and they were deposited in the Academy of Sciences. At the end of eighteen months one of these toads was dead, but the other two were still living. Nobody could doubt the authenticity of this fact, yet the experiments were severely criticized, as well as the observations which they seemed to confirm. It was contended that the air must have come to these animals through some imperceptible hole which escaped the notice of the observer. Some probability, however, was given to this circumstance by some researches of Dr. Edwards, published in 1817. He has observed that toads, shut up totally in plaster, and absolutely deprived of air, lived for a great number of days, and much longer than those which were forced to remain under water. This certainly is one of the most extraordinary phenomena which the history of reptiles can furnish. It appears an exception to the necessity of air, which is regarded as indispensable to the life of all animals, and seems to break the chain which united them under the most interesting relations of existence. It appears, however, that the air evidently penetrated through the plaster, as Dr. Edwards proved, for the toads perished as soon as the plaster which enclosed them was placed under water. The opponents of Herissant were therefore justified to some degree in their scepticism. Still the fact of animals existing so long under such circumstances, even with a little air, is most surprising, and calculated to produce very strange reflections. If these reptiles lived in this manner longer than they would have done in the open. dry air, the reason is that they lost less by transpiration, and if they died much later than they would have done in water, it was because the air certainly had some access to them.

We also find, on observation, that toads, shut up for a long time in solid masses, have the mouth filled with a sort of mucous membrane. M. Schneider has remarked that frogs, during their hybernation, submerged in mud, had the same part stuffed with mucus and mud.

With the toads the feet are seldom used in walking. They all creep, and when surprised, do not seek safety in flight, but stop suddenly, swell their body, render it hard and elastic, distil from the tubercles on the skin a white and fetid humour, shoot a peculiar fluid from the anus, and attempt to bite. But their bite occasions no great inconvenience, merely producing at times a slight inflammation.

The liquid ejaculated from the anus is not urine. It has been supposed, but very erroneously, to be venomous. That which oozes from the tubercles is equally harmless. It has been pretended that when these liquids were deposited on vegetables, fruits, mushrooms, &c., that they produced vomiting. It appears certain, however, that when these fluids have been swallowed, they have produced violent nausea, and other unpleasant stomachic affections. This is not very surprising, and may perhaps, be partly explained by the association of ideas. M. Bosc, however, tells us, that if during the heat of summer, any one, after having handled the common toad, puts his hand to his nose, he will be troubled with the same unpleasant symptoms. Schelhammer relates an anecdote of a child, who had a severe pustulous eruption, in consequence of another child having held a toad for some minutes before its mouth.

The cutaneous fluid of the toads is yellowish, of an oily consistence, capable of concretion by exposure to the air, and of a very bitter, acrid, and caustic taste. Tincture of corona solis it will turn red, of a deep hue, and it forms an emulsion with water. It appears to contain an acid partly free, and partly combined to a basis, which is a very bitter and fatty substance, and an animal matter which has some analogy with gelatine.

In the countries where the temperature is cold, the toads pass the winter in the holes of rocks, often united in great numbers. In the United States, they have been frequently found thus benumbed by cold, in the same holes which were occupied by rattlesnakes.

As soon as the warmth of spring begins to be felt, the toads repair in crowds to the waters in their neighbourhood for the purpose of engendering. The union of the male and female subsists for a longer or shorter time according to the season, from two, to upwards of twenty days. They then croak perpetually, and the male repulses other males with his hinder feet. When there are more males than females in one marsh, they unite in numbers around a couple, and wait until the female lays her eggs. When the eggs appear they are fecundated by the male. They are then, in general, abandoned in the water. They form two chaplets, which being joined together, would sometimes exceed forty feet in length. In ten or twelve days after being laid, they acquire a double volume. The small tadpoles issue from them towards the twentieth day, and they acquire their gills in two or three after.

It is commonly believed that the tadpoles of the toad, live on the detritus of vegetables in the water. But M. Bosc, from a series of observations, rather believes that they feed on infusory animalculæ, and the larvæ of insects, &c.

The toads continue to reproduce only until the fourth year. They probably live a long time, but nothing positive is known on this subject. Some are observed to acquire enormous dimensions.

They are capable of being tamed. Pennant relates that at the residence of M. d'Arscolt, there was a toad, which had established its domicile under a staircase, and which had grown so familiar, that every evening as soon as it perceived light in the house, it raised its head, and seemed to ask to be placed on a table, where it found its supper prepared, and consisting of worms, flies, beetles, and other insects. It lived in this manner six and thirty years, and finally died only in consequence of an accident. It was of an enormous size.

The toads have been the subject of a great number of fables, both ancient and modern. To their glance has been attributed the power of charming both men and animals. Toads have formed an ingredient in many magical compositions, employed by persons in the country, who combine in themselves the most happy union of ignorance, wickedness, and superstition. These reptiles, however, are so far from being hurtful and offensive, that they are incapable of self-defence, and constantly become the prey of serpents, of pikes, of storks, vultures, and a number of other animals.

Toads die very quickly, when they are powdered with salt or tobacco. It is also said that gardeners drive them from their gardens, by burning old leather there.

There are few persons who would knowingly eat the flesh of toads. Nevertheless, even at Paris, the thighs of these animals are constantly sold for the thighs of frogs. In Africa and America, they are habitually eaten by the negroes.

The old physicians employed this disgusting being in a variety of pharmaceutical preparations. Dried and powdered, the flesh was considered diuretic and diaphoretic. It was prescribed living, as a topical application in cases of cephalagia and epigastralgia. Macerated in oil, it was considered detersive and anodyne. Ettmuller, Joël, Vallesnieri, and many others, have left us curious details on this subject. It is fortunate that we are now pretty well rid of all the rubbish of remedies of this kind, which, like an useless and troublesome scaffolding, so long obstructed the entrance into the sanctuary of medicine.

The *common toad* (*rana bufo*) is too well known to need any external description, even if that had been omitted or left deficient in the text. It usually sojourns in obscure and sheltered places, and passes the winter in holes, which it hollows for itself. It is found throughout Europe, is common in this country, and abounds greatly in the neighbourhood of Paris, where it is frequent in gardens. In March and April it reproduces in the water, or otherwise; the female proceeds to the water, drawing the male with her. She produces small and innumerable eggs, united, by transparent jelly, into two cords, often a hundred and thirty feet in length, which the male draws after him with his hind feet. The tadpole is blackish, and is remarkably small when it loses its tail and gets its feet. The branchial aperture is on the left side. Daudin tells us, in opposition to the opinion of almost all naturalists, and, indeed, to common observation, that this toad shuns the water, and lays its eggs in places near subterraneous sources.

The common toad walks slowly, and leaps but little. It lives in general about fifteen years, and produces at four years of age. Its cry has some relation to the barking of a dog. Sometimes, during the summer, it makes a feeble croaking at the entrance of its hole, which it gives over as soon as any one approaches.

The *bufo calamita* lives in the temperate regions of Europe, and particularly in the mountains. It is common enough in France. It undergoes all its metamorphoses in the water, and afterwards inhabits dry places, the clefts of walls, the holes of rocks, and passes the winter there in a lethargic state, and sometimes in small numbers. In Saxony it is common in the houses.

It lives on land, never leaps, but runs tolerably swift. It climbs on walls and trees, to conceal itself in their holes; and to assist it in this sort of movement, it has two small osseous tubercles under the palm of the hand. It never goes to the water but for the purpose of reproduction, which takes place in spring.

The cry of the male resembles that of the *hyla viridis*, and is produced by means of a vesicle placed at the entrance of the throat.

This batracian is said to shed a very strong odour of gunpowder.

The *brown toad* (*rana bombina*), which leaps tolerably well, sojourns, by preference, in the neighbourhood of fresh and stagnant waters in the south of Europe. It sheds a very strong odour of garlic when disturbed. The croaking of the male resembles that of the green frog, and the female utters a feeble grunting noise. Its eggs come from the body in a single cord, but thicker than the two cords on which the eggs of the common toad are strung. The tadpole has but one branchial aperture on the left side. It is slow in passing to the perfect state, and grows very large before it loses its tail and gets its hinder feet; so that when it quits the tadpole envelope, it appears somewhat diminished in size. It is eaten in some places, as if it were a fish.

M. de Lacépède considers as analogous to this species the *rana ridibunda* of Pallas, which inhabits the waters of the Volga and the Dural, near the Caspian Sea. Its cry resembles a laugh, and its weight is sometimes half a pound.

The *bufo obstetricans* is found throughout all France. It was first figured and described by M. A. Brogniart. It is never seen in the water, not even at the time of reproduction. The male assists the female in getting rid of her eggs, which are tolerably large, and about sixty in number. He attaches them in packets on both his thighs by means of certain threads of a glutinous matter. He carries them every where with him, taking all the necessary care for their preservation. This affords a rare example of paternal attention in this class of animals.

At the end of some time, the eyes of the tadpole which these eggs contain, are visible through their membranes, the albuminous matter of which is more transparent and solid than in the other species. When the young are about to be excluded, the toad seeks out some stagnant water, and deposits them there. They soon open, and the young animal comes forth and swims immediately.

A species, called *bufo spinosus* by Daudin, in consequence of its having certain tubercles terminated by an obtuse corneous sort of spine, is found in France, and particularly in mountainous situations. It seems probable that to it may be referred all the observations made upon monstrous toads in Europe.

This toad is never met with on the surface of the soil; it can only be procured by means of the plough, and the country people are persuaded that it never quits its retreat voluntarily. Daudin suspects that it lays its eggs in the earth, in humid places, near the subterraneous sources of water.

The *bufo musicus* of Daudin is met with in Carolina, where it lives in holes in the ground, never coming out until towards evening or after rain. Its croaking is far from being *musical*, as has been pretended—it is weak and disagreeable. Bertram, however, tells us that in the early spring, when the toads assemble together in great numbers, in ponds and canals, they make a loud noise, which is not altogether devoid of harmony. They leave the water after laying, and spread themselves over elevated grounds. Their young, when they have undergone all their metamorphoses, are scarcely larger than a grasshopper, and they proceed immediately from the water to hop and walk on the dry land. They live on various insects.

For the species figured it may be sufficient to refer to the text and table.

The genus PIPA, after the example of Laurenti, the Baron, and M. Dumeril, is now constantly distinguished from the toads by modern naturalists. The *pipa of Surinam*, which is its type, is assuredly a most hideous reptile; it is from six to eight inches in length, and four or five in breadth; its skin is of a sombre brown, and thickly sown with reddish tubercles; on its thighs and sides, small warts are visible, and its skin is folded and wrinkled.

The female, whose back is hollowed with a great number of small cells, is more voluminous than the male; the latter has an enormous larynx made like an osseous triangular line, within which are two moveable bones, which have the power of closing the entrance of the branchiæ.

The pipa lives in the fresh waters of South America, and sometimes in the obscure parts of houses, at Cayenne and Surinam, whence it is named *tedo* and *curucu*. According to Seba, and Mademoiselle Merian, the negroes of the colonies employ its flesh as food.

This reptile has been long celebrated for the mode in which it perpetuates its species, and in this respect has become the object of research to some of the most distinguished naturalists of the last century.

It it now demonstrated that the female lays its eggs after the manner of toads, but that the male, fastened on her back, fecundates them, and then places them on the back of the mother ; she then repairs to the water, where her skin swells, and forms rounded alveoli, in which these eggs are lodged, to be subsequently disclosed. At the moment of their birth, the young have a membranous tail, before the fall of which they do not quit their cell. They then have feet, and when they have departed, the female, having rubbed the epidermis from her back against some hard body, returns to land.

We now arrive at a genus, whose name has been celebrated from antiquity, and embellished with the tints of fable in all ages. It was on the fortunate soil of ancient Greece, in the bosom of a wise and warlike nation, whose imagination, favoured by a happy climate, exaggerated even the wonders of creative power, that the reputation of the SALAMANDER originated, and that an immortal and generally adopted name was employed to characterize an obscure reptile, which has usurped the most universal celebrity, and is even still one of the objects of the curiosity of man.

This animal, which the rude inhabitants of other countries regard as an object of terror, and, as a malevolent being, abhor and proscribe, as an object not less disgusting than dangerous, has formerly passed, and still passes in the eyes of many persons, as being able to brave the violence of fire, the most active of the elements, to escape from the force of its action, and not only to come safe and sound out of the flames, but even to extinguish them. However, after having furnished so many emblems to the poet, more brilliant than faithful, this little oviparous quadruped, once so highly privileged,

THE PIPA. *R. PIPA. Lin.*

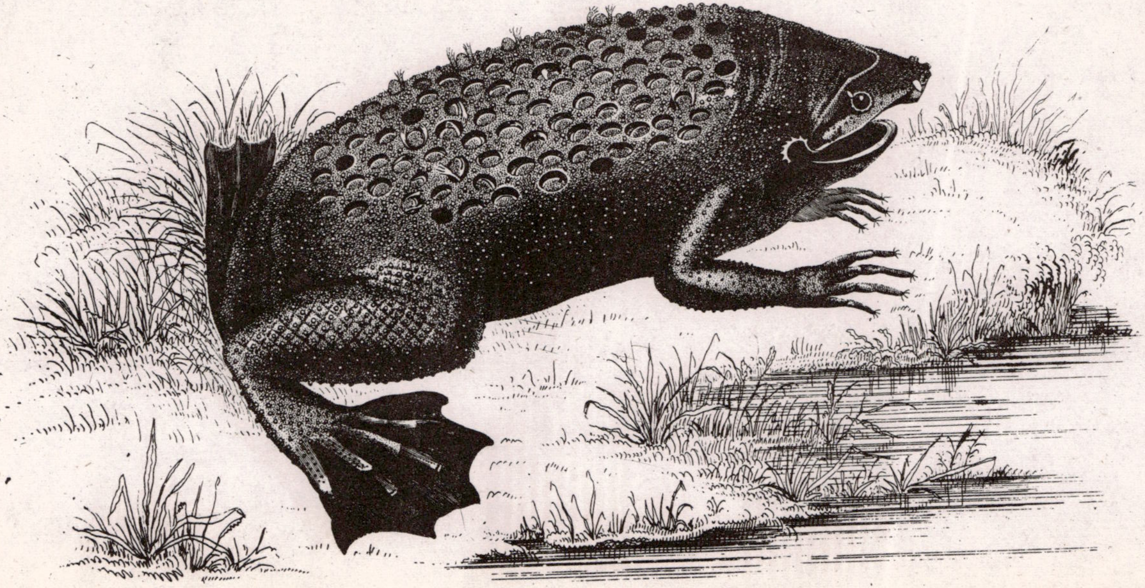

has fallen into oblivion and contempt, so much so that the interest which it really deserves to excite, is no longer taken in it, since it has been despoiled of those attributes in which it was so gratuitously invested. This is an evident proof of the immense influence of the light of truth directed against objects of arbitrary belief, or absurd conceptions. The salamander has ceased to furnish similitudes, even to the poets, and to be the emblem of deathless love, and death-despising valour. This daughter of fire, with a body of ice, whose origin was not less surprising than its power, which owed its existence to the purest of elements, by which it could not be consumed, which mountebanks had vaunted as capable of arresting the progress of the most violent conflagrations, and authors have assumed as the basis of so many interesting allegories, has dwindled down into a simple batracian reptile of the second family of its order, and constituting the type of a distinct genus. Such is the sad work which our modern naturalists, sworn enemies of all fictions but their own, have made with the ingenious and brilliant fabrications of antiquity.

However, if time has dissipated the false glory of the salamander, and refuted all its chimerical pretensions, it has on the other hand accumulated very important facts in its history, and advantageously substituted for a futile romance, a series of interesting facts founded in truth.

The salamander was placed by Linnæus among the lizards, to which it has no relation, being destitute of claws, and a heart with two auricles. Among the species of this genus, which remain in the water only during their tadpole state, or when they reproduce, and whose eggs exclude the young in the oviductus, we shall notice the following—

The *Common Salamander*, whose external description is in the text, exhibits very important facts in its anatomy, which cannot be passed over in silence, as they are highly interesting in the study of the philosophy of nature.

The composition of the osseous cranium, resembles that of frogs in the hinder and under part. But it differs from it singularly in other respects. For example: it has no bone forming a cincture in its anterior part; as in the other batracians, there are but two lateral occipital bones, but each of them are intimately united to the part analogous to the *os petrosum*.

The cranium, almost cylindrical, is widened towards the face, which represents a semi-circle, and behind by two branches disposed crosswise, and containing the internal ears.

As in the frogs, the vomeres are two in number. They each give out a slender process, which like them supports the palatine teeth. At the anterior and internal paries of the orbit is a large membranous space.

There are two frontal bones, in the terrestrial Salamander, which are articulated in front with the proper nasal bones, and laterally with the anterior frontals.

The parietals flatted and wider behind, are also two in number. The two occipital condyles are very much separated from each other, and placed on each side of the occipital foramen.

The *os petrosum* and the lateral occipital, are represented by a single bone, to which are attached the pterygoidian, the jugal, and the tympanic bones.

The external apertures of the nostrils are very much separated, which is in consequence of the breadth of the ascending processes of the intermaxillaries. The canal of the nasal fossæ is very short. The dentary portion of the maxillary bones is carried backwards, but does not join the pterygoid or jugal bone. The latter placed transversely on the pterygoid, is united only by a ligament to the posterior point of the maxillary, and presents a facet for the articulation of the jaw. The lachrymal bone is very small, and placed at the external angle of the anterior frontal. The bones of the nose form a vault above the nasal fossæ, which have no lower cornets.

The lower jaw, which is of a parabolical figure, exhibits a true dentary bone, forming the symphysis with its congener, and bearing teeth pretty nearly the same as in the majority of the lizards. The remainder is composed in the adult Salamanders of a single piece, which is double the length of the preceding, at the lower half of the internal face, gives out a coronoïd crest behind, and bears the articular tubercle which is intimately united to it by synostosis.

Both jaws are armed with numerous and small teeth.

There are fourteen vertebræ from the head to the sacrum, and twenty-six to the tail, according to the Baron in the Ossemens Fossiles; about forty-two, according to MM. Carus and Funk. The atlas is articulated to the head by two concave faces, and with the axis by the lower face of its body, concave also. But all the following vertebræ have the posterior face of their body convex.

The articular processes of the dorsal vertebræ are horizontal, and united on each side by a crest, like a rectangular roof with the lateral edges a little re-entering. Their spinous processes are nothing but a simple ridge. The transverse processes have at their summit two tubercles, which carry the vestiges of ribs.

In the attachment of the pelvis to the spine, there are numerous individual differences. Sometimes it is the fifteenth vertebra, and sometimes the sixteenth, which supports this part of the skeleton.

In front of the symphysis of the pubis, is a cartilage formed like a Y.; it is embodied in the muscles, and seems in some measure to represent the marsupial bones of the dildelphis.

Under the body of the caudal vertebræ, from the thirteenth, is observable a small transverse lamina directed obliquely backwards, and pierced with a hole at its base.

There is but a mere vestige of the sternum, and it is more membranous than cartilaginous. The ribs are so short, that they seem to be mere transverse processes of the vertebræ, having but a single point of articulation on which they are but triflingly moveable. These rudimentary bones are twelve in number on each side.

The shoulder is very curious, from the prompt synostosis of its three bones into a single one, which bears the glenoïd fossa on its posterior edge, sends towards the spine a quadrilateral lobe, widened above, which is the omoplate, and furnishes to the breast a rounded disk, composed of the clavicle and coracoïd bone, separated by a suture.

This disk is constantly pierced by a small hole and surrounded with a large cartilaginous plate of a crescent form, which, under the breast, crosses with its congener.

The spinal edge of the omoplate is surmounted by a cartilaginous appendage.

The scapulary head of the humerus is rounded. Below it may be observed a compressed and obtuse tuberosity, and a thick pointed process; the first turned forward, and the second backward. The two bones of the fore-arm are situated one above the other.

The carpus is composed of five bones, and two cartilages, all flat and angular. The metacarpus has four short bones, flat, and contracted in their centre. The toes are only four: the first has but one ossified phalanx, the third has three, and the second and fourth have two each.

The head of the femur is oval: at the internal face of the neck of this bone, is a pointed process, in place of the trochanter. Its tibial extremity is wide and flatted. On the whole, it differs but little from the humerus. The tibia, very thick above, has a slender stem, tolerably long, and descends lower than the fibula, which is equally thick. There are nine bones in the tarsus, and five pieces in the metatarsus.

The muscles of the salamander have been described with peculiar care by Dr. Funk; but, to enter into any detailed myology of this reptile in this place, would be totally beside our purpose.

The brain of the salamander is so small, that it does not equal the spinal marrow in diameter. This last is composed of two nervous cords, enveloped in a tight membrane, and from which the spinal nerves spring out by roots, a little more voluminous. The ganglionic nervous system is not very accurately known.

The globe of the eye is pisciform. It is fixed in the orbit by the optic nerve, and a peculiar muscle. The skin so completely covers it, that the cornea alone is visible. This, according to the expression of Wurffbain, gives to the animal *mirum torvum et obtusum vultum.*

There is neither lachrymal gland, nor *viæ lachrymales.* The sclerotica is irregular, and unequal in its thickness. The cornea is very transparent. The chorioid is black. The ciliary ligament is small and narrow. The iris and pupil exhibit no remarkable peculiarity. The chrystallina is large and compressed, and has a hard and spherical nucleus for its centre, as in fishes.

The organ of touch does not appear to possess any great delicacy. That of smell, on the contrary, is greatly developed. The olfactory nerves are spread in a mucous, grey, vascular membrane, which lines the conical nostrils.

The tongue of the salamander is short and thick. It is

fixed on a particular bone, which does not allow it much power of moving. It is supplied from certain mucous cells, with an abundant viscosity.

The apparatus of hearing seems to differ but little from that of chondropterygian and branchiostegous fishes. Under the skin and muscles is found, in the temporal region, a cartilaginous operculum, not rounded, but rhomboidal, and encased in a separate osseous piece. More deeply is a cavity, invested with a greyish pulp, in which is a kind of lodge for a rudimentary small bone of the tympanum. This is white, soft and chalky, and effervesces with acids. Still more deeply is situated the first cavity of the vestibulum, which communicates with the mastoidian cell, then come the three semicircular canals, and, finally, two labyrinthian cavities, one of which is internal and elliptical.

The heart of this reptile is enclosed in a peculiar pericardium. It is more or less globular, and has but a single auricle, and a single ventricle. Its colour is red. A large vena cava, which receives the blood of the pulmonary veins, opens into the auricle, the parietes of which are much less thick than those of the ventricle, which gives rise to the aorta. The lymphatic system is but little known.

The lungs are convergent in front, and divergent behind. They are composed of small membranaceous cellular sacs, which are often secondarily divided by incomplete partitions, communicating one with another. They receive the air through a short and narrow trachea.

The liver is situated towards the middle of the body, not far from the point of the heart. It almost covers the lungs. It is convex externally, and concave within, and its figure is that of an irregular trapezoid. It is suspended in the thoraco-abdominal cavity, by a double fold of the peritoneum.

The bile is green, acrid, and bitter. The kidneys, situated on each side of the vertebral column, are narrow in front, but more voluminous behind. Their colour is a deep red. There appear to be no ureters.

The skin of the terrestrial salamander is coriaceous, firm, and yet fine and smooth, and covered with a semi-transparent epithelium. Some lenticular cells pour upon its surface a milky bitter, and acrid humour, which would seem to have given rise to the fable of the salamander being able to resist fire. According to Lacepede, its taste is so caustic, that a drop of it coming in contact with the tongue, produces a sensation of burning.

There is no œsophagus, properly speaking, in the salamander. The pharynx degenerates, insensibly, into a fusiform stomach, whose cavity is smooth, and provided with cells for the secretion of mucus.

Certain masses of an adipose, and, as it were, oily tissue, of a yellow colour, have been found in the body of this reptile, and have been supposed to minister to the nutrition of the animal during its winter slumbers. But this is very far from being demonstrated. Reproduction in the salamander seems to resemble that of fishes.

The terrestrial salamander is found in France, in Germany, and even sometimes in very high latitudes. It is also found in the southern parts of Europe.

It takes up its abode in the humid earth, in the tufted woods of high mountains, in ditches and shady places, under stones and the roots of trees, in hedges, by the banks of streams, in subterraneous caverns, and ruined buildings. Though generally feared, it is by no means dangerous. The milky fluid which exudes from its skin, and which it sometimes shoots to the distance of several inches, though nauseous, acrid, and, according to Gesner, even depilatory, is fatal only to very small animals. This humour, however, doubtless was the cause of a general proscription of the salamander. According to Pliny, by infecting with its poison all the vegetables of a vast extent of territory, this reptile could produce death to entire nations!

It is almost unnecessary to repeat now, that there is not the slightest foundation for the story of this animal being able to resist the action of fire.

If the salamander be struck, it raises its tail, and seems affected by catalepsy. It seldom quits the hole where it makes its habitual residence. It passes its life in general under ground. During summer, it dreads the heat of the sun, and seldom ventures forth, except in rainy seasons, or by night. Its walk is slow and heavy. It is stupid, and totally destitute of courage, never braving danger, as has been pretended. It is true, indeed, that it does not seem to perceive the approach of peril, against which it advances blindly, without deviating from its route; but this is mere stupidity, not courage.

It lives on flies, worms, young snails, scarabei, earth-worms, &c. It also eats humus.

Though very tenacious of life, it falls rapidly into convulsions, if it be steeped in vinegar, or sprinkled with salt.

The perceptive powers of this reptile seem to be remarkably dull. It shews no dread of the presence of man, or of animals stronger than itself. Other animals, however, seem to have an instinctive horror of it. Its bite is perfectly harmless, though Matthioli has declared it to be equally mortal with that of the viper—an atrocious absurdity.

The salamander utters no cry. On being thrown into the water, it tries immediately to get out again, and comes every moment to the surface to respire. When on the ground, it frequently rolls itself into a spiral.

It appears, according to the authority of Gesner, that in countries too much elevated in latitude, the salamanders pass the winter in a sort of burrow under ground, where numbers of them are to be found, assembled, and intertwisted together.

The salamander, like the viper, is ovoviparous. The eggs open in the oviducts, and the young come forth fully formed. The latter, whose tail is compressed vertically, are folded in two, to the number of from eight to twenty in each of the five oviducts, where they are nourished by a peculiar fluid, and from which they do not come until they have gone through all their metamorphoses, that is, have lost their gills, and acquired their feet. Then they are deposited near marshes, to the number of forty, and even sometimes fifty at a time. Their colour is an uniform black.

Nothing is more erroneous than the opinion that the terrestrial salamander is destitute of sex, and that each individual is capable of self-reproduction.

There are some varieties of the terrestrial salamander. One remarkable one was described by Thunburg in the Memoirs of the Academy of Stockholm, in 1787, under the name of *Lizard* of Japan. It inhabits the island of Riphon, one of the largest of that empire. Its colour is black, varied with irregular white spots. The natives attribute to it the same medical properties as to the skink, and consider its flesh as a powerful stimulant and an energetic remedy. The shops in the neighbourhood of Jeddo are consequently filled with salamanders of this kind, suspended to the ceilings, and dried.

With respect to the divisions of the Aquatic Salamanders, we have but few observations to make. They are externally distinguished from the land salamanders by having a compressed, not a rounded tail; but in all the main points of anatomical conformation, they agree. They have been rendered celebrated by the experiments of Spallanzani on their astonishing faculty of reproducing parts which have been removed, and those, too, with all their peculiar bones, muscles, vessels, &c. They have been caught in the ice, and remained there a long time without perishing. The eggs are fecundated by the males in the water, and come forth in long chaplets. The young preserve their gills for a longer or or shorter time, and the colours of these animals change according to age, sex, and season.

The *S. Marmorata*, Latreille, attains to the length of eight or nine inches. It is found abundantly in the southern parts of France, the neighbourhood of Montpellier, Fontainbleau, &c. It lives habitually in the water, but sometimes quits it in the evening, or when the weather is hot or stormy. It sheds rather a fœtid odour, and passes the winter in the holes of rotten trees.

There are many more of these salamanders; but it would be wholly unproductive of interest to our readers to take up any more time concerning them.

The genus AMPHIUMA has been re-established by our author. It cannot be passed over in this place without a few observations.

In 1822, Dr. Mitchill sent to the administration of the Museum of Natural History in Paris a very exact description of the animal which forms the type of this genus. In the course of the same year, another description was given of it in a number of "The Medical Recorder," under the name of *chrysodonta larvæformis*. Its external character and conformation, have also been very correctly given by Dr. Harlan in the third volume of the Journal of the Academy of Sciences of Philadelphia, and in the number for June, 1825, of the Annals of the Lyceum of Natural History of New York. The Doctor also gives two excellent figures of this animal.

From all these descriptions, and from the osteological researches of the Baron on the siren and amphiuma, we must be convinced how erroneous is the opinion that these two reptiles are only individuals of different ages, but of the same species.

The *amphiuma means* of Garden, to which our author

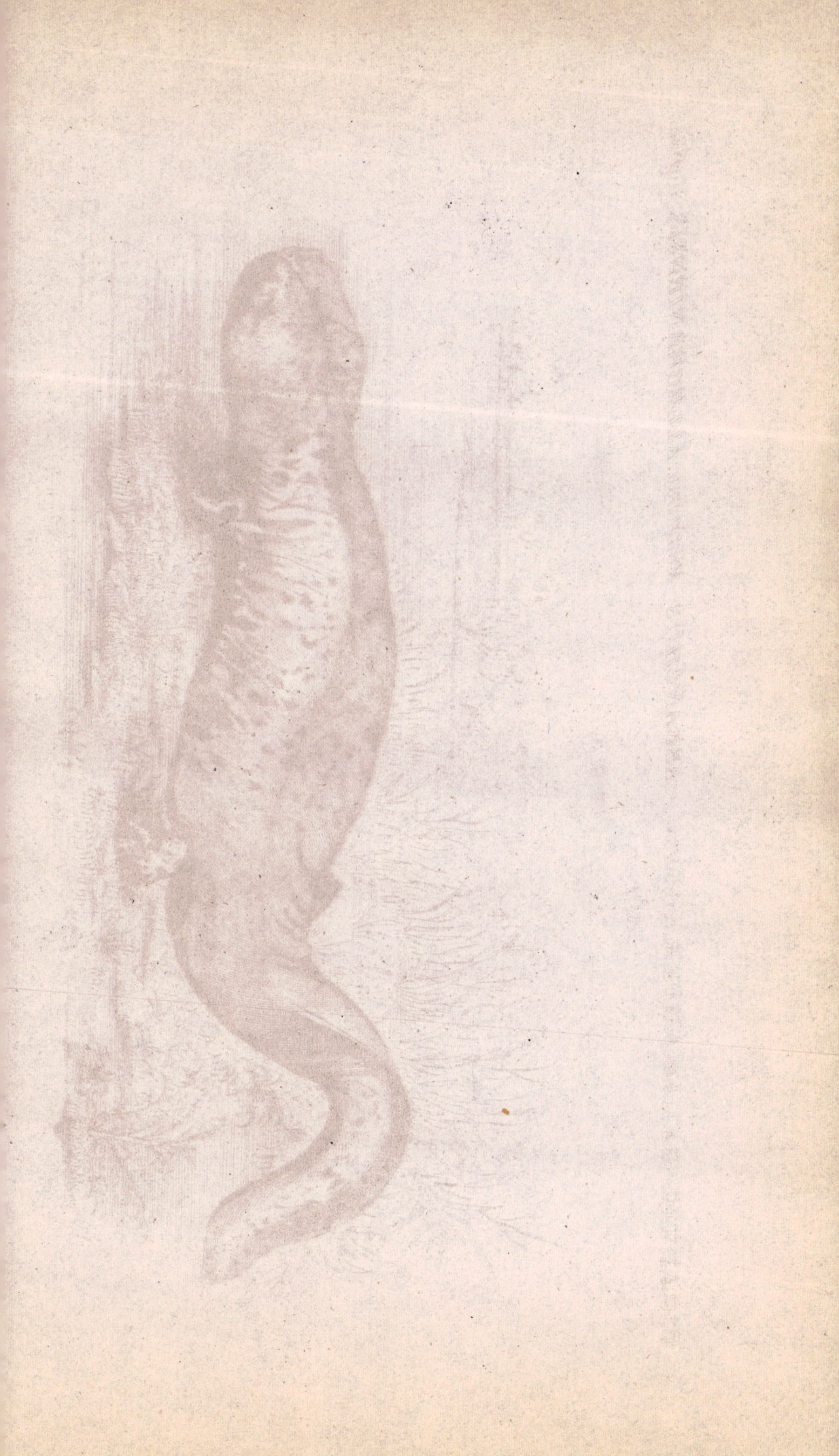

GIGANTIC SALAMANDER, (of Barton.) ABRANCHUS. *Menopoma.* ALLEGHANIENSIS. *Harlan.*

HARLAN'S MENOBRANCHUS. *MENOBRANCHUS LATERALIS. Harlan.*

AXOLOTL OF THE MEXICANS.

SIREN PISCIFORMIS. Shaw.

would apply the name *amphiuma didactylum*, has the body elongated and cylindrical; the head depressed and obtuse; the tail compressed, pointed, trenchant above, and rounded underneath. The nostrils are pierced at the end of the muzzle; the eyes lateral, round, small, and without lids; the lips slender. The teeth are conical, pointed, a little arched, and crowded one against the other; the tongue is not very apparent. The fore feet are formed like tentacula; the toes are only two in number on all the feet.

This whole animal is covered with a smooth skin, presenting no other inequalities than the folds of the sides, and some granulations on the head. It is of a blackish grey above, and pale underneath, without spot or stripe. It varies in length from six inches to two feet.

This reptile inhabits ponds in the neighbourhood of New Orleans, in Florida, Georgia, and South Carolina. It is sometimes found sunk in mud, two or three feet deep, and concealed like an earthworm. A great number of individuals were thus found in hollowing a ditch near Pensacola. It can also exist for some time on land. The negroes of these colonies call it the *Serpent of Congo*, and dread it, but without reason, as being venomous.

The *amphiuma tridactylum* of our author is a new species, differing from the preceding only in the number of its toes.

We insert figures of Dr. Harlan's two genera, Menopoma without external gills, and Menobranchus with them, as also of the Axolotl, for the specific descriptions of which we refer to the text and table.

The animal which has given rise to the establishment of the genus Proteus, is a being of a very extraordinary kind, having much analogy with the larvæ of the salamanders, while they are yet provided with their gills. The knowledge of it is owing to the Baron de Zoïs, a gentleman of Carniola, a country where the proteus is sometimes seen during the exundations of the subterraneous lakes, a phenomenon of a very peculiar character in that region. From the individuals collected by that enlightened cultivator of natural science, Laurenti and Scopoli drew the descriptions which first made this reptile known, without, however, giving entire satisfaction to naturalists by their details.

The *proteus anguinus* of Laurenti, *siren anguina* of Schneider, is about a foot in length, and of the thickness of one's finger. Its tail is compressed vertically. Besides the internal lungs, it has three gills on each side of a coral red colour, formed like feathery tufts, and which it seems to preserve all its life. It has teeth in both jaws. The eye, which is excessively small, is concealed by teguments, like that of the asphalax. Its ear, covered by flesh, has in that respect much analogy with that of the salamanders. Its skin is smooth, whitish, and mucous.

For a long time it was believed that the lakes of the environs of Sittich, in Lower Carniola, constituted the sole habitat of the proteus. But within these some years past, discoveries have been made in the grotto of Adelsberg, on the high road from Trieste to Vienna, which have set naturalists to examine this subject anew with assiduity and success.

Hermann, Schneider, and some other writers have imagined that the proteus was only a reptile in the larva state, but in the entire district which it inhabits no salamander is known from which it could possibly proceed, nor any with the true larva of which we are not perfectly acquainted.

The proteus walks but slowly, but swims extremely well, and utters a feeble sort of sound, like the noise made by the piston of a syringe. It possesses some vestige of a larynx. Between its gills there are holes which penetrate into the back part of the mouth. The liver is divided into three lobes, and proceeds from the thorax to the pelvis. The gall bladder is very ample. In the stomach of one of these animals, which is very thick and coriaceous, was found a small shell-molluscum, which indicates the species of nutriment on which this reptile lived, although it could not be brought to eat any thing in a state of captivity. There succeeds to the stomach a narrow intestine, which makes three folds before it terminates at the rectum. The heart, situated between the fore-feet, has but a single ventricle and auricle, and the lungs similar to those of the salamanders, have the form of simple and slender tubes, each terminated by a vesiculary dilatation. The spleen and pancreas are long and narrow, and the kidneys very long and narrow in front, widen towards the anus, into which they open. M. de Schreibers thinks that he has recognized some traces of ovaria.

The skeleton of the proteus resembles that of the salamanders, except that it has many more vertebræ, and fewer rudiments of ribs. But the osseous head is totally different from theirs in its general conformation, and approximates to that of the siren in a very marked manner. It is provided only with pterygoidean bones, and is without crest, and altogether more depressed than the cranium of the siren. The parietals also advance less to the side of the frontals, which occupy a longer and wider space in proportion. The orbitals and the ossa petrosa are also much less elevated. The nasal bones are mere rudiments, and slide between the intermaxillaries, which have long ascending processes, and the edge of which is armed with a range of eight or ten teeth.

Behind these intermaxillary teeth a parallel range is observed, which may be supposed to belong to the vomeres. These last have each twenty-four small teeth, and are continued back, with an osseous branch also furnished with some teeth. This proceeds to attach itself to the internal edge of the tympanic bone, having a void between it and the basis of the cranium. There are neither maxillary nor palatine bones.

All the under part of the cranium is flat, and formed by a single sphenoïd. There are but two tympanic bones, two petrous portions, and two occipitals, and the fenestra ovalis is entirely pierced in the petrosum.

The nostrils, without any osseous envelope, externally or underneath, penetrate into the mouth under the lower lip.

The under jaw has its dentary edge furnished with teeth. Its coronoïd apophysis strongly marked, gives attachment to a crotaphite muscle, which passes over the pterygoidean bone, and produces the swelled appearance of the head.

Between the head and pelvis are thirty vertebræ, and twenty-five from the pelvis to the end of the tail. All, except the very last, are well ossified, and articulated as in the first by concave faces filled with cartilage.

On each side, from the second vertebra, are seven rudiments of ribs, very small, and the head of which is not divided.

Except the neck of the omoplate, all the rest of the shoulder is cartilaginous. The pericardium is enveloped by a cartilage, which might almost pass for the remains of a sternum.

The pelvis is still less ossified than the shoulder, and the bones of the feet which are very small and slender, have cartilaginous extremities.

We shall conclude this account of the reptiles with a few brief remarks, on the generic peculiaries of the Siren.

This animal, of which the species *lacertina* attains the length of more than three feet, is of the number of those beings, which seem peculiarly formed to set classification at defiance, and which are distinguished in the Animal Kingdom, for the anomalies of their organization. It inhabits the marshes of Carolina, and especially those devoted to the culture of rice, where it lives on earth-worms, insects, young mollusca, &c., at least according to the report of professor Barton, who denies it the faculty of devouring serpents. It was in 1765-6, that Dr. Garden first introduced the Siren to the notice of the scientific world, having sent both description and individuals to Linnæus, and our countryman Ellis.

The learned Swede, believing with Garden, that this animal never changed its form, created for it a peculiar order of amphibia, which he termed *meantes.* But many other naturalists of note even to a recent period, maintained that the *lacertina* of Linnæus, was not a perfect animal, but merely the larva of some batracian reptile, more or less resembling an unknown salamander, which, in the course of age, would necessarily lose the external gills by which it was characterized. Such was the opinion of Pallas, of Hermann, of Schneider, and of Lacépède. Camper, in which he was followed by Gmelin, went even so far as to make a fish of it, of the eel genus.

Our author has established on anatomical grounds, that the Siren is the type of a separate genus, whose osseous frame differs totally from that of the salamanders; that this reptile could never have hind feet or lose its gills; that it was consequently a true *amphibium*, respiring as it pleased, for its whole life, either in the water, with gills, or in the air with lungs. These deductions, first published in 1807, have been fully confirmed by time.

From the correspondence of Garden, with Linnæus, and with Ellis, it appeared that the American physician had seen sirens, whose size varied from four inches to three feet and a half, all equally provided with gills, and propagating, without losing them.

All travellers, and all naturalists of the New World, particularly Barton, have confirmed the facts announced by Garden; Messrs. Say, Harlan, Mitchill, and Green, have published interesting notes on the siren, or on the singular reptiles which approximate to it. Many sirens of all sizes have been sent into Europe, always with gills and without any appearance of hinder feet; yet M. Rusconi, a learned physician of Milan, has raised doubts concerning all these testimonies, and thinks that the siren undergoes metamorphosis, because he was informed by a German traveller, that he saw at the Museum of the College of Surgeons in this metropolis, a siren with its four feet, and without any gills.

This assertion certainly deserves the reproach of levity; this pretended adult siren was known a long time ago, and had not escaped the investigating eye of Garden, and was in fact, the *amphiuma means*, which we have already described.

The simultaneous existence of a larynx, and trachea, with a branchial apparatus not only permanent, but perfectly ossified in most of its parts, is a peculiarity of high importance in comparative anatomy. This is exhibited in the siren. It also contributes to prove what has been advanced by M. Cuvier with respect to frogs and salamanders, namely, that the branchial apparatus is nothing more than a mere complicated hyoïd bone, and not a combination of pieces from the sternum and larynx.

Though it greatly approaches the salamanders and the proteus, the siren nevertheless differs from them in a great number of internal characters. Its head has neither the same general conformation, nor the same proportions between its parts; the muzzle is very narrow in front, in consequence of the extreme reduction of the maxillaries, which consist only in a very small osseous nucleus; behind is observable an occipital crest which predominates over the petrous portions and the parietals.

The pieces which compose the lower jaw, are not transverse, but are directed obliquely forwards. The intermaxillary bones have no teeth, but their trenchant edge, as well as that of the lower jaw, is furnished with an almost corneous sheath, which is easily detached from the gum.

The cavity of the nasal fossa is covered underneath, with

a simple fibrous membrane; their internal orifice is on each side, near the commissure of the lips between the lip and the palatine teeth; there is neither mastoidean, pterygoidean, jugal, upper occipital, nor basilary bone.

In the palate under the anterior and lateral portions of the sphenoïd and the orbital bones, are seen two thin plates all bristling with crooked teeth, and which might be taken for vestiges of vomeres or palatines, or of palatines and pterygoideans. The first of these plates has from six to seven ranges of these teeth, while the smallest has but four.

The hyoïd bone of the siren is the same as that of the larva of the salamander, but very much ossified in many of its parts.

The bones of the carpus are cartilaginous.

The vertebræ are more numerous than those of the salamander, and differently figured; their bodies, which correspond by concave faces, are united by cartilages formed like a double cone, as in fish. There are but eight pairs of ribs. The eye is very small, and the ear concealed.

END OF REPTILES.

A

SYNOPSIS OF THE SPECIES

OF THE

CLASS REPTILIA.

BY J. E. GRAY, ESQ. F.G.S., &c. &c.

N.B.—In the following List those species alone are included, which have either been seen by the author, or which have been well figured or described.*

REPTILES.

Vertebrated animals, respiring by lungs, having warm red blood, heart with one ventricle and two auricles and the skin covered with scales.

* In forming this list, the collection of reptiles of the British Museum, the College of Surgeons, and of Mr. Bell, in London, of the Gardens of Plants and Ecole de Medicine at Paris, of the Royal Museum of Leyden and Berlin, and the Free Town of Francfort, have been studied with attention through the kindness of their several keepers.—J. E. Gray. Oct. 1830.

Section I. CATAPHRACTA.

Body covered with two shields. The vent longitudinal or circular: the ossa quadrata, and the pterygoid processes forming part of the skull; the organs of generation simple and single in both sexes.

Order I. TESTUDINATA, OR TORTOISES.

Ribs, vertebræ and sternum united together in a bony case, enclosing the body, and protecting the head and limbs; the jaws toothless.

The case is usually covered with horny shields, which are granular in the newly-hatched animal, and which increase in size by the addition of layers on the under side of the edge, and having the first formed granular part marked on their surface; this part is called the areola. The shields of the upper shell are called, from their position, discal, consisting of the vertebræ or central series, and costal, composed of the two lateral series and marginal plates, which have often an additional small narrow plate placed on the back of the neck, called the nuchal plate.

The under shell is usually covered with six pair of shields, one called the gular (which have sometimes an additional intergular plate between them), the second the humeral, the third the pectoral, the fourth the abdominal, the fifth the femoral, and the sixth the anal plates; the suture between the upper and lower shell is generally covered by the end of the pectoral and abdominal plates, and there is a small plate placed at each end of it, the front called the axillary, and the hinder the inguinal plate.

Genus I. TESTUDO.

Feet club-shaped, claws five before, four behind blunt; jaws horny. Shells hemispherical, solid, covered with horny plates; the hinder marginal plate broad, incurved.

I. *Shell and sternum immoveable, sternal plates in the gular pair being separate.* Testudo.

Indian Tortoise. Test. Indica, Lin. *Test. Retusa,* Merrem. *Test. Elephantopus,* Harlan. *Test. Nigra,* and *Test. Californica,* Gaimard. *Test. Gigantea,* Schw. and *Test. Dussumieri,* Schegel. Shaw, t. 3. Frey. Voy. t.

Shell black, nuchal plate, often deficient; pectoral plates short. India, naturalized in California, Isle of France and Galapago Isles. Length three feet.

Hercules Tortoise. Testudo Hercules, Spix. Braz. t. 14. *Test. Carbonaria,* Spix. t. 16.

Shell subquadrate, depressed, contracted in the sides, black, areola yellow, rarely exceeds three quarters of an inch in width. Nuchal plate none. Sternum behind roundly lobed, leg red spotted, length twenty-four inches. South America, Var. *Truncata.* Shield and bones beneath elevated, conical, truncated.

Tabular Tortoise, Shaw. *Test. Tabulata,* Walb. Schoepf. t. 12. f. 2. t. 13. *T. Sculpta,* Spix. t. 15. *T. Cagado,* Spix. t. 17. Young, *T. denticulata,* Lin.

Shell oblong, depressed, deeply grooved; black brown areola large; the sternum behind rather acutely lobed. Brazils.

Schweiger Tortoise. Test. Schweigeri, n.

Shell oblong, depressed, pale brown, obscurely brown dotted; beneath brown rayed; areola small central; sternum before acutely cut. Hab. —— ?

Gopher Tortoise. Test. Polyphemus, Bartram. *T. depressa,* Lesueur, Mss. Guerin. Icon. t. 1. f. 1. *T. Carolina,* Leconte Mus. Paris.

Shell oblong, depressed pale brown grooved; nuchal plate

broad, square; front of sternum projecting; tail obsolete. North America.

Bell-shaped Tortoise. Test. Marginata. Schoepf. t. 11, 12, f. 1. *T. Graji.* Herm. *T. Gracæ*, Lacep. *T. Campanulata*, Walb.

Shell oblong, ventricose; hinder margin flattened, expanded toothed; shields grooved; blackish yellow varied; areola yellow, contracted; nuchal plate, very slender, long. Egypt?

Radiated Tortoise, Shaw. *Test. Radiata*, Shaw. Zool. t. 2. *J. Coui*, Daud. *T. Calcarata*, Merrem.

Shell hemispherical; black, yellow rayed, areola sunk; nuchal plate none. Madagascar.

Sulcated Tortoise. Test. Sulcata, Shaw, Miller, Cym. Phys. t. *Test. Calcarata*, Merr. *Test. radiata Var.* Gray, Syn. Rept.

Shell sub-globular, yellow; shield rather convex, deeply grooved; areola small, superior; nuchal plate none. Senegal. Dongola. Dr. Ruppell.

Leopard Tortoise. Test. Pardalis, Bell. Zool. Jour. t. 15. *T. armata*, Boie Mss. *Test. Bipunctata*, Cuv. R.A.

Shell hemispherical; yellowish, black spotted; nuchal plate none. Cape of Good Hope.

Starred Tortoise. Test. Stellata, Test. Actinodes, Bell. Z. Jour. t. 23. *Test. Elegans*, Shaw, t. 6. f. 1. and Schoepf. t. 26. *T. Stellata*, Schw.

Shell globular; shields convex, grooved; black, yellow-rayed; areola yellow, large; nuchal plate none. India. Ceylon.

Geometrical Tortoise, Shaw. *Test. Geometrica*, Lin. Shaw. t. 1. f. 1. Jun. *T. Luteola*, Daud.

Shell globular oblong, black; shields yellow-rayed; nuchal plate long; β nuchal plate short. *Test. Tentoria*, Bell. Z. Jour. t. 26. Cape of Good Hope, Madagascar.

Greek Tortoise. Test. Græca, Lin. *T. Carolina*, Herm. *T. Geometrica*, Brunn. *Test. Hermanni*, Schn. Edw. t. 204. Shaw. t. 1.

Shell oblong, globose; black, yellow-rayed; nuchal plate, slender; caudal plate incurved. Var. β hinder margin expanded, Schoepf. t. 9. Inhab. South of Europe, Var. no nuchal plate. *Test. Zolhafa*, Forsk.

Areolated Tortoise, Test. Areolata, Thumb. Schoepf. t. 23.

Shell oblong, rather depressed; shields yellowish, deeply grooved; areola brown; sutures deep; nuchal plate, narrow. Var. β greenish; hinder lobe of sternum mobile. Cape of Good Hope, variable in number of dorsal and marginal plates.

Marked Tortoise. Test. Signata, Walb. Schoeff. t. 28. *T. Cafra?* and *T. Juvencella?* Daud.

Shell rather depressed; dorsal shields flattish; yellowish, black-rayed; areola blackish; nuchal shields narrow. Africa?

II. *Dorsal shell and sternum both solid; sternal plates eleven, the gular pair produced united into one.* Chersina. *Gray.*

Angular Tortoise, T. (*Chersina*) *Angulata*, Dumeril. *Test. Bellii*, Gray Spec. Zool. t. 3, f. 1. *Test. Pusilla*, Lin. *T. miniata*, Lacep.

Shell oblong, ventricose; black, shields grooved; areola yellow; nuchal plate very narrow. Cape of Good Hope.

III. *Hinder part of dorsal shell mobile, united to the anterior by a carious ligamentous suture; sternum solid, shields twelve.* Kinyxis. *Bell.*

Home's Tortoise. Test. (*Kinyxis*) *Homeana.* Bell. Lin. Trans. xiv. t. f. 2.

Shell brown; hinder edge reflexed; upper edge of fifth vertebral plates prominent; nuchal shield long, narrow; front of sternum expanded. Demerara. Guadaloupe.

Bells Tortoise. Test. (*Kinyxis*) *Belliana*, Gray.

Shell brown, edge hardly expanded; centre of hinder vertebral plates convex; nuchal shield long, narrow; front of sternum, narrow. Inhab. ?—Mus. Gray.

Worn Tortoise. Test. (*Kinyxis*) *Erosa.*

Shell brown; hinder edge reflexed, denticulated; centre of fifth vertebral plates prominent; nuchal shield none. Young, back rounded. *K. Castanea.* Bell. l. c. t. f. 1. *Test. Denticulata*, Shaw. Zool. t. 13. *Test. Erosa*, Schw. Inhab.

IV. *Dorsal shell solid; front lobe of sternum mobile.* Pyxis. *Bell.*

Radiated Box Tortoise. Test. (*Pyxis*) *Aranoides*, Bell. Lin. Trans. xv. t. 16.

Shell hemispherical, varied black and yellow; length 6, breadth 4 inches. Inhab.

Genus II. Emys.

Feet palmated; claws five—four. sharp; nostrils pervious; jaws horny; shell solid, covered with horny plates, marginal plates twenty-three or twenty-five, hinder pair free; sternal shields eleven or twelve; neck retractile; pelvis only attached to the vertebra.

I. *Shell hemispherical; sternum rounded before, divided by a transverse suture; both lobes mobile, and united to the back shell by a cartilaginous suture; sternal plates twelve.* Cistuda, Gray.

American Box Terrapin, E. (*Cistuda*) *Carolinæ*, *Test. Carolinæ*, Lin. *Test. Brevicaudata*, Lacep. *Test. Clausa*, and *T. Irregulata*, Daud. *Terrapene Carolinæ*, *Ter. Nebulosa*, and *Ter. Guttata*, Bell. Edw. t. 205.

Shell hemispherical, brown yellow varied, convex, lowly keeled; marginal plates twenty-six or twenty-seven. North America; Var. *β. fusca*, brown, not spotted.

Eastern Box Terrapin. E. (*Cistuda*) *Amboinensis*, Illust. Ind. Zool, t. f. 1. *Emys Couro*, Schw. *Terrapene Bicolor*, Bell. Zool. Jour. ij. t. 14. *Test. Amboinensis*, Daud.

Shell hemispherical, slightly tri-keeled, blackish, margin expanded, broad; nuchal plate linear; sternum yellow and black varied; head with two yellow lines on each side. Java and Penang, Var. *Leveriana*. Shell oblong, with back flattened, not keeled.

Three-banded Box Terrapin. E. (*Cistuda*) *Trifasciata*, Gray. *Sternothærus Trifasciatus*, Bell. Zool. Jour. iij. t. 13.

Shell oval, keeled, dull yellow, varied with black and reddish, with three dorsal streaks; sternum behind nicked; head yellow, with two brown bands on each side. Inhab.—

European Box Terrapin. E. (*Cistuda*) *Europea*. *Testudo Europea*, Schn. *Test. Flava*, Lacep. *T. Punctata*, Gotw. *T. Orbicularis*, Linn.? *T. Lutaria*, Merrem. *T. Meleagris*, Shaw, Zool. t. 5. Jun. *T. Pulchella*, Schoeff. t. 26.

Shell depressed, oval, yellow spotted; sternum behind nicked

7

before rounded; body and head black and yellow spotted; tail long. South of Europe. These are the only Water Tortoises known which have concave sternums.

II. *Sternum united to the back shell by a horny suture; solid, before and behind lobed; sternal plates twelve.* Emys.

A. Margin of shell acutely toothed; sides of sternum rounded. Found in the Old World.

Spinous Terrapin. Emys Spinosa, Gray, Illust. Ind. Zool. t.

Shell suborbicular, pale brown; edge sharply toothed; areola with a central spine. Penang.

Dhor Terrapin. Emys Dhor, Gray, Illust. Ind. Zool. t.

Shell suborbicular, hinder edged, acutely toothed; areola punctate; brown, black spotted. India.

Spengler Terrapin. Emys Spengleri. Test. Spengleri, Walb. t. 3. *Test. Serrata*, Shaw, t. 9. *T. Tricarinata*, Bory, Atlas. t.

Shell oblong, pale brown; tri-keeled; hinder margin deeply toothed. Africa.

Thick-necked Terrapin. Emys Crassicollis, Bell, Gray, Illust. Ind. Zool. t.

Shell ovate, black, tri-keeled; hinder edge toothed. Sumatra.

Thuryi Terrapin. Emys Thuryi, Gray, Illust. Ind. Zool. t.

Shell oblong, black, slightly tri-keeled; hinder edge slightly toothed; head blackish, eyebrow and chin yellow lined. Bengal.

Hamilton's Terrapin. *Emys Hamiltonii,* Gray, Illust. Ind. Zool. t.

Shell oblong, black, yellow-rayed; hinder edge toothed; head, and body, and limbs yellow spotted. India.

Eyebearing Terrapin. *Emys Occilifera,* Kuhl. Beytr.

Shell hemispherical; behind toothed; above with blackish eyed rings, beneath yellow; a doubtful species.

B. Shell margin entire; sternum with the sides keeled. Living in the Old World.

Tented Terrapin. *Emys Tecta,* Bell, Gray, Illust. Ind. Zool. t.

Shell ovate, oblong, solid; olive, beneath varied brown and red; animal red and yellow lined.

Lined Terrapin. *Emys Lineata,* Gray, Illust. Ind. Zool. t.

Shell oblong, olive; head blueish ash; chin and cheeks yellow varied; nape and eyebrow scarlet lined. India.

Batagur Terrapin. *Emys Batagur,* Gray, Illust. Ind. Zool. Shell suborbicular, depressed, pale olive; margin dilated; body ash; chin and lips yellowish. Var. *β Test. Baska.* Shell oblong, ovate.

Common Terrapin, or Mud Tortoise. *Emys vulgaris.* *Test. Lutaria,* Lin.?

Shell ovate, olive, black dotted, clouded and ringed with blackish and orange, head and limbs orange lined. Adult smooth, young rugose. South of Europe. Spain.

Caspian Terrapin. *Emys Caspia,* Ruppell. *Test. Caspia,* Gmel. Trav. t. bad.

Shell oblong, slightly contracted on the sides, olive yellow netted; beneath black, irregular yellow spotted, head olive brown, with narrow sub-concentric yellow lines. Caspian Sea. Dr. Ruppell.

C. Shell margin sub-entire; sternum sides rounded. Living in the New World.

Rough Terrapin. Emys Scabra, Lacep. t. 8. f. 2. *Emys Dorsata*, Schoepft.

Shell ovate oblong, acutely keeled; head with a fork band over each eye. South America.

Dotted Terrapin. Emys Punctularia, Schweiger, (not figured.)

Shell oblong, convex, brown; shield blackish edged; head with red lines and spots. South America.

Marbled Terrapin. Emys Marmorea, Spix, t. 10.

Shell oval, greenish, varied with black and yellow; shields yellow edged; head with yellow lines. Brazils.

Muhlenberg's Terrapin. Emys Muhlenbergii, Schw. *Emys Bipunctata*, Say. Schoeff, t. 31.

Shell oval, low, side contracted; shields grooved, varied with yellow; head blackish, with two yellow spots. North America.

Spotted Terrapin. Emys Guttata, Schw. *Test. Punctata*. Bosc. Schoeff, t. 5.

Shell low, oval, smooth, black, yellow dotted; head yellow spotted. North America.

Painted Terrapin. Emys Picta, Schw. Schoeff, t. 4.

Shell oblong, flattish, smooth, olive; shields pale edged; head and neck yellow lined. Young. *Cinereous Tortoise*, Brown, Illust. 48. North America.

Specious Terrapin. Emys Speciosa, Gray, Syn. Rept.

Shell oval, flattish, keeled; shields grooved, black, yellow spotted; beneath yellow with black areola. North America?

Concentric Terrapin. Emys Concentrica, Schw. Schoeff. t. 15. *Test. Palustris*, Gmel. *T. Centrata*, Bosc.

Shell oval, side edges recurved; shield, brown ringed; head and limbs ash, black speckled. North America. *Var. polita* shell, black polished.

Reticulated Terrapin. Emys Reticulata, Gray. *Test. Reticulata*, Latr. Daud. t. 22, f. 3.

Shell oblong, olive, pale netted; beneath yellow, with four round black spots on side of the margin, and one at each end of the sternocostal suture. North America.

Banded Terrapin. Emys Vittata, Gray, Syn. Rept.

Shell oblong, olive; behind doubly toothed, varied with unequal yellow lines; beneath yellowish, with squarish eyed spots on the sutures of the marginal plates, and roundish ones on the end of the symphysis; head and feet yellow lined. North America?

Furrowed Terrapin. Emys Decussata, Gray.

Shell oblong, pale brown, shields rugulose; beneath yellowish, with subocellated spots, on the sutures of the marginal plates; animal greenish. North America.

Lettered Terrapin. Emys Scripta, Gray. *Test. Serrata*, Daud. t. 21. f. 1. Schoeff. t. 3. f. 5. Jun.

Shell oblong, longitudinally rugose, brown, sides yellow banded; beneath yellow, with a series of roundish spots on the middle of each marginal plate, or two on the gular plates, and four on each sternocostal symphysis. North America.

Serrated Terrapin. Emys Serrata, Gray.

Shell oblong, longitudinally rugulose, irregularly cross-banded; beneath yellowish, with a series of subocellated spots on the sutures of the marginal plates, and dark-edged pale lines on the symphysis. North America.

11

Ornamented Terrapin. *Emys Ornata*, Gray.

Shell oblong, longitudinally rugose olive; vertebral shields, irregularly ringed; beneath, pale yellow, with an irregular black band down the centre, and on each side, and squarish-eyed spots on the suture of the marginal shields. South America. Young, beautifully orange and green-ringed.

Rugose Terrapin. *Emys Rugosa.* *Test. Rugosa*, Shaw, Zool. t.

Shell oblong, rugulose; black, with yellow lines and dots; beneath yellow with black dots and lines; Var. β shell livid, black dotted. North America.

Lake Erie Terrapin. *Emys Lesueurii*, Gray, Syn. Rept.

Shell oval, smooth, tuberculately keeled; behind, toothed; olive, with anastamosing black-edged pale lines; beneath, yellow, shields, black edged, margin, black ringed; head, yellow lined. β *Emys Geographica*, Lesueur, Jour. Acad. N. Philad. t. North America.

Bell's Terrapin. *Emys Bellii*, Gray, Syn. Rept.

Shell oblong; dorsal line, depressed; olive, netted with green and black dots; beneath blackish, yellow dotted, surrounded by an irregular yellow margin.

White Spotted Terrapin. *Emys Kinosternoides*, Gray, Syn. Rept.

Shell oblong, pale brown, with black-edged white keel, and side-spots, and with a dentated white edge; beneath yellow spotted; sternum rounded at both ends. (—— ?)

Ring-bearing Terrapin. *Emys Annulifera*, Gray, Syn. Rept.

Shell oblong, depressed; pale brown, yellow lined with

yellow ring placed across the sutures of the vertebral plates; beneath pale brown, varied with darker brown. (—— ?)

III. *Sternum united to the back shell by a horny suture; before rounded, behind rounded, or slightly nicked, divided by one or two cross sutures; sternal plates, eleven gular pair united.* Kinosternon. *Spix.*

Three-keeled Kincsternon. E. (*Kinosternon*) *Scorpoides. Test. Scorpoides,* Lacep. *Test. Tricarinata,* Daud. *Kinosternon Shavianum,* Bell. *Kin. Longicaudatum,* Spix, t. 12? Shaw Zool. t. 15. Jun. *Test. Retzii,* Daud. Shaw, t. 11.

Shell oblong, olive polished, rather compressed three-keeled, keels continued, dorsal shields long, hexangular, sub-imbricate; sternum as broad as the opening of the shell. South America.

Three Ridged Kinosternon. E. (*Kinosternon*) *Triporcata,* Gray. *Terrapena Triporcata,* Wiegmann, Mus. Berl.

Shell, oblong, olive, with three very high compressed ridges; sternum cross shaped, very narrow; the largest species of the genus.

Pennsylvanian Kinosternon. E. (*Kinosternon*) *Pennsylvanica. Test. Pennsylvanica,* Bosc. *Test. Subrufa,* Lacep. Edw. t. 287. Var. *Test. Glutinata,* Daud. *Test. Boscii,* Merrem.

Shell oval, brown; dorsal shields, flattened; sternum, rather narrower than the opening of the back shell; behind nicked. North America.

Musky Kinosternon, *E.* (*Kinosternon*) *Odoratum. Test. Odorata,* Bosc. *Cistuda Odorata,* Say. *Sternotheros Odoratus,* Bell.

Shell ovate, keeled; sides declivate; sternum very narrow;

before acute, behind acutely nicked; head brown, yellow lined on each side. North America.

IV. *Sternum very narrow, cross-shaped, rounded at both ends, united to the back shell by very long bony processes. Sternal plates five pair, and a pair over the symphysis.* Chelydra Schw.

Serrated Alligator Tortoise. Emys (Chelydra) Serpentina. Test. Serrata, Penn. *Chelydra Lacertina,* Schw. Young. *Test. Serpentina.* Lin. Shaw, t. 29.

Shell oblong; vertebral shields flat; upper edge of costal plates, prominent; behind bluntly six toothed. Young sharply three-keeled and toothed. North America.

III.—Chelys, Gray.

Feet palmated; claws five—four, sharp; nostrils sub-tubular; shell solid, covered with horny plates; hinder pair of marginal plates separate; sternal plates thirteen, with an intergular one; neck contractile on the side; pelvis attached to the sternum and vertebræ.

I. *Lips horny; sternum front, and sometimes hinder lobes mobile; intergular plate marginal.* Sternotherus, Bell (part.) Gray.*

Chestnut Chelys. Ch. (Sternotherus) Castaneus, Gray. *Emys Castaneus,* Schw. *Testudo Subniger,* β Daud. *Sternotherus Castaneus,* Bell. Zool. Jour. t. 14.

Shell oval, sharply keeled, convex, chestnut; shields radiately rugose; areola black. Var. β hinder lobe, mobile.

* I have used as the name of the sub-genera those which have been given by authors to what they considered genera in this family; but I have generally restricted their signification, so as to render them more definite.

Test. Subniger, Lacep. 2. O. t. 7. f. r. *Test. Nigricans*, Merrem.

II. *Jaws horny; sternum solid; intergular plates six sided between the gular and pectoral plates.* Chelodina.

Long-necked Chelys. Ch. (*Chelodina*) *Longicollis*, Gray. *Testudo Longicollis*, Shaw. New Holland. t. 7.

Shell oblong flat; sternum yellow, shields brown edged.

III. *Jaws horny; sternum solid, intergular plate small, marginal; nostrils short, tubular.* Hydraspis. *Bell.*

Species of the Old World or its Islands.

Cape Chelys. Ch. (*Hydraspis*) *Subrufa*. n. *Test. Subrufa*, Lacep. Q. O. t. 12. *Test. Indica*, Daud.—Jun. *Test. Galeata*, Schoeff. t. 3. f. 1. *Test. Scabra*, Retz. *T. Senegalensis*, Daud.; and *Emys Olivacea*, Schw.

Shell oblong, low, pale brown, smooth; nuchal plate, none; vertebral plate depressed. Cape of Good Hope. Senegal.

Adamson's Chelys. Ch. (*Hydraspis*) *Adamsonii*, *Emys Adamsonii.* Schw.

Shell low, oval, very broad behind, yellow, black spotted; nuchal plate none; vertebral plates keeled, first urn-shaped, hinder triangular. Africa. Nigritia.

Macquary's Chelys. Ch. (*Hydraspis*) *Macquarii. Emys Macquarii.* Cuv. R. A.

Shell oval, depressed, contracted in front, behind subdentate, dark olive; shields rugulose; dorsal line sunk; nuchal plate narrow. New Holland.

15

Species of America.

Flat Headed Chelys. Ch. (Hydraspis) Planiceps. Test. Planiceps, or *T. Platycephala.* Schn. Berb. Naturf. t. 16. *T. Martinella,* Cuv. R. A. *Emys Discolor,* Thumb. *E. Caniculata?* Spix, t. 8. *E. Caruncnlata,* Cuv. R. A. *Emys Geoffroyana,* Schw.

Shell oblong, flattened, black brown, centre concave; nuchal plate narrow; sternum yellow edged. *E. Aspera,* Cuv. R.A. is a variety with larger warts in the neck.

Depressed Chelys. Ch. (Hydraspis) Depressa, Gray. *Test. Depressus.* Pr. Max. t. not Spix.

Shell elliptical, pale brown, black rayed; nuchal plates narrow; head and neck black spotted. Brazils.

Radiated Chelys. Ch. (Hydraspis) Radiolata, Gray. *Emys. Radiolata,* Mikan. Pr. Max. *T. Depressa,* Spix, t. 3. f. 12.

Shell oval, slightly narrowed before, yellowish, black rayed; nuchal plate distinct; neck beneath yellow, eyelids streaked. Brazils.

Red Footed Chelys. Ch. (Hydraspis) Rufipes, Gray, Spix, t. 6, Jun. *Emys Stenops,* Spix, t. 9, f. 3, 4.

Shell elliptical, keeled in front, brown, beneath yellowish; nuchal shield linear; head and neck thick, olive brown, beneath yellowish. South America.

Green Chelys. Ch. (Hydraspis) Viridis, Gray. *Emys Viridis,* Spix, t. 2, f. 4.; t. 3, f. 1.

Shell elliptical, olive green, brown dotted; nuchal shield linear; intergular shield large, cordate. Brazils.

16

Expanded Chelys. Ch. (*Hydrapsis*) *Expansa*, Gray. *Emys Expansa*, Schw. *E. Amazonica*, Spix, t. 1. *E. Tracaxa.* Spix, t. 5, f. 1. 2.

Shell ovate, depressed, brown, black spotted; margin very much expanded; nuchal shield none; nose longitudinally grooved, Var. β *Emys Erythrocephala*, Spix, t. 7, is perhaps the same. It is from a stuffed specimen. Brazils.

Dumeril's Chelys. Ch (*Hydraspis*) *Dumerilliana*, Gray. *Emys Icterocephala*, Spix, t. 4.

Shell ovate, black, discal shields flat, hinder margin expanded; head globose; nose convex, smooth. Brazils.

Cayenne Chelys. Ch. (*Hydraspis*) *Cayennensis*, Gray, *Emys Cayennensis*, Schw.

Shell ovate, convex, tuberculately keeled; smooth, yellow green, black varied; nuchal shield none; head brown, with two yellow spots. β *Emys Gibba*, Schw. Shell black. Guiana.

Demerara Chelys. Ch. (*Hydraspis*) *Lata*, Bell MSS.

Shell suborbicular, depressed, black; head and neck black orange spotted. Demerara. *Bell.*

Consult also, 1, *Emys Bitentaculata*, Cuv. MSS.—2, *Emys Barbatula*, Gravenhorst, Delic. t. 5. f. 3. 4. - 3. *Emys Constricta*, Cuv. MSS.—4. *E. Maximiliana*, Michan, and—5, *Hydraspis Pachyura*, Boie MSS.

IV. *The beak very broad, depressed, covered with soft lips; the nostrils long, tubular; intergular plate marginal;* Chelys, *Dumeril.*

Brazilian Matamata. Chelys Matamata, Test. Matamata, Brong. Jour. H. N. 7, t. 13. *Ch. Fimbriata*, Schn.

Shell oblong; shields elevated acute, forming three keels. South America.

IV.—Trionyx. Geoff.

Feet palmated, claws three, sharp; the shell covered with a soft skin expanded on the edge into a flexible margin. *Living in fresh water in tropical climates.*

I. *The margin of the shields cartilaginous and the sternum narrow.* Trionyx, Gray.

Fierce Trionyx. Trionyx Ferox, Merrem. Pen. Phil. Trans. lxi. t. 10. *Trionyx Georgicus*, Geoff.

Shell rather convex, obscurely keeled: front and hinder margin wanting; sternum callosities four, two hinder, large, united into one. *Trionyx Spiniferus*, Lesueur, Mem. Mus. xi. appears to be the young, as may also be *Test. Brongniartii*, Sch. and *T. Carinatus*, Geoffroy.

Armless Trionyx. Trionyx Muticus, Lesueur, Mem. Mus. xv. t.

Shell elliptical, confounded with the neck, and armless in the front; sternal callosities four, two hinder large, united together. North America.

Egyptian Trionyx. Trionyx Ægyptiacus, Geoff. Rept. Egypt. t. 1. *Test. Triunguis*, Forsk.

Green, white spotted; shell convex, slightly keeled; sternum callosities four, the hinder ones triangular, separate. Egypt.

Indian Trionyx. Trionyx Indicus, Gray, Illust. Ind. Zool. t.

Olive green, with black-edged irregular pale tortuous and forked streaks; sternal callosities four, the hinder ones rounded triangular. India.

Hurum Trionyx. Trionyx Hurum, Gray, Illust. Ind. Zool. *Trionyx Gangeticus*, Cuv. Oss. Fos.

Dull brown; head green, with two yellow spots on each side

over the eyes; sternum with four callosities, the lateral truncate on the inner hinder angles, and the hinder ovate-triangular. India.

Javan Trionyx. *Trionyx Javanicus*, Geoff. Ann. Mus. xiv. t. Illust. Ind. Zool. t.

Dull green, with numerous narrow lines of minute white granular specks. Head green, with five or six diverging black lines, and a streak between the eyes. Sternal callosities two, linear, lateral. When young, back black ocellated. Java and India.

Flat Trionyx. *Trionyx Subplanus*, Geoff. Ann. Mus. xiv. Illust. Ind. Zool. t.

Shield nearly flat; front edge smooth; sternum without any callosities. India.

Euphratic Trionyx. *Trionyx Euphraticus*, Geoff. *Test. Rascht*, Oliv. Voy. t. 41. This needs a better description. Euphrates.

Margin of the shield with a series af small bones in front and behind, limbs covered, when withdrawn, by the valves on the side of the sternum. Emyda. *Gray.*

Punctuated Trionyx. *Tri. (Emyda) Punctatus*, Lacep. Q. O. t. 7. f. 1. Schoepf. t. 30. A. B. *Trionyx Coromandelicus*, Geoff. Gray, Illust. Ind. Zool. t.

Deep green, occiput white spotted; sternum with seven large callosities, the hinder of them united. India.

N. B.—M. Cuvier informed me on his late visit to this country that there has been lately received at the Paris Museum a species of this genus, with 4 *claws* on *each foot*; it will form a section (*Tetraonyx*), and it may be indicated by the name of *Trionyx* (*Tetraonyx*) *Cuvieri.*

5. Chelonia, Brongn.

Feet fin-shaped, compressed; shell with a bony margin all round, and a ring-like sternum.

Living in the sea; eating algæ and mollusca.

I. *Shell covered with a leathery skin. Sphargis*, Merrem. *Coriaceous Turtle, Ch.* (*Sphargis*) *Coriacea. Testudo Coriacea.* Lin. Pen. Brit. Zool. 4. t. 1.

Shell with three longitudinal dorsal ridges; when young the skin granular, and fins very long. See Schoepft. t. 29. Mediterranean Ocean.

II. *Shell covered with horny scales.* Chelonia.

A. shields depressed. Chelonia.

Green Turtle. Chelonia Mydas. Testudo, Lin. Lacep. 2. O. t. 1. f. 1

Shell cordate, greenish brown, slightly keeled, shelving on both sides; lower jaw deeply denticulated. Varies first in the number of the dorsal shields, and thus *Test. Atra*, Lin., and *Test. Multiscutata*, Kuhl. Second in the number of sternal shields, thus *Test. Cepediana*, Daud. Sometimes the shields are variously coloured, and thus *Chelonia Virgata*, Dumeril, and *Chelonia Radiata, Ch. Maculosa* and *Ch. Lachrymata*, Cuv., R. A. *Testudo Japonica*, Thumberg, appears to be only a variety with the shields so thin, that he mistook the sutures of the bones for the edge of the scales.

Loggerhead Turtle, Chelonia Caretta, Testudo Caretta, Lin. Schoepf. t. 16.

Shell convex, vertebral shields gibbous; head very large; when young the nose produced, *Testudo Nasicornis*, Lacep. Edw. t. 206, Var. Vertebral shields sometimes seven. *Chelonia Olivacea*, Eschscholtz Atlas, t. 3. Mediterranean and other seas.

20

B. *Shields Imbricate.*

4. *Imbricated Turtle.* *Chelonia Imbricata*, Schw. *Testudo*, Schoepf. t. 18, *a.*

Shell elliptically keeled; shields yellow, spotted and rayed. In the young, the ends of the shields are obliquely truncated, and scarcely imbricate, *Caretta Nasicornis*, Merrem. Schoepf. t. 17, f. 1. Indian and Atlantic Ocean.

Ord. 3. Emydosauri.

Vertebræ of the back and the ribs free, mobile; the mouth toothed; the feet digitate, webbed; above and beneath covered with imbeded squarish plates, with small scales on the sides.

Gen. 1. Crocodilus.

Toes 4, 5, claws 3, 3; tail compressed, above keeled, serrated; teeth conical in a single series; tongue fleshy.

1. *The head depressed; the canine teeth of the lower jaw received in a pit in the upper jaw.* Alligator.

* *Feet digitate.* South America.

Spectacled Alligator. Croc. (Alligator) Sclerops, Schw. Seba, 1, t. 104, f. 10, Spix. Braz.

With a cross rib between the orbits; nape with a band of shields. South America and Brazils. The frontal cross rib varies in shape.

Eye-browed Alligator. *Croc.* (*Alligator*) *Palpebrosus*, Cuv.

Eyelids bony; nape with a band of shields, each bearing a crest. Var. the nuchal shields are sometimes irregular. *Croc. Trigonatus*, Seba, 1, t. 108, f. 3. Guiana.

† *Feet, subpalmated.* North America.

Pike-muzzled Alligator. Cayman Croc. (Alligator) Lucius. Cuv. *Croc. Cuvieri*, Leach Zool. Misc. t

Head depressed, parabolic, nuchal shields 4. North America.

21

II. *The head oblong; canine teeth of the lower jaw received in a notch in the edge of the upper; feet palmated.* Crocodilus.

Common Crocodile. *Crocodilus Vulgaris*, Ann. Mus. x. t. 3.

The head equal, broad; nuchal plates six; cervical six, subequal; the dorsal shields quadrate, six rowed. Africa and India.

Ruppell's Crocodile. *Croc. Octophractus*, Ruppell, MSS.

The head equal, long; nuchal plates four, cervical eight, unequal; the dorsal shields slightly keeled. Africa, Dongola, and Senegal.

Double-crested or Indian Crocodile. *Crocodilus Biporcatus*, Cuv.

The forehead with two nearly parallel longitudinal ridges; nuchal shields six; dorsal shields ovate, eight rowed. India and its islands.

Rhombic Crocodile. *Crocodilus Rhombifer*, Cuv.

The forehead convex, with two diverging ridges; nuchal shields six; dorsal scales quadrate, six rowed; scales of the limbs thick, keeled.

Siam. Crocodile. *Crocodilus Galeatus*, Cuv. Hist. Anim. Par. 6, a.

Crown of the head with a two-toothed elevated crest; nuchal shields six. India. Siam.

Two-shielded Crocodile. *Crocodilus Biscutatus*, Cuv.

Nuchal shields two; the middle dorsal scales quadrate, the outer irregular, scattered.

West Indian Crocodile. *Crocodilus Acutus*, Cuv. Geoff. Ann. Mus. Par. II. t. 37.

Head produced, convex at the base; middle dorsal scales

quadrate, outer irregularly scattered; nuchal scales six. West Indies.

Armed Crocodile. Crocodilus Cataphractus, Cuv. Oss. Foss. f.

Head produced; nape with four bands of bony shields like the Alligators. Mus. Col. Surgs.

Flat-headed Crocodile. Croc. Planirostris, Graves.

Head, base, flat; shields tubercular; nuchal plates six, distant. Africa?

Intermediate Crocodile. Crocodilus Intermedius, Graves.

Head produced, sub-cylindrical: dorsal shields, six-rowed, nuchal, six. America?

III. *Head long, produced, muzzle cylindrical; teeth and feet like Crocodiles.* Gavialis.

Common Gavial. Croc. (Gavialis) Gangeticus.

The nuchal shield united to the dorsal plates; the crown and orbits appear to dilate with age. See Faugas. Hist. Mont. St. Petri, t. 46, 48; from whence *Croc. Tenuirostris*, Cuv. Ganges.

Order III.? Enaliosauri. *Conybeare.*

Vertebræ of the back and the ribs free, mobile; the mouth toothed; the feet fin-shaped. Only found in the fossil state; perhaps not of this section.

Gen. 1. Ichthysaurus, Koenig

Head large; teeth conical, placed in a groove; neck short.

For the anatomy and species of this genus, consult Home Phil. Trans. Conybear. Geolog. Trans. and Cuv. Os. Fos. vol 5.

Ichthyosaurus Communis. Conybeare. Cuv. Os. Fos. t. 29. f. 1, 9

Ichthyosaurus Platyodon, Conybeare, teeth compressed. Cuv. Os. Fos. t. 28. f. 4, 5.

Ichthyosaurus Tenuirostris, Conybeare, Cuv. Os. Fos. t. 29. f. 8, 9.

Ichthyosaurus Intermedius, Cuv. Os. Fos. t. 29. f. 2, 5.

Ichthyosaurus Latifrons, Koenig. Icon. Sect. t. f. ined. Mus. Brit.

Ichthyosaurus Grandipes, Sharp. Geol. Trans. 1830.

Gen. 2. Plesiosaurus, Conybeare.

The head small: neck very long, of many vertebræ; the teeth placed in pits.

For the anatomy of this genus, consult Conybeare, Geolog. Trans. and Cuv. Os. Fos. v.

Plesiosarus Dolichodeirus, Conybeare, Geolog. Trans. Cuv. Os. Fos. v. t. 31 and 32. f. 1, 2, &c. Lyme Regis.

Plesiosaurus Recentior, Conybeare, Kimmeridge.

Plesiosaurus Carinatus, Cuv. Os. Fos. v. 486.

Plesiosaurus Aucois, Cuv. L. c. v. 486.

Plesiosaurus Pentagonus, Cuv.

Plesiosaurus Trigonus, Cuv.

Gen. 3. Saurocephalus, Harlan. *Saurodon*, Hay.

Teeth like incisors, placed in pits with a regular hole on the inner edge of the Alveola.

Saurocephalus lanciformis, Hay. Amer. Phil. Trans. y. t. 16. *Saurocephalus*, *Leanus*, l. c. t. 16.

Section II. Squamata.

Tongue free, organs of generation double, ossa quadrata separated from the skull.

Order I. Sauri.

Mouth not dilatatile, skin covered with various unequal scales.

Division I.

*Tongue long, deeply bifurcated, teeth placed on the inner edge of the jaws.**

Gen. 1. MONITOR.

Tongue retractile, head covered with small polygonal shields. *Inhabiting the old world, living near water.*

1. *Tail compressed, nostrils subapical, or medial, open;* Monitor, nob.

Two Banded Monitor, Mon. bivittatus, nob. *Tupinambis bivittatus*. Kuhl.

Black, with five cross rows of white spots or rings, back of head spotted, above banded; cheeks, with two black streaks; nostrils subapical. India. Jun., more spotted. *Tup. exilis.* Reinw. Mss. Mus. Brit.

Laced Monitor, Mon. Varia, n. *Lacerta Varia*, Shaw. White. Voy. t. p. 253. Phil. Voy. t. at p. 279.

Nostrils one-third from the tip of the nose; toes short; subequal. New Holland. *Tup. Marmoratus*, Oppel. of Manilla, appears to be a variety. Mus. Brit.

Yellowish Monitor, Mon. Flavescens, nob. *Uran. Russelii*, Schegel.

Yellowish, brown netted, scales very large; toes very short, nearly equal, India. Mus. Ruppel.

Eyed Monitor, Mon. Ocellatus, Von Heyden Ruppel Atlas, t.

Yellowish brown, with pale eyed spots, dorsal scales oblong,

* All the species of the genera of this section require a complete reexamination, the following is only an attempt at noting the species which I have seen named in collections.

surrounded by small granular scales, ventral scales square. Dongola, Senegal. Mus. Ruppel.

Green Spotted Monitor, Mon. Chlorostigma, nob.

Black; head, body, and tail, white spotted; scales, rather large; toes very unequal, nostrils one-third from the tip. Young, belly whitish, black spotted. Rawack, Mus. Paris.

Timor Monitor, Mon. Timorensis, nob. *Tup. viridomaculatus*, Daud.

Like T. Bengalensis, but the scales are larger, especially those of the tail, toes longer, nostrils more than one-fourth from the tip. Timor. Mus. Paris.

Green Monitor. Mon. Viridis.

Nostrils medial, scales large, yellow, with large dark spots, Mus. Leyd.

Elegant Monitor. Mon. Exanthematica, n. *Lac. Exanthematica*, Bosc. t. 5, f. 3. *Tup. Cepedianus*, Daud. t. 29. *Lac. Argus*, Daud. from Seba 1, t. 85. f. 3. *Uran Elegans*, Mer.

Nostrils medial, with a black streak from the eye; head with black edged white spots, throat black spotted, belly black banded. Africa.

Brown Spotted Monitor. Mon. Bengalensis, n. *Tup. Bengalensis*, Daud. *Uran Punctatus*, Mer.

Nostrils medial, head black, dotted; back, black dotted and slightly eyed; throat black dotted; scales rather large; toes short. India. *Lac. Dracæna*, Lin. from. Seba. 1, t. 101, f. 1, now in Mus. Par. The *Tup. Caudivertus*, Daud. has the nostrils medial, and is very like this species.

Clouded Monitor. Mon. Nebulosus.

Nostrils medial, above green brown, white dotted, loins and base of tail, with bands of eyed spots ; beneath white, black netted, toes unequal, rather long, head long, angular. Java. Mus. Par.

Ornamented Monitor. Mon. Ornatus, Tup. Ornatus, Daud. Ann. Mus. II. t. 48 *Lac. Capensis,* Sparrm.

Nostrils medial, above black, beneath white ; throat, with nine black bands, nape, with four or five curved lines, and with seven bands of round white spots. Africa. Mus. Leyd.

Nilotic Monitor. Mon. Niloticus, Lac. Niloticus, Hasselq. Geoff. Rept. Egypt, t. 3, f. 1.

Pale, varied and eyed with brown ; toes very unequal. North Africa and India. Mus. Brit.

Heraldic Monitor. Mon. Heraldicus.

Nostril submedial, rather nearer the eyes ; head with regular and variously placed white lines ; nape and back with a ring of eyed white spots Toes short, subequal. Bengal. Mus. Paris.

Varied Monitor. Mon. Pulcher, Leach. *Uran Elegans,* Merrem. *Lac. Monitor,* Linn. *Tup. Stellatus,* Daud. n. t. 31 ? Seba, 1. t. 94, f. 1, 2. y. t. 30. f. 2. y. 68. f. 2.

Nostrils submedial, brown ; head and neck with concentric white rings of dots ; back, white dotted, with nine bands of round white dots ; beneath white, brown dotted. Central and South Africa. Mus. Brit.

II. *Tail round, nostrils large, valvular, placed near the orbits.* Psammosaurus.

Grey Ouran. Mon. (P.) Scincus, Licht. *Tup. Grisseus,* Daud. Geoff. Rept. Egypt. t. 3.f. 2. t. 4. f. 14, 15.

Grey, dotted with brownish scales, nearly square, edged with small grains. North Africa. Mus. Brit.

White throated Ouran. Mon. P. Albogularis. Tup. Albogularis, Daud. iij. t. 32.

Head, and neck, beneath, and sides, whitish, brown, dotted with two white lines, from the eye to the nape. Inhab.

N.B. The *Dragon* figured by Klein Tent Herpet, t. is an animal of this genus which has had the head united to the skin of the body, just behind the fore legs, so that it appears only to have two legs.

Gen. II. Holoderma. Wiedeman.

Tongue contractile? Head broad, depressed, covered with crowded, irregular, many sided, convex, tubercular shields. Palate not toothed; body squamulose with parallel transverse ridges of larger distant tuberiform long scales; abdominal scales, four sides, smooth; femoral pores none; tail round.

Horrid Monoxillo. Holoderma Horridum. Wied. *Tachydermun Horridum*, Wagler, II. t. 18.

Brown, yellow spotted, tail with close yellow rings. New Spain.

Gen. III. Teius.

Tongue contractile, head shielded, palate mostly toothed. Femoral pores distinct. Found only in America.

I. *Abdominal scales square, as long as broad, smooth; dorsal scales keeled. Tail compressed, keeled.* Crocodilurus, Spix. Ada Gray.

Great Dragon, Lacépède. *Common Ada, Teius (A.) Crocodilus,* Merrem. Lacep. t. 9. Daud. t. 28.

Back with scattered keeled scales. Guiane.

28

Double Crested Ada. Teius (A.) Bicarinata, Lac. Bicarinata, Gm. *Tup. Lacertinus*, Daud. *Croc. Amazonicus*, Spix, t. 21.

Back with equal keeled scales, belly with eight series of scales. South America. Young, side of body spotted. *Croc. Ocellatus*, Spix, t. 22. f. 1.

II. *Abdominal scales smooth, longer than broad; dorsal scales smooth; tail round; collar small.* Teguixin.

Variegated Safeguard. Teius. (T.) Monitor, Merrem. *Lacerta Teguixin*, Lin. Seba, i. t. 96. f. 23. t. 97. f. 1, 5. t. 99. f. 1. *Tup. Nigropunctatus*, Spix, t. 20. Pr. Max. Beytr. t.

Black, with bands of yellow spots; beneath yellow; tail, black and yellow-banded. *Tup. Maculatus*, Daud. *Tup. Monitor*, Spix, t. 19., appears to be the young.

According to Mr. Caup's account of his genus *Exypuestes*, which he says, has the scales of the Monitor, and the tongue, teeth, head, plates, and country of *Teius*, it must belong to this subgenus. He described the back as finely and equally scaled, and the palate toothless.

III. *Abdominal scales smooth, broader than long; tail round; dorsal scales smooth; collar none. The young streaked with black on the sides.* Ameiva.

The Ameiva. Teius Ameiva Vulgaris, Licht. Spix, t. 23. *Lac Ameiva*, Lin. *Ameiva Argus*, Fitz.

Green, black spotted, sides with four bands of black-edged white spots. Young *Teius Ocellifer*, Spix.

Lettered Ameiva. Teius (A.) Litterata, Daud. Seba, i. 29

t. 83. and *Lac. Gutturosa,* Daud. from Seba. ij. t. 103, f. 3, 4.

Blue green, varied with oblong black dots; sides with white-eyed black cross-bands; neck beneath plaited. South America, not Germany.

Blue-headed Ameiva. *Teius* (*A.*) *Cæruleocephalus*, Seba. i. t. 91, f. 3.

Head blue; dorsal line white, with two yellow lines along each side; thighs white spotted.

Blue Ameiva. *Teius* (*A.*) *Cyaneus*, Lacep. i. t. 31, Seba. ij. t. 105, f. 2.

Bluish, sides with roundish white spots.

Side-streaked Ameiva. *Teius* (*A.*) *Lateristriga*, Cuv. Seba, i. t. 90, f. 7. (Spix, t. 24, f. 1 ?)

With a dark white-edged line on each side.

Striped Ameiva. *Teius* (*A.*) *Lemniscatus*, *Lac. Lemniscatus*, Gmel. Seba, i. t. 92, f. 54. Daud. ij. t. 36, f. 1.

Dusky blue, back with eight white lines, tail round, limbs white spotted, with three white lines.

Three-streaked Ameiva. *Teius* (*A.*) *Tritæneatus,* Spix. t. 24, f. 2.

Checkered Ameiva. *Teius* (*A.*) *Tesselata,* Say, Long. Exp. ij. 50.

Black, with nine or ten longitudinal, and eighteen or twenty cross lines; dorsal scales small, beneath whitish. Tail long, brownish olive, black spotted. North America, Say.

Blue and black Ameiva. *Teius* (*A.*) *Cyanomelas*, Pr. Max. Beytr. v. t.

Head short; tail and back black, with a central broad band, and the sides with two narrow whitish blue streaks. Brazils.

30

Green Ameiva. Teius (A.) Viridis, Merrem. *Lacerta. Teyou*, Daud. *Ameiva Teyu*, Lecht.

Top of head and dorsal line green; body olive violet, with six white longitudinal lines; throat and belly silvery white.

Collared Ameiva, Teius (A.) Collaris. Agama Collaris, Say, Long. Exp.

Olive, with five or six alternate broad dusky and narrow yellow or grey spotted fibrous bands; sides greenish yellow; sides of neck fibrous, varied with red and black bands, beneath pale. North America, Arkansa Territory.

IV. *Abdominal scales lanceolate, keeled; tail round; collar distinct.* Kentropyx, Spix.

Intermediate Centropyx. Teius (C.) Intermedius, Schlegel. Back with three pale lines; dorsal scales minute. Surinam, Mus. Leyd.

Spurred Centropyx. Teius (C.) Calearatus, Spix. t. 22, f. 2.

Scales of the back and throat granular. Males with two small spine-like scales on each side of the vent; scales of the belly, legs, and tail, keeled, green; sides of the back blackey. Brazils.

Striated Centropyx. Teius (K.) Striata. Lacerta Striata, Daud. *Pseudo-Ameiva Striata*, Fitz. Merrem. Wetter. Ann. i. t. 1, Pr. Max. Beytr. xiii.

Grey, sides blue, with two longitudinal brown lines; abdominal scales twelve rows; dorsal scales keeled.

Genus IV. Lacerta.

Tongue contractile; head shielded; palate toothed; lateral line, none; the bones of the skull advanced over the temples

31

and orbits; femoral pores distinct. *Inhab. temperate parts of Old World.*

1. *Back with granular scales; belly with a collar of large plates; femoral pores numerous.* Lacerta.

A. collar separate the whole length from the chest-plates by small granular scales; the frontal plate nearly as broad before as behind.

Eyed Lizard. Lac. Ocellata, Edw. t. 202.

Belly with eight or ten rows of plates; occipital plate large, Young. *Lac. Lepida*, Daud. iij. t. 37. f. 1. South of Europe. *Lac. Rhombica*, Merrem, *Lac. Jamaicensis*, Daud., are from Edwards' figure.

Green Lizard. Lac. Viridis, Lin. Daud. t. 34., and *Lac. Bilineata*, Daud. t. 55. f. 1.

Scales of back slender, keeled; of tail, sharply keeled; collar free, serrated; occipital plate rudimentary; abdominal plates six-rowed; hinder legs not reaching the armpits. Europe. Length, twelve inches.

Wall Lizard. Lac. Muralis, Merr. Daud. iij. t. 8. f. 1. *Lac. Vivipara*, Jacq. Act. Helv. j. t. 1.

Scales of back and sides smooth; of the tail, octant above, slightly keeled; collar entire, adnate; belly, scales six-rowed; hinder legs reaching the armpits; length five inches.

Schreber's Lizard. Lac. Schrebersiana, Edw. An. Sci. t. 5. f. 5. *Lac. Fusca*, Daud.

Occipital plate rudimentary; abdominal plates six-rowed; temples covered with granular scales and a cheek-plate; hind legs long. Mus. Par.

Cape Lizard. Lac. Lalandii, Edw. An. Sci. t. 5. f. 6, and t. 8. f. 5.

Occipital plate rudimentary; abdominal plates six rowed;

temples covered with granular scales only; hind legs short, ith medial scales before the vent. Cape. Length thirteen inches.

Duges Lizard. Lac. Dugesii, Edw. t. 6. f. 2.

Occipital plate rudimentary; belly plates, six-rowed; temples with granular plates only; hind legs long; one medial plate before the vent. Madeira.

Edwards' Lizard. Lac. Edwardsiana, Duges.

Abdominal plates eight-rowed; scales of back imbricate, pointed; four sub-maxillary plates on each side; limbs slender; thighs cylindrical. Shore of the Mediterranean.

B. Collar free the whole length; the frontal plate small, and narrowed behind.

Olivier's Lizard. Lac. Olivieri, Ardouin. Sav. Rept. Egypt. t. 2. f. 3.

Abdominal plate six or eight-rowed; one large medial scale before the vent. North Africa.

Savigny's Lizard. Lac. Savignii, Ardouin. Sav. Rept. Egypt. t. 1. f. 8.

Abdominal plates six or eight-rowed; three larger medial plates before the vent. Egypt.

Shielded Lizard. Lac. Scutellata, Ardouin. Sav. Rept. Egypt. t. 1. f. 2.

Abdominal plates twelve or fourteen-rowed; interparretal plate rudimentary; three large medial scales before the vent; the hinder larger than the side ones. Egypt.

Dumerill's Lizard. Lac. Dumerillii, Edw.

Abdominal plates twelve or fourteen-rowed; interparretal plates rudimentary; three larger medial scales before the vent, the hinder smaller than the side ones. Senegal. Mus. Par.

Knox's Lizard. Lac. Knoxii, Edw. t. 6. f. 8.

Abdominal plates twelve or fourteen-rowed; interparietal plates large. South Africa.

Variable Lizard. Lac. Variabilis, Licht. *L. Variabilis, L. Arguta, L. Velox,* and *L. Cruenta,* Pallas.

Abdominal plates fourteen or sixteen-rowed; head rather acute; scales small, rather triangular or roundish, smooth; caudal scales closer keeled. Perhaps *L. Deserti,* Lepech, and *L. Corunea,* Merrem. Tartary.

C. Collar united in the middle to the abdominal plates, free on the sides.

Panther Lizard. Lac. Pardalis, Licht.

Abdominal scales ten rowed; head rather acute; scales of the back smooth; of the tail rhombic above, keeled; tail little longer than the body. Egypt.

Bosc's Lizard. Lac. Boscii, Daud. iij. t. 36. f. 2. *Lac. Velox,* Edw. Savig. Rept. t. 1. f. 9.

Abdominal plates ten-rowed; scales keeled; cervical ones small; dorsal large, rhombic, blunt; tail twice as long as the body. Egypt.

Red-spotted Lizard. Lac. Rubropunctata, Licht. Savign. Rept. Egypt. t. 1. f. 1?

Abdominal plates ten-rowed; head acute; scales of the back small, smooth; of the tail large, sub-quadrate, olive, keeled; tail twice as long as the body. Egypt.

Spotted Lizard. Lac Guttulata, Licht. Sav. t. 1. f. 8?

Abdominal plates eight-rowed; head rather acute; scales of

back very small; of tail sub-rhombic, above keeled; tail long. Egypt.

Grammic Lizard. Lac. Grammica, Licht. Sav. t. 1. f. 7.

Abdominal plates fourteen or twenty-rowed; head acute; scales of the back small, scarcely keeled; of tail rhombic, keeled; collar sometimes obsolete; tail half as long again as the body. Egypt.

Indian Lizard. Lac. Leschenaultii, Edw. t. 6. f. 9.

Abdominal plates six-rowed. Inhab. Coromandel.

II. *Back with granular scales; belly with small plates; collar indistinct.* Psammodromus. Fitz.

Spanish Lizard. L. (Psammodromus) Hispanicus, Fitz MSS. See also Var. of *Lac. grammica,* Licht.

Blackish blue, lighter underneath. Spain.

III. *Back and tail with keeled scales, belly with smooth imbricate scales, collar none, femoral pores numerous.* Algyra Cuv.

Common Algyra. Lac. (Algyra) Cuvieri, Lac. Algyra, Lin.

Tail long, round, brown, beneath yellowish, with two yellow streaks on each side; length four inches. Spain.

IV. *Back and tail with lanceolate imbricate keeled scales, abdomen with smooth imbricate scales, collar none, femoral pores none.* Tropidosaurus, *Boie.*

Mountain Lizard, Lac. (*Tropidosaurus*) *Montanus,* Boie.

Olive dark metallic green, blackish on the sides, and with a narrow black dorsal, and two narrow white lines on each side. Java. I have seen this genus both at Paris and Leyden, it has much more the appearance of a scinct than a lizard, but I could not see its tongue, which I suppose caused Cuvier to place it in this group.

Gen. V. TACHYDROMUS. Oppel.

Tongue contractile; head shielded; back, belly, and tail, with square keeled scales; lateral line distinct, with small scales; femoral pores one on each side the vent; body very long; feet very far apart and small.

Chinese Tachydrome. *Tachyodromus Sexlineatus*, Daud. iij. t. 39.

Silvery, with six brown bands. Tail three or four times as long as the body. Java and China. Common in insect boxes from China. The other species, *Quadrilineatus*, Daud. in the Paris Museum, is in too bad a state to distinguish it from the former. Is this *Lac. Sept.* of Linne?

Section II.

Tongue short, contractile, end slightly lobed.

Gen. I. IGUANA.

Teeth three lobed or toothed, placed in the inner edge of the jaw. Body and head compressed, palate mostly toothed.

A. Ribs simple; throat dilatile; head short; back crested; palate toothed; femoral pores numerous. Iguana, nob.

I. *Tail equally scaly; toes unequal; head shields flatish; dewlap denticulated.* Iguana.

Common Guana. *Iguana Tuberculata*, Laur. *Squamosa*, t. 5, I. *Viridis*, t. 6, and I. *Emarginata*, t. 8, Spix.

Sides of neck with convex scales; front edge of dewlap toothed; sides of lower jaw with orbicular plates. Var. Nose plates prominent, hornlike. *Iguana Cornuta*, Em. t. 4. f. 4. Young, dorsal spines lower. *Iguana Cærulea*, Daud. and *I. Lophyroides*, Spix, t. 9. South America, Mus. Brit.

Smooth-necked Guana. *Iguana Delesatissima*, Laur. *1 Nudicollis*, Merrem. Mus. Bresl. t. 13. f. 1.

Sides of neck smooth; front of dewlap slightly toothed; sides of lower jaw with small plates. Brazils and Gaudaloupe.

2. *Tail equally scaly; toes unequal; head shield, flatish; dewlap entire.* Brachylophus. (*Les Brachylophes.*) Cuv.

Banded Guana. *Iguana* (*Brachylophus.*) *Fasciatus*, Brongn. Bul. Soc. Philom. t. 61. f. 1.

Green, brown banded. South America, not Java, as said by Brongniart. Mus. Brit.

4. *Tail compressed, with rings of spinous scales; toes nearly equal; head shields convex.* Amblyrhynchus, Bell.

Rough headed Guana. *Iguana* (*A.*) *Cristatus*, n. Bell. Zool. Jour. i. t. 12.

Brown, sides and belly yellow? Galapagos.

Black Guana. *Iguana* (*A.*) *Ater*, nob.

Nearly black; beneath dusky. Galapagos.

B. Ribs simple, throat with a cross fold.

I. *Head long, scutellated; back crested; tail with rings of spinous scales; back with square scales; femoral pores distinct.* Cyclura nob.

Palate toothed. Ctenosaura. *Weigmann.*

Cycluroid Guana. Iguana. (Ctens.) Cyclurcides. n. *Ctenosaura Cycluroides.* Wiegmann, Iris xxi.

Dorsal crest continued; scales of back and sides, obsoletely keeled; of sacrum mucronate. Palatine teeth many, small; tail very long and round, crested with a series of spines. Mus. Berl.

Spiny Tailed Guana. Iguana. (Ctenosaura) Acanthura. Lac. Acanthura. Shaw. Zool. iij. 326. *Cyclura Shawii*, Gray Ann. Phil.

Pale brown; dorsal crest interrupted over the sacrum; tail cylindrical, with a ring of depressed scales between each ring of spiny ones; palatine teeth few, large. Mus. Brit.

Armed Iguana. Iguana (Ctenosaura) Armata.

Head long, pyramidical; tail with rings of spines, having two rings of depressed scales between them at the base; teeth three or five lobed; palatine teeth small, in one row. Mus. Bell.

Allied Iguana. Iguana (Ctenosaura) Similis.

Grey, black dotted, body with four oblique dark bands; occiput forming a concave band behind; dorsal crest low but continued over the sacrum. Teeth blunt, three lobed; palatine on two raised lines on each side. Head two, body nine inches. Mus. Bell.

Lancet Toothed Iguana. Iguana (Ctenosaura) Lanceolata.

Grey, black dotted. Tail black banded, base black spotted; head scales rather large, convex; dorsal crest, completely interrupted over the sacrum. Palatine teeth, in two bunches; the teeth of the jaw, long lanceolate sharp-edged, two or three lobed; femoral pores six. Length of body and head fourteen, of tail twenty-four inches. Mus. Bell.

Bell's Iguana. Iguana (Ctenosaura) Bellii. Gray MS.

Grey, black netted; dorsal and caudal crest of short, broad scales, interrupted over the sacrum. Scales of the occiput and sides of the head, convex ; scales of the mastoid ridge, small, conical. Palatine teeth in two bunches; tail, base smooth, middle and end with distant rings of slightly elevated scales, having four rings of small scales between each ; toes short, thick. Mus. Bell. Length of body eighteen inches.

Iguana Cyclura of Cuvier may be probably one of these species, but his account of it is too general to distinguish it.

Palate toothless. Cyclura. *Harlan.*

Keeled Guana. Iguana (Cyclura) Carinata. Cyclura Carinata. Harlan. Acad. Nat. Sci. Phil. iv. t. 15.

Deep dirty brown ; dorsal crest interrupted over the shoulders and loins ; scales of body uniform ; legs and feet minutely square. Tail above keeled, with three rings of depressed scales between each ring of spiny scales. Inhab. Turk's Islands, Bahama.

Round Guana. Iguana (Cyclura) Teres, n. *Cyclura Teres.* Harlan. l. c. t. 16.

Dark green, teeth small, uniform pointed ; dorsal crest interrupted over the loins ; scales of the sides square, of the thighs and legs bristly. Tail cylindrical, with two rings of depressed scales between each rings of spines. Tampico.

Clouded Guana. Iguana (Cyclura) Nubila. Gray.

Blue with oblique bands of roundish spots ; tail compressed, with four rows of small depressed scales between each ring of spines. South America ? Brit. Mus. Figured here under the name of *Clouded Lizard. L. Nebulosa.*

II. *Head covered with convex scales, and often some superciliary plates ; back, covered with small scales slightly crested, femoral pores none.* Ophyessa, Boie.

1. *Tail, simple, compressed, dorsal scales, minute,* (Xiphura, nob.)

Sword Tailed Ophyessa. Oph. Superciliosa, Lacerta, Lin. from Seba, 1. t. 199. f. 4. *Lophyrus Xiphurus,* Spix, t. 10, good.

Head with sub-equal, keeled scales, occiput with cross rows of convex tubercles. Back with small, belly and limbs and tail with larger keeled scales. Brazils. Mus. Brit.

Pearly Ophyessa. Oph. Margaritaceus, Lophyrus, Spix, t. 12, f. 2.

Back and tail slightly crested, supercilliary scales, small. Brazils. *Loph. Auronitens,* Spix, t. 13, scarcely appears to differ from this except in age.

Lozenge Ophyessa. Oph. Rhombifer, Lophyrus, Spix, Braz. t. 11, adult. *Loph. Albomaxillaris,* Spix, t. 13, f. 2. Young, *Agama Catenata,* Pr. Max. t.

Brown, beneath paler, limbs pale banded, back with two sinuous, brown bands, forming a series of rhombic spots, super-orbital scales small. Brazils.

9. *Tail round, slender, with imbricate scales.* Plica, nob.

Brazilian Ophyessa. Oph. Braziliensis.

Head with rather convex, keeled scales, superciliary ones rather smaller, bluish; chest, inside of legs, belly, and vent, whitish. Scales of back and throat granular, of belly rather larger, of limbs and tail largest, keeled; tail round; back and loins with a small toothed crest. Brazils.

Painted Ophyessa. Ophyessa Picta, Agama, Pr. Max. t. *Lophyrus Ochrocollaris,* Spix, t. 12, f. 2. *L. Panthera.* Spix, Braz. t. 13, f. 1. Young.

Brown; body, face and limbs darker banded; beneath pale, Superciliary scales oblong; occipital plate, distinct; back slightly crested. Brazils.

40

Umber Ophyessa. *Ophyessa Plica*. *Lac*. *Plica*, Lin. *Agama Plica*, and *A*. *Umbra*, Daud. *Loph*. *Agamoides*, Gray. Ann. Phil.

Back with minute scales, sides of neck behind the ear with four bundles of spines.

3. *Tail round, tapering, with spinose, whorled scale; ears spinose in front, back and belly scaly*. Tropidurus, *Pr. Max*.

Palate toothed, Oplurus, Cuv.

Collared Tropidurus. *Trop*. *Torquatus*, Pr. Max. t. *Agama Taraguira*, Licht.

Green; black stripe on each side the neck; eyelid black rayed, back with lines of spot when young. Mus. Brit. See *Agama Nigricollis*, Spix, t. 16, f. 2, but Spix says it has no palatine teeth. This may be, *Ephimotes*, Cuvier 47, but certainly it has no pores.

Tubercular Tropidurus. *Trop*. *Tuberculata*, Gray. *Agama Tub*. Spix, t. 15, f. 1, ♁ 2, ♂

Green, black spotted; streak on side of neck black; scales of back and feet keeled; the dorsal ones the largest. Brazils.

Cuvier's Tropidurus. *Trop*. *Cuvieri*, Gray. *Oplurus Torquatus*, Cuv. ij. 48. Seba. j. t. 94. f. 4. Cop. Shaw Zool t. 69.

Dorsal scales minute, keeled; caudal scales large, keeled; with a black half collar on each side the neck. Braz.

*** "*Palate toothless*," Spix.

Half-banded Tropidurus. *Trop*. *Semitæniatus*. *Agama sem*., Spix, t. 15. f. 1.

Green-brown, smooth; dorsal line blue; dorsal scales minute; abdominal rather larger; caudal largest, keeled.

Cycluroid Tropidurus. Trop. Cyclurus. Agama Cyclurus. Spix, t. 17. f. 1.

Olive brown, longitudinally keeled; back and tail black-banded; beneath brown. See also Seba, t. 97. f. 4. Pr. Max. Bon. Trans. xiv. t. 15.

4. *Tail tapering, with large verticillate spinose scales; head and body with minute granular scales.* Uranocentron, *Caup.* Doryphorus, *Cuv.*

Long-tailed Doryphorus. Oph. (*Uran*) *Azureus. Agama Azureus,* Daud. iv. t. 46.

Bright blue, spotless; tail long. Inhab. Surinam.

Short-tailed Doryphorus. Oph. Uran. Brevicaudatus. Agama Brevi, Daud. iv. t. 47. *Lac. Azurea,* A. Gmel.

Blue, black-banded; tail short, depressed. - Mus. Col. Surg.

III. *Head with numerous small regular shields; back and tail with large keeled spines, converging towards the back, and forming a crest; femoral pores none.* Leiocephalus, Gray.

Keeled Leiocephalus. Leiocephalus Carinatus, Gray. Ann. Phil.

Green, blackish-banded; dorsal scales large, broad, keeled; keels forming oblique ridges; the vertebral keel the highest. ——— ? Mus. Brit.

IV. *Head covered with small shields; back and tail with large keeled scales; belly with smooth scales; tail long, with imbricate keeled scales; femoral pores large.* America. Tropidolepis, *Cuv.* Scelophorus, *Wiegmann.*

Northern Tropidolepis. *Tropidolepis Undulatus*, Gray. *Agama Und.*, Daud. *Lacerta Hyacenthina*, and *Lac. Faciata*, Green.

Ash, with irregular brown cross-bands; head with regular flat plates. North America. Mus. Brit.

Thorny Tropidolepis. *Tropidolepis Aculeatus*, Gray.

Head with convex irregular scales: eyebrow plates band-like, back with a crest of short compressed scales; scales of body and limbs dagger pointed. Martinique.

Cuvier places *Agama Nigricollis*, and *A. Cyclurus*, Spix, in this genus; but Spix says they have no femoral pores, they are more like *Ophyessa*.

Collared Tropidolepis. *Trop. Torquatus*, Gray. *Scelophorus*, Weigmann. Iris xxi. *Tecoixin* Hernandez.

Olive brown, with a black collar, in an orange band; dorsal scales large, keeled; rhombic, smooth, tip denticulated. Mexico. Mus. Brit.

Spinous Tropidolepis. *Trop. Spinosus*, Gray. *Scelophorus*, Weigmann.

Grey brown, back with a series of brown spots; scales rhombic, tooth-edged; apex longly pointed, acute; abdominal ones, smooth truncate. Mexico.

Side Streaked Tropidolepis. *Trop. Pleurostictus*, Gray. *Scelophorus*, Wiegmann.

Olive grey; sides black and yellow spotted; scales rhombic, keeled, tooth-edged, mucronate; those in base of tail largest. Mexico.

Green Tropidolepis. *Trop. Grammicus.* Gray. *Scelerophorus.* Weigmann.

Shining coppery olive, with black flexuous cross lines;

scales rhombic keeled, mucronate. Variety, scales smaller Mexico. Mus. Brit.

Æneous. Tropidolepis. Trop. Æneus, Gray. *Sclerophorus,* Wiegmann.

Shining coppery; scales ovate, lanceolate, keeled, toothed. Mexico.

Banded Tropidolepis. Trop Scalaris, Gray. *Sclerophorus,* Wiegmann.

Grey brown; side band white, with white edged brown semi-lunar spots; scales ovate, keeled entire. Mexico. Mus. Brit.

Bell's Tropidolepis. Trop. Bellii, Gray.

Metallic green, scales of the back and upper part of the body, and tail long, strongly keeled, dagger pointed, the keels forming 14-16 ridges; of the belly broad, blunt, those of the limbs and tail smaller, becoming larger near the end; length 10 inches. Mus. Bell.

Banded Tropidolepis. Trop. Fasciatus, Gray.

Pale brown; back with wavy cross bands; head varied, pale and dark brown; scales of back tail and limbs moderate keeled, keels 18-20; continued, oblique, entire; of the limbs and tail smaller. Mus. Bell.

V. *Head, short; round, dilated behind, covered with convex scales, and crowned with spines. Body and tail short, depressed, covered with irregular keeled spines; belly with larger scales, fringed on each side; pores distinct, palate toothless.* Phrynosoma. Weigmann. America.

Douglass' Toad Lizard. Phrynosoma Douglassi, Gray. *Agama* Bell. Lin. Trans. good.

Yellow; brown spotted. Mus. Brit.

Orbicular Lizard, or Horned Toad Lizard. ***Phrynosoma Cornuta.*** Gray. ***Agama cornuta.*** Harlan. Journ. Acad. N. S. Phil. good.

Brown ; with two large hornlike spines behind the head. Mus. Brit.

Common Toad Lizard. ***Phrynosoma Buffonium,*** Weigm. Isis, xxi. Seba, 1. t. 83. f. 1, 2.

Scales of the belly keeled ; spines on side of body two rowed, short, straight, three sided. Surinam. Mus. Berl.

Orbicular Toad Lizard. ***Phrynosoma Orbiculare,*** Weigmann.

Abdominal scales smooth ; spines on side of body one rowed, large, recurved, rather compressed. Mexico. Mus. Berl.

C. Ribs forming complete circles ; head scutellated ; throat extensile.

I. *Head hooded ; back crested ; pores o. ; toes margined on the side.* Basilicus, nob. (part Cuvier.)

Mitred Basilisc. ***Bas. Americanus,*** Laur. *Lac. Basilicus,* Lim. Seb. i. t. 100. f. 1. Guiana. Mus. Brit.

Banded Basilisc. ***Basiliscus Vittatus,*** Boie.

Blue, with two series of spots down the sides of the back, and white streak along each side of the dorsal ridge, another edging the two lips and extended to the shoulders ; beneath white ; throat brown marbled. Dorsal ridge scarcely elevated ; hood small. Length about nine inches. Mus. Berl.

II. *Head squarish, occiput compressed, produced into a crest ; back lowly crested ; pores none.* Chamœleopsis, Wiedemann, MSS. Mus. Berl.

Hernandy's Chameleopsis. ***Chamœleopsis Hernandesii,*** Wiedemann, MSS. Gray. Mus. Brit.

Black, white speckled ; scales minute, thin. Mexico. Mus. Brit.

III. *Head elongate, simple ; toes, last joint but one dilated, pear-shaped beneath.* Anolis, Merrem. Anolius, Cuv.

Equestrian Anolis. *Anolis Equestris*, Merrem. Cuv. R.A. t. 5. f. 2.

Purple ash, scapular band white ; dorsal line toothed.

Edwards Anolis. *Anolis Edwardsii*, Merrem. Edw. Glean. i. t. 245. f. 21.

Uniform slate colour, and nearly allied to Equestris.

Two-spotted Anolis. *Anolis Bimaculata*, Daud. *Lac. Bimaculata*, Sparmann, Swed. Acad. v. t. 4. f. 1. bad.

Cuvier's Anolis. *Anolis Cuvieri*, Merrem. *Anol. Velifer*, Cuvier, R.A. t. 5. f. 1.

Lined Anolis. *Anolis Lineatus*, Daud. Rept. t. 48. f. 1. *Lac. Strumosa*, Lin. Seba, ij. t. 24. f. 4 ?

Dotted Anolis. *Anolis Punctatus*, Daud. t. 40. f. 2.

Gouty Anolis. *Anolis Podargicus*, Daud. Catesby Car. t. 3.

Violet Anolis. *Anolis Violaceus*, Spix, Braz. t. 17. f. 2. Brazils.

Green Anolis. *Anolis Viridis*, Pr. Max. Good. Brazils.

All these species are so very much alike that it is impossible to distinguish them without long descriptions ; indeed the genus requires a complete revision, which I hope shortly to be able to publish : the following is very distinct.

Slender Anolis. *Anolis Gracilis*, Pr. Max.

Head elongate, slender, two keeled in front ; back slightly keeled. Dull brown, with seven cross-bands of white spots ; throat orange. Brazils.

IV. *Head angular; body with smooth scales, larger beneath; back simple; tail round; toes simple; femoral pores distinct.* Polychrus, Cuv.

Common Marble. Polychrus Marmoratus, Lac. *Lac. Marmorata,* Lin. Spix, Wagler, t. 21.

Eyelids black rayed; scales above, small, blunt, beneath lanceolate, keeled, of the tail largest. *P. Virescens* and *P. Strigiventer,* Wagler, and *P. Acutirotris,* Spix, t. 14. appear to be scarcely distinct, though they are said to have no femoral pores. *Polychrus Fasciatus,* La Porte, if from Molucca, must be an *Agama.* See also *Agama Molinaii,* Lesson. Bul. Sci. from Chili.

Genus III.—Gecko.

Head and body depressed; scales small, tubercular, usually larger beneath; toes five, sub-equal, generally furnished with transverse scales beneath; claws retractile; throat simple; palate toothless; eyes large, nocturnal.

Toes with a single row of broad cross scales beneath; last joint compressed, attached. Platydactylus nob.

A. Toes free, clawless, dilated; thumb small. Phelsuma.

Clawless Platydactile. Gecko Inunguis, Cuv. R. A. t. f. 1.

Femoral pores none. Violet; beneath white, with a black streak on each side. Isle of France.

Eyed Platydactyle. Plat. Ocellatus. Gecko, Cuv. R. A. t. f. 4.

Femoral pores none. Grey, with white-eyed brown spots. Isle of France.

Cepedian Platydactyle. Plat. Cepedianus. Gecko, Merr. Cuv. R. A. t. f. 5.

Orange, marked with blue, and a white streak on the side femoral pore distinct. Isle of France.

Ornamented Platydactyle. Plat. Ornatus. ***Phelsuma***, Gray. Ann. Phil.

Brown; back with six rows of red oval spots. New Holland.

B. Toes, first and fourth clawed, free, dilated; pores none. Tarentola, *Gray.*

Fasciculated Gecko. Plat. Fascicularis. Gecko, Daud.

Grey; head rough; back with twelve series of groups of three or four small acute spines. South of Europe.

Annulated Gecko. Plat. Ægyptiacus. Gecko, Cuv. *Gecko Annulaire*, Geoff. Egypt. t. 5. f. 7.

Grey; scales granular, with rows of large, flat, simple, round tubercules. Egypt.

American Gecko. Platydactylus Americanus, Gray.

Pale, lined and varied with brown; scales granular, with twelve or fourteen close rows of blunt, simple, equal tubercles. New York. Mus. Par.

C. Toes clawed, thumb clawless, subanal pores distinct. Gecko.

Toes free, dilated the whole length. Gecko, Gray.

Common Gecko. Gecko Guttatus, Daud. t. 49. *Lac. Gecko*, Lin.

Body black, with white spots; scales minute, with low rounded simple tubercles. Java. Mus. Brit.

Chinese Gecko. Gecko Reevesii.

Black, with cross band of white spots, and some obscure rather larger tubercular scales. China.

Madagascar Gecko. Gecko Madagascariensis. Gray.

Scales smooth, with large tubercles on the sides; femoral and subanal pores in two straight diverging lines; young, with a black and white lateral line. Madagascar.

Banded Gecko. Gecko vittatus, Houttyn. Act Ulys. ix. t. 11. f. 2. Daud. iv. t. 50.

Brown, with a white dorsal band, forked over the head and the base of the tail. Java.

D. Toes free, scarcely dilated. Eublepharis, Gray. Stenodactylus, Part. Licht.

Hardwicke's Gecko. Eublepharis, Hardwickii, Gray.

Grey, brown banded; back tubercular. Java.

Ascolobates Stenodactylus. Licht. Savigny, Egypt. *Phyllodactylus Marmoratus.* Mus. Franc. *St. guttatus*, Cuv.

Scales above small; white spotted; tail, black, with irregular white rings. Egypt.

E. Toes dilated, webbed, sides of body and tail margined. Pteroplura, Gray. Ptychozoon, *Kuhl.*

Horsfield's Gecko. Pte. Horsfeldii, Gray. *Lac. Homalocephala*, Creicht.

Brown, with darker cross bands; beneath pale; tail, when reproduced, roundish, not pinnated.

E. Toes and thumb clawed, subwebbed, sides sub-finned.
Leache's Gecko. Gecko Leachianus, Cuv.

Grey, with large white spots; scales uniform, smooth.

II. *Toes, with a series of cross scales, divided by a groove, beneath, last joint compressed, sheathed.* Thecodactylus.

a. Toes dilated to base, scales under toes transverse, thumb clawless, femoral pores, none. Thecodactylus, Cuv.

Smooth Sheath claw. Thecadactylus lævis, Daud.

Tail round, conical; grey, brown marbled; tail when reproduced, turnip shaped. Gecko rapicauda, Daud. iv. t. 31.

b. Toes, base, slender; scales 2 or 3 at the tips. Phyllodactylus, Gray.

Beautiful sheath claw. Phyllodactylus pulchellus, Gray, Spic. Zool.

c. Tail round, femoral pores, none.

Porphyry sheath claw. Thec. porphyreus, Gecko, Daud.

Pale reddish, marbled with many roundish pale spots, Java.

d. Scales under the toes radiating, toes, slender, tail round, slender, femoral pores, distinct. Ptyodactylus, Cuv.

Egyptian sheath claw. Thecodactylus Lobatus, Gecko, Geoff. Rept. Eypt. t. 3, f. 5. *Gecko Ascolobates*, Merrem.

Pale brown, with larger scattered scales. Egypt. *Ptyodactylus Guttatus*, Ruppel. t. 4, appears scarcely distinct.

e. Toes bare, webbed; tail, body, and limbs, margined. Uroplates.

Fimbriated sheath claw. Thecodactylus fimbriatus, Gecko, Daud. iv. t. 52.

Pale brown, scales minute, with scattered larger ones. Madagascar, Mus. Brit. *Gecko Sarroube*, appears to differ very little, if at all, and *Lacerta Caudiverbera*, Seba, ij. t. 103, f. 2, is only known from La Feuille's account.

III. *Toes, base dilated, with a double row of oblique scales beneath, last joint compressed.* Hemidactylus.

Warty Gecko. Gecko Verrucosus, Cuv. *Lac. Turcica.* Lin. Edw. t. 204?

Reddish grey; back scattered with roundish tubercles; subanal pores only.

Mabuia Gecko. *Gecko Mabuia*, Cuv.

Grey, varied with brown; back with about six or seven rows of small scattered tubercles on each side. South America.

It is very difficult to determine the species of this genus as the individuals vary in the size of the larger tubercular scales, and in the presence or absence of the femoral pores, which have hitherto been considered good specific characters. They are however named as follows. *Thecodactylus Policaris*, Spix, t. 18. f. 2. *Gecko Aculeatus*, Spix, t. 18. f. 2, 3. *Gecko Spinicauda*, Houttyn, Act. Ulys. ix. t. *Gecko Trihedrus*, Daud. *Stellio Platyurus*, Schn. *Stellio Argyropus*, Tiles. Mem. Acad. Petersb. viii. t. ii. a., and *Hemidactylus Gronosus*, Ruppell. Atlas. t. 5. f. 1. The two last are said to have no pores.

Margined Hemidactyle. *Hemidactylus Marginatus*, Cuv.

Sides of body and tail, hinder edge of thighs and legs, slightly fringed. Bengal. Mus. Ind. Comp.

IV. *Toes slender, compressed, free, and fringed on the edge, end compressed, bent and arched.* Cyrtodactylus, Gray.

a. Tail round; pores none. Cyrtodactylus, *Gray.* Gonyodactylus, *Kuhl.* Stenodactylus, *Licht.* Gymnodactylus, *Spix.*

Beautiful Cyrtodactyle. *Cyrtodactylus Pulchellus*, Gray, Zool. Jour.

Pale, with chocolate cross-bands. India.

Marbled Cyrtodactyle. *Marmoratus Gonyodactylus*, Kuhl. MSS.

Ash brown, marbled with darker spots, beneath ash. Java.

Eyed Cyrtodactyle. *Cyrtodactylus Ocellatus*, Gray.

Pale brown, with ocellated yellowish spots. Mus. Brit.

Brasilian Cyrtodactyle. Cyrtodactylus Spixii, Gray. *Gymnodactylus Geckoides*, Spix, t. 18. f. 1.

Brown, back with six series of small angular warts. Brazils.

Piping Cyrtodactyle. Cyrtodactylus? Pipiens. Stenodactylus, Licht. *Lac. Pipiens*, Pallas.

Tail, not annulated.

b. Tail depressed, subanal pores distinct. Phyllurus, Cuv.

Flat-tailed Cyrtodactyle. Cyrtodactylus Platura. Lacerta, Shaw. *Agama*, Merrem. *Phyllurus Cuvierii*, Bory. White Voy. t. 32. f. 2.

Tail cordate; grey, brown marbled, scales, with scattered pointed tubercles.

Nilius's Cyrtodactyle. Cyrtodactylus Nilii. Phyllurus, Bory. Dict. Class. t.

Tail bluntish, spathuliform, above brick-red with a black half collar and three bands.

c. *Toes free, ending in a small simple round disk, without any plates beneath; claws retractile.* Sphœriodactylus, *Gray.*

Banded Spheriodactyle. Spheriodactylus Sputator. Gecko, Lacep. 2. O. t. 28. f. 1.

Reddish, with brown cross-bands.

See also *Lac. Sputator Var*. Lacep. Rept. i. t. 28. f. 2. *Lac. Sputator*, Sparmann. Nov. Act. Stock. 1784. t. 4. f. 1, 3.

Gen. IV.—Chameleon.

Teeth placed on the upper edge of the jaws; toes united into two groups to the claws; tail prehensile; body compressed, covered with squarish scales. Africa or India.

a. With a series of spines along the belly and chin.

Common Chameleon. Chameleo Vulgaris. Lac. Africana.

Occiput keeled, arched; scales equal. Grey, banded and streaked with yellow. Africa and India.

Beaded Chameleon. C. Verrucosus, Cuv. *Ch. Monilifer*, Boie.

Occiput keeled, arched; with larger scales on the sides.

Senegal Chameleon. Chameleo Senegalensis. Ch. planiceps. Mer. Seba, 1. t. f. 2. *Ch. Gymnocephalus*, Lacep.

Occiput flat; side parallel edges not margined. Senegal. *Cham. Galeoratus* appears to be a variety of this species.

Eared Chameleon. C. Dilepis. Leach. Gray. Spic. Zool. t. f. 4. *Ch. Planiceps.* Merrem. *C. Bilobus.* Kuhl.

Occiput flat, rather wider behind, under edged with two flaps; a white band on each side. Interior of Africa.

Streaked-sided Chameleon. Chameleo Lateralis, Gray.

Lead colour, sides white streaked; occiput compressed, slightly keeled; scales granular, with a series on the back.

b. With a series of processes on throat, none on belly or tail.

Panther Chameleon. C. Tigris. Kuhl. Gray. Spic.

Scales minute, regular; brown, with minute black specks.

Cape Chameleon. Chameleo Pumillus. Daud. iv. t. 53. *Ch. Margaritaceus.* Mer. Seba, i. t. 32. f. 4, 5.

Scales minute, larger on the sides; olive varied. Var. With larger processes on the head. *Ch. Pardalis*, Cuv. *C. Fimbriatus.* Wied. Cape of Good Hope.

Brookes Chameleon. Chameleo Brookesii, Gray. Spic. Zool. t. 3. *Cham. Superciliaris*, Kuhl.

Scales minute; back with a series of processes on each side; eyebrows produced into horns. Mus. Brit.

c. Without any spines on the chest, belly, or tail.

Two Horned Chameleon. Cham. Bifidus, Brongn. Daud. t. 54. *Ch. Bifurcus*, Kuhl.

Scales square; occiput flat, dilated behind; the sides with groups of white scales; male with two compressed horns over the nostrils; female hornless. Isle of France.

Hooded Chameleon. Cham. Cucullatus, Gray.

Scales oval, unequal; occiput compressed, flattened, with 2 flat processes on each side of its hinder edge, nose produced. Madagascar. Mus. Brit.

Three Horned Chameleon. Cham. Owenii, Gray.

Scales square, small; head short, with three long, conical, curved horns over the nostrils. Fernando Po.

Parson's Chameleon. Cham. Parsonii, Cuv. Fos. v. f. 30, 31. Phil. Trans. viii. t. 8.

Scales ovate, equal; occiput flat, truncated behind; eyebrow crest, produced and elevated on each side the muzzle into a long irregular lobe. Mus. Bell., and Col. Surg.

Gen. V. Agama.

Teeth placed on the edge of the jaws, toes free, long, head and body depressed, covered with imbricate scales. Old World.

A. *Femoral pores, none.*

a. Head, lyrate, back and tail crested. Lyriocephalus.

Scutated Lizard. Lac. Scutata, Lin. Seba, 1, t. 109.

Nose with two rounded tubercles, body with cross bands of larger scales, Mus. Fort. Pitt. Chatham.

b. Head, lyrate, ears distinct, back and tail crested, Agama.
Gariocephalus, Boie.

Agama Tigrina, Merrem. Shaw, t. 68. *Lophyrus gigas,* Boie.

Pale brown, with transverse spotted bands.

Lophyrus Kuhlii, Boie, MS.

Crest less elevated, back with five or six oblique white bands.

c. Cytophanes.

Cyrtophanus Cristatus, Boie. *Agama Cristata,* Merrem. Seba, 1, t. 94, f. 4.

Scales of body, smooth, of limbs and tail, larger and keeled.

d. Back scales large, shelving upwards. Calotes, Boie.

Common Calotes. Calotes Ophiomachus, Kaup. *Lac. Calotes,* Lin. Seba, 1, t. 93, f. 2.

Blue, sides white banded, with two rows of compressed spines behind the ears, scales large, rounded, vertical on back.

Indian Calotes. Calotes Tiedemani, Kaup. Isis. xx. t. 8. Young. *Agama Versicolor,* Daud. t. 44. *Ag. Flavigularis,* Daud. and *Agama Indica,* Gray.

Brown, varied with spots, with two or three groups of small conical spines over the ears. India.

Smaller Calotes. Agama Minor, Gray, Zool, Jour.

Tail short, back brown, with larger brown spots.

e. Back scales, small, horizontal. Bronchocela, Kaup.

Blue Calotes, Agama Cristatella, Kuhl. Seba, 1, t. 89, f. 1.

Blue, not banded, nuchal crest, very small, low, scales of body, keeled, smaller than on tail and limbs.

Long-legged Calotes. Agama Vultuosa, Harlan. Jour. Acad. Phil. iv. t. 19. *Agama Calotes*, Kuhl. Seba, 1, t. 89, f. 2. *Calotes Gutturosa*, Cuv. Guerin, Icon. t. 7, f. 3.

Blue; crest short, scales, all keeled. India. *Le Geleote*, Lacep. 1, t. 12, is perhaps this species. *Agama Gutturosa*, according to Merrem and Lichterstein, is an American species, and is not this. Seba, 1, t. 89, f. 1, is an *Iguana*. *Col. Gutturosa*, Mus. Berl. with a blue neck, is white banded.

Kuhl's Calotes. Calotes Tympanistra, Kuhl.

Blue; crest, very small, scales of belly subequal, keeled. Mus. Berl.

f. Acanthosaura.

Armed Calotes. Agama Armata, Gray, Zool. Jour. When young, with very thin scales. *Calotes Lepidogaster*, Cuv. R. A. Tropidogaster. Mus. Paris.

Head, with long cylindrical spines over eyes, a dorsal crest.

g. Agama.

Tubercular Agama. Agama Tuberculata, Gray Zool. Jour.

Head moderate; scales of back rather smaller than on limbs and tail; of neck and nape with ridges of rather larger triangular scales; under side of legs with short strong keeled scales. India. Mus. Brit.

Obsoletely Crested Agama. Agama Dorsalis, Gray.

Head large, dilated behind; scales rather small, smooth; nape, and back, and tail with very obscure crest of rather larger scales. India. Mus. Brit.

Occipital Agama. Agama Occipitalis, Gray.

Head moderate; scales of back broad, ovate, lanceolate, keeled, spinose, of head smooth and thin; nape with a very obscure crest, and bundle of short three-sided spines. Africa. Mus. Brit.

Common Agama. Agama Spinosa. Lac. Agama, Lin. ?

Brown; head moderate; scales of back large, smooth; of limbs larger, keeled; of belly moderate, nearly smooth; nape with a slight crest; back of head and sides of neck with blundle of long subulate scales. Africa. Brit. Mus.

Sitana. *Cuv.* Semiphorus. *Wag.*

Pondichery Sitana. Agama Pondicerana, Nob.

Body and limbs with keeled imbricate scales; fulvous with two series of large rhombic brown spots. Male with very large pouch extending to between the front legs. Hence the genus *Sitana,* Cuvier, called *Tropidosaura* in Mus. Paris.

b. Head rounded, depressed; ears distinct; back with band of larger scales; tail with whorled bands of large spinous scales. Stellio.

Common Stellio. Lac. Stellio, Lin. *Stellio Vulgaris,* Daud.

Olive spotted with black. Rup. Atlas. f. 2.

c. Head ovate lyrate; ears distant; back and tail not crested; tail tapering, with imbricate scales. Trapelus.

Cape Agama. Trapelus hispidus, Kaup. Iris. xx. t. 7. *Agama Orbicularis,* Merr. ? Seba, j. t. 83. f. 1, 2.

Scales unequal, with trihedral spinous scales. Cape.

Yellow-striped Agama. Ag. Ater, Daud. *Ag. Subspinosa,* Gray, Ann. Phil. 1827. *Trapelus Subhispidus,* Kaup.

Brown, striped with a yellow dorsal line and red spot on each side; scales small; of occiput large, ovate, smooth.

Mutable Agama. *Ag. Ruderata*, Oliv. Voy. t. 29. *Trapelus Mutabilis*, Cuv. *T. Ægyptiacus*, Geof. Rept. Egypt. t. 5. *Ag. Orbicularis*, Daud. t. 45. *Ag. Deserta*, Licht. ?

Tail one-half as long again as the body ; scales of the head convex ; of the body unequal, irregular. Egypt. Mus. Brit. *Trap. Savignii*, Ardouin. Savigny Rept. Egypt. f. 3 and 4, scarcely appears to differ.

Ruppel's Agama. *Agama Sinaita*, Ruppel. t. 3. *Agama Straminea.* Mus. Berl.

Brownish grey, with paler spots ; dorsal scales unequal, slightly keeled, margined ; edge of ear one-spined ; a scaleless pit before the shoulder. Egypt. Perhaps same as former.

Sand Agama. *Agama Arenaria*, Van Heyden.

Yellowish grey, spotless ; dorsal scales equal, slightly keeled, not margined ; edge of ears one-spined ; but pit before the shoulder with scales. Egypt.

Aral Agama. *Ag. Aralensis*, Licht. *Lac. Sanguinolenta*, Pall. ?

Body dirty straw colour ; reddish, spotted ; scales equal, semicircular, keeled, ending in a spine ; tail black ringed.

Slender Agama. *Agama Agilis*, Oliver. Voy. Savigny Rept. Egypt. f. 5. *Agama Isodactyli.* Mus. Berl.

Scales small, equal, nearly smooth ; tail long, slender, black banded ; legs, especially the hinder, long and slender. Egypt

d. Head round depressed ; ears covered ; claws sharp ; back and tail with irregular scales. Phrynocephalus.

* *Margin of lips produced ; toes fringed ; claws long.*

Eared Phrynocephale. *Ph. Auritus*, Kaup. *Lac. Aurita*, and *Lac. Mystacea*, Pallas. *Lac. Lobata.* Shaw.

Brown with a small lobe on each side the neck.

** *Margin of lips simple, claws moderate.*

Pallas's Phrynocephale. *P. Caudevolvula. Lac. Caudevolvula*, Pallas. *Lac. Guttata*, Lepech. Reis, t. 22.

Scales of body very minute, smooth; tail long, smooth.

Eyed Phrynocephale. Ph. Ocellatus. Agama. Licht.

With unequal sized black-eyed white spots.

Ural Toad Lizard. P. Uralensis, Kaup. *Lac. Uralensis*, Lepech. Rees, t. 22. *Lac. Helioscopea*, Pallas?

Scales of back unequal. Buchara.

e. Head roundish; throat three pouched; sides with wing-like expansions supported by the ribs. Draco, Lin.

Short-Pouched Dragon. Draco Abbreviata, Gray, Zool. Jour. *Draco Fimbriatus*, Kuhl?

Scales large; thighs fringed behind; throat pouches short, central reaching to the thorax. Wings spotted beneath.

Banded Dragon. Draco 5 Fasciatus, Gray, Zool. Jour. *Draco Viridis*, and *Draco Fuscus*, Daud. t. 41?

Wings ash, with four continued blackish bands.

Lined Dragon. Draco Lineatus, Daud.

Back ash; wings brown, with numerous white lines; scales of tail large, blunt, and uniform.

Timor Dragon. Draco Timorensis, Kuhl.

Middle dorsal scales larger, keeled. Brownish; wings and body glaucous, and brown marbled; pouch black. Kuhl.

Black Pouched Dragon. Draco Hæmatopogon, Boie. Is peculiar for a black spot on each side of the throat pouch.

B. *Femoral Pores distinct.*

f. Head long; eyes large; tail round; back and tail with irregular imbricate scales. Gemmatophora.

Bearded Gemmatophore. Gem. Barbata. Agama, Cuv.

Back with cross bands of larger scales; scales of throat, sides, and back of ears elongated, bristle pointed. New Holland, Mus. Bell. and Col. Surgeons.

Muricated Gemmatophore. G. Muricata, Lac. *Muricata*, Shaw. Zool. *A. Grandoculis*, Lac. and *A. Jacksoniensis*, Kuhl.

Black brown, with a series of pale spots; back with longitudinal series of elevated scales. New Holland.

g. Head long, having a large pleated frill on each side of the neck behind the ears. Back and tail with irregular scales. Clamydosaurus.

Frilled Lizard. Clamydosaurus Kingii, Gray, Kings. Voy. Pale brown, scales keeled, of back limbs and tail larger, of belly small, and of sides smallest. Frill with large keeled scales on both sides. New Holland. Mus. Brit. Mr. Frazer.

h. Head squarish; back and tail compressed, crested. Lophura.

Amboina Lophura. Lophura Amboinensis. Lac. Javanica, Hornsted. *Lac. Lophura*, Shaw, Zool. t. 62.

Tail above compressed, fin-shaped.

Cuvier's Lophura. Lophura Cuvieri, Gray, Mus. Paris.

Brown green spotted; body with four or five oblique bands of blue-eyed spots; scales of body and tail small subequal; back and end of tail with a series of distant, short, compressed spines. Tail compressed, interrupted by distant rings, upper and lower edge two keeled. Cochin China.

Lesueur's Lophura. L. Lesueurii, Gray, Mus. Paris.

Dark brown, varied with pale netted lines; scales of head conical, with scattered acute conical tubercles on the head and neck ; back with a crest of short compressed spines, forming a double crest about two-thirds the length of the back, sides with eight or nine cross bands of oval larger keeled scales, and others similar across the base of the tail and limbs. Paramatta, Lesueur.

Beautiful Lophura. Lophura Concinna, Physignathus Concinnus, Cuv. R. A. t. f. *Ph. Iguanoides.* Mus. Paris.

A crest of conical compressed scales the whole length of the back ; head with granular scales. Body and tail with small squarish scales, the scales of the tail becoming larger near the end ; tail much compressed ; young with scarcely any crest. Dark blue, with some oblique white bands on the side, and whitish beneath, with some rather larger compressed scales on the side of the chin.

i. Head short, arched ; back, with minute scales, not crested; tail with whorled scales. Uromastyx.

* *Scales of tail large, spinose,* Uromastyx. Cuv. Mastigura. Flern.

Common Uromastyx. Uromastyx Spinipes, Merrem. *U. Acanthinurus*, Bell. Zool. Jour. *Stellio Spinipes*, Daud. Geoff. Egypt, t. 2.

Olive dull greenish brown, subcaudal segment, with subdentate scales, placed in two or three series. North Africa.

Two Coloured Uromastyx. Uromastyx Dispar.

Tail like the former, male, black brown ; female, ash yellow.

Ornamented Uromastyx. Uromastyx Ocellatus, Licht. *Uromastyx Ornatus*, Heyden, Ruppel, Trav.

Dull green, body brown, ringed subcaudal segment consists of a single series of armless scales. Africa. Dongola.

Black Uromastyx. *Uromastyx Niger*, Merrem. *Stellio Niger*, Daud.

Tail twice as long as the body, with twenty-seven rings of spines; toes long, sub-depressed; scales small, rhombic.

Hardwick's Uromastyx. *Uromastyx Hardwickii*, Gray, Zool. Jour. *U. Reticulatus*, Cuv. R.A.

Greenish, marbled with black dots; a large black spot in front of each thigh. Hindostan.

Cuvier indicates *Uromastyx Grisseus*, from New Holland,

** *Tail with rings of small armless spines. Lecolepis*, Cuv. *Cynosaura*, Schegel.

Bell's Uromastyx. *Uromastyx Bellii*, Gray, Zool. Jour. *Lecolepis Guttatus*, Cuv. Guerin. Icon, t. 7. *L. Guttato Lineatus*, Mus. Paris. *Cyn. Punctatus*, Schegel. Mus. Leyd.

Olive; back, with three lines alternating with black edged white spots; limbs white-eyed; sides black spotted.

Reeves Uromastyx. *Uromastyx Revesii*, Gray, Mus. Brit. Olive, with a series of bright red spots down each side.

Spotted Uromastyx. *Uromastyx Maculatus*, Gray. *Lecolepis Maculatus*, Cuv. Mus. Par.

Pale brown; head, with a black streak from the eye to the neck; back, legs and tail with six rows of oblong black spots, sides of the legs black netted.

Order II. Ophiosauri.

Mouth not dilatable, skin covered with regular equal scales.

Section I.

Body with equal similar scales, above and below, and a compressed line of small scales on each side; tongue short, contractile, two cut. Ptygopleura.

Gen. I. Zonurus.

Teeth placed on the side of the jaws; legs four, moderate; ears, exposed; vent medial.

1. *Toes 5, 5; femoral pores distinct; tail with spinose whorled scales.* Africa. Zonurus.

Common Zonurus. Lac. Cordylus, Linn. *Cordylus Griseus*, Seba, i. t. 48. f. 4.

Brown with a yellow line down the back; the scales nearly equal. *Cor. Niger*, Cuv., is from a badly preserved specimen, and *Cor. Dorsalis*, from a good coloured specimen. Cape.

Armed Zonurus. Zonurus Cataphractus. Bonn. Trans.

Head with large scales; scales of the body large and hard, of the sides three toothed; of the thighs, neck, and tail, ending in a trigonal spine.

Small scaled Zonurus. Zonurus Microlepidotus. Gray.

Dorsal scales about half the size of ventral, bluntly keeled.

2. *Toes 5, 5, femoral pores distinct; tail unarmed.* Cicigna, Gray, 1815. Gerrhosaurus, Wiegmann.

Common Cicigna. Cicigna Sepiformis. Scincus, Merrem. *Lac. Seps*, Lin. *Gerrhosaurus Flavigularis*, Weigmann.

Olive brown, beneath white, chin, throat, and upper lateral scales, yellow; fore, middle, and hinder toes longest.

Smooth Cicigna. Cordylus Lævigatus, Cuv.

Brown, with four black and yellow lines on each side the back, and two series of black and yellow spots; scales obliquely four sided. keeled. Perhaps the same as former.

Madagascar Cicigna. Cicigna Madagascariensis. Gray. Green, with yellow lines on each side the back; back and sides brown spotted; scales smooth; when young, scales keeled; variety. *β Ornata.* Back with five yellow and six black lines, and sides black, yellow spotted. Madagascar. Mus. Brit.

C. Toes 5, 5; femoral pores, none; tail with whorled, unarmed scales. America. Gerrhonotus, Wiedman.

Burnet's Gerrhonote. Gerrhonotus Burnettii, Gray.

Scales of back, and sides of body and tail, obliquely four sided, slightly keeled; dark brown; sides, with dark band and cross rows of white spots beneath.

Depp's Gerrhonote. Gerr. Deppii, Wiegm. Isis. xxi.

Dorsal scales four angular, smooth; olive black, irregularly white spotted, beneath white; tail round, white ringed. Mexico.

Retired Gerrhonote. Gerrhonotus Tæniatus. Wiegm.

Dorsal scales four angular, smooth; bluish, with black angular cross bands; tail black, ringed.

Blue Gerrhonote. Gerrhonotus Cæruleus, Wiegm.

Dorsal scales, four angular; olive bluish, with three series of black spots; tail round, hemeolate. Brazils.

Rough-necked Gerrhonote. Gerrhonotus Rudicollis, Wiegm.

Head rough, with elevated shields; scales four, angular; placed in cross series. Grey green, base of tail, four angular.

Imbricate Gerrhonote. Gerrhonotus Imbricatus, Wiegm.

Head shields elevated; dorsal scales keeled, four angular, olive grey, tail, hemiolate. Mexico. Brit. Mus.

Smooth-headed Gerrhenote. Gerr. Leiocephalus, Weigm. Head smooth, scales four angular, in cross series, of middle of back keeled, of nape and sides smooth; grey green, bluish beneath, with nine longitudinal lines. Mexico.

Toes 4, 4, femoral pores—Saurophis *Fitz.* Tetradactylus, *Merrem.*

New Holland Saurophis. Saurophis Lacepedii, Lac. Tetradactylus, Lacep. Ann. Mus. ij. t. 59. f. 2. *Tetradactylus Chalcidices,* Merrem.

Gen. II. Ophisaurus.

Teeth on the side of the jaw; legs rudimentary; ears distinct; lateral line distinct; vent medial.

a. Hinder extremities only rudimentary, on the side of the vent, undivided. Pseudopus, *Merrem.*

Pallas's Scheltopusik. Pseu. Serpentinus, Merrem. *Ps. Pallasii,* Cuv. *Lac. apoda,* Pall. Nov. Com. xix. t. 9.

Scales smooth; caudal scales square; front feet rudimentary, length two feet. Russia.

Durville's Pseudopus. Pseutopus Durvillii, Cuv.

Brown; dorsal and caudal scales keeled, rough. Brit. Mus.

b. Legs none. Ophisaurus, *Daud.* Hyalinus, *Merrem.*

Common Glass Snake. O. Ventralis, Daud. *Anguis,* Linn.

Yellow green, black spotted. Sometimes the spots form continued lines. *Oph. Striatalus* and *Oph. Punctatus,* Cuv. MSS. differ very little, if at all, from this species.

Gen. III. Chalides.

Teeth on the side of the jaws? Legs four, distinct; ears hid under the skin; vent medial, linear.

a. Toes 5, 5. Chalcides, *Fitz.*

Common Chalcides.

Ferruginous with six dorsal lines.

b. Toes 4, 5; *femoral pores distinct.* Heterodactylus.

Spix's Heterodactyle. Het. Imbricatus, Spix, Braz. t. 27.

Olive brown; sides of body and tail yellow-lined; tail long; central dorsal scales acutely keeled. Brazils.

c. Toes 4, 4. Brachypus, *Fitzinger.*

Common Brachypus. Bra. Abdominalis, Thunb.

Brown, with stripes on the belly.

d. Toes very small, rudimentary, 5 before, 3 behind. Chalchis and *Colobus*, Merrem. *Cophias*, Fitz.

Annulated Chalchides, Shaw. *Cophias Flavescens*, Gray. *Le Challide*, Lacep. t. 32. *Chamæsaura Cophias*, Schn.—*Chalchide Menodactylus*, Daud. *Chalchydes Flavescens*, Bonnat. *Chalchide Tridactylus*, Daud. *Colobus Daudini*, Merrem. Guiana.

Gen. IV. Amphisbœna.

Teeth in sockets; ears hid; vent roundish, subterminal.

a. Legs two, very small; subanal pores distinct Chirotes.

Worm-like Bimane. Lacerta Lumbricoides, Shaw. *Chamasaurus Propus*, Schn. Lacep. t. 41. Mexico.

b. Legs none, subanal pores distinct; thoracic rings regular. Amphisbœna, *Lin.*

White Amphisbœna. Am. Alba, Lin. Lacep. ij. t. 21. f. 1. Eyes small; occiput smooth. Perhaps same as next.

Sooty Amphisbœna. Amph. Fuliginosa, Linn. Seba, t. 18. f. 2. t. 100. f. 3. t. 83. f. 4.

Eyes small; occiput grooved; black, white-varied.

See *Amp. flavescens.* Pr. Max. t. *Amp. pathyura*, Wolf.—*Amp. punctata*, Bell.—*Amp. vermicularis*, Wagler.—*Ampt. Cinerea*, Vandelle. *Amp. cxyura*, Wagler, Braz. t. 25. f. 1. *Blanus*, Wagler.

Blind Amphisbæna. *Amph. Cæca*, Cuv.

Eyes none. Martinique.

c. Legs and subanal pores none ; thoracic rings irregular ; the plates united together in front. Leposternon, *Spix.*

Dotted Leposternon- *Leposternon Microcephalus*, Spix *Amph. Punctata*, Pr. Max.

Blue gray, with numerous black dots.

Sharp-nosed Leposternon. *Leposternon Oxyrhychus.* Dekay Siliman Jour. xiv. 907.

Yellowish ; back with three longitudinal grooves ; nose pointed, not mucronate.

11. *Body covered with imbricate scales.*

Gen. V. Scincus.

Body covered with uniform imbricate scales ; head long ; tongue short, contractile, two-cut.

A. Legs four ; ears distinct.

a. Toes 5, 5, *margined ; muzzle acute, produced.* Scincus.

Shop. Scinc. *Scincus Officinalis*, Linn. Egypt.

b. Toes 5, 5 ; *muzzle rounded ; body moderate ; scales very hard, bony ; tail broad, depressed.* Trachydosaurus.

Rugose. Trachydosaurus. Tra. Rugosa, Gray. King's Voy. *Scincus Pachyurus.* Peron MSS.

c. Toes 5, 5 ; *muzzle rounded ; body moderate ; scales thin, silvery ; tail conical* Tiliqua, *Gray.*

New Holland Tiliqua. *Tiliqua Whitii.* *Lacerta Scincoides*, Shaw. White Jour. at p. 242. *Cyclodus*, Wagler. *Sc. Tuberculatus*, Merrem.

Large ; pale ; scales large, whitish ; six series on the back of neck. New Holland. Brit. Mus. *Cyclodeus Flavigularis*, Wagler, t. 6, appears to differ little from this.

Black and yellow Tiliqua. Tiliqua Nigroluteus. Scincus, Quoy and Gaim. Frey. Voy. t. 41. New Holland.

New Holland Tiliqua. Tiliqua Crotaphomelas. New Holland.

Indian Tiliqua. Tiliqua carinatus, Schn. Gray, Zool, Jour. *Scincus rufescens*, Cuv. *S. multicarinatus, Sc. lineatus, and Sc. nigro fasciatus*, Kuhl. Mus. Par.

Greenish, with a yellow line along each side, scales, three keeled. India, Java, &c.

Cape Tiliqua. Tiliqua Capensis, Gray. *Sc. trivittatus*, Cuv.

Brown, with three longitudinal pale lines, on the back and tail, with black spot between the lines. Cape.

Three Streaked Tiliqua. T. trivittatus, Gray, Zool, Jour.

Pale brown, with three broad black edged yellow streaks on back and tail; sides, pale black spotted. India.

Ribbon Galley-Wasp. T. tæniolata, Lac. tæniolata, White, Jour. t. ap. p. 245. *Sc. octolineatus*, Daud. *Sc. decemlineatus*, Lacep. *Sc. undecemstriatus*, Kuhl.

Brown, back with ten or twelve white bands, with black sides, the two dorsal ones on each side uniting into one over the eyes, scales thin polished. New Holland.

Many scaled Galley-Wasp. Tiliqua multiscutatus Sc. Cuv. *Anolis parcé*, Geoff. Rept. Egypt, t. 4. f. 4.

Large Galley-Wasp. Tiliqua Cyprinus Sc. Cuv. *Anolis gigantesque*, Geoff. Rept. Egypt, t. 3. f. 3.

Scales smooth, tail longer than the body, brown, with a pale line on each side. Egypt.

Eyed Galley-Wasp. Tiliqua ocellatus Sc. Schn. *Sc. variegatus*, Cuv. Daud. t. 56. *Anolis marbré*, Geoff.

Rept. Egypt, t. 5. f. 1. Savigny, Rept. Egypt, t. 2. f. 2. *Sc. Tiligugus*, Merrem. Oliv. Voy. 1. t. 16.

Pale brown, with semilunar, black spot divided in the centre, by a short whitish band. South of Europe, and Egypt. See also *Sc. vittatus*, Oliver, Voy. t. 29. f. 1. Varied with a pale line on each side the back. Brit. Mus.

Common Galley-Wasp. Tiliqua occidua, Lacerta occidua, Shaw, Sloane, Jam. y. t. 273. f. 9. *Sc. fossar.* Merrem.

Large; pale brown; scales moderate, about twelve series of scales on the back of the neck. Brit. Mus. Jamaica.

Lacepedes Galley-Wasp. Sc. Mabouya, Sh. Lacep. t. 24.

Smooth, greenish brown, back black dotted, and a brown band from the temple over the shoulder.

Doubled Galley-Wasp. Tiliqua bistriatus Sc. Spix, t. 26. f. 1.

Coppery brown with a broad black streak from the nostrils to the nose, edged with white on each side. Brazils.

Five lined Galley-Wasp. Tiliqua quinquelineatus, Lacerta quinquelineatus, Lin.

Greenish blue, with five dorsal lines, the central one forked over the head. Tail reproduced, blue. *Lacerta fasciata*, Lin. Catesby, Car. t. 67.

Seven Striped Galley-Wasp. Tiliqua homolocephalus, Wiegmann.

Olive brown, with seven black longitudinal streaks.

Black spotted Galley-Wasp. Tiliqua nigro-punctatus Sc. Spix, t. 26. f. 2.

Black brown, golden, with an unconspicuous black band along each side to the thigh, back, black punctate, tail short.

Coppery Galley-Wasp, Tiliqua ænea, Gray.

Golden green; back with five lines of black spots, sides from the eyes, with a dark brown band, beneath pale, sides of belly black spotted. Brazils. Brit. Mus.

Sloan's Galley-Wasp. Tiliqua Sloanii, Sc. Daud. t. 55. f. 2. Pr. Max. x. 11. t. 1. f. 1.

Golden green, with four longitudinal black streaks extended over the base of the tail, paler beneath, scales thin, smooth, Toes short thick. Brit. Mus.

Red-headed Galley Wasp. Ti. Erythrocephala. Sc. Erythrocephalus, Gilliams, Jour. Acad. N.S. Phil. i.

Reddish brown cupreous, beneath whitish, head above red.

Two Coloured Galley Wasp. Ti. Bicolor. Sc. Harlan. Acad. N. S. Phil. iv. t. 18. f. 1.

Dusky brown, darkest on the head, beneath silvery, two longitudinal whitish lines on each side the body; tail tapering with two obsolete lines on the hinder part of the thighs.

Lateral Galley Wasp. Tiliqua Lateralis. Scincus, Say *Sc. unicolor*, Harlan.

Light brown, with blackish lateral lines, beneath greenish white; a cross row of scales behind the plates larger than the remaining cervical scales.

Bell's Galley Wasp. Tiliqua Bellii, Gray.

Pale brown, with irregular cross bands of small white scales, with central black bands, sides varied with dark brown, scales small, smooth, about four series on back of neck.

Double-streaked Galley Wasp. Tiliqua bistriatus, Gray.

Brown, sides with two narrow streaks, back with five series of black spots, with a central yellow triangle in each.

Thin Scaled Galley Wasp. *Tiliqua Tenuis*, Gray.

Pale brown, back with irregular black spots, forming a band on each side the back, scales thin polished.

Fine Scaled Galley Wasp. *Tiliqua Microlepis*, Gray.

Pale, with numerous irregular transverse brown bands ; head very depressed, forehead flat, eyes large, tail slender, toes very unequal. Scales very thin, closely adpressed keeled.

Banded Galley Wasp. *Tiliqua Fasciatus.*

Silvery blue, head with three, neck with one, back and tail with six broad black bands, toes, hinder rather unequal, five last in Brit. Mus.

Spix has placed his genus *Lepidosoma* near the Scinks. In one of my notes made on the animal at Paris I have remarked that it is nearly allied to *Polychrus*, but has no pores or keeled scales.

d. Toes 4, 5, *rest like Tiliqua, but no eyelids.* Gymnotholamus, *Merrem.*

Four-lined Galley Wasp. *Gym. quadrilineatus*, Merren. *Lacerta lineata*, Lin. Seba, y. t. 41. f. 6. *Lac. quadrilineatus,* Lin. Pr. Max. Beytr. xiii. f. 2. *Sc. Cyanurus*, Sching.

Blackish, with four yellowish dorsal bands.

e. Toes 5, 5, *muzzle rounded, body very long slender, feet small, far apart, ears distinct.* Lygosoma, Gray.

Short-footed Lygosoma. Scincus Brachypus. Schn. Geoff. Rept. Egypt, t. f. 9, 10. *Spœnops Capistrata,* Wagler.

Hinder toes unequal, long ; pale brown, with narrow brown longitudinal band, and a black streak through each eye.

Dotted Lygosoma. Lygosoma Punctata, Gray.

Hind toes unequal, long ; pale brown, with many series of minute dark spots ; head brown ; tail long, dotted.

Scincoid Lygosoma. *Lygosoma Scincoides, Seps.* Cuv.

Ears large; hind toes unequal, long. Pale brown, with black slender bands formed of spots on the centre of each scale. Mus. Paris.

Lygosoma Serpens, Lacerta Serpens, Gmel. Blo. Naturfosh. y. t. 2.

Hinder toes subequal short.

Golden Ligosoma. *Ly. Aurata,* Gray. *L. Serpens,* Gray.

Golden brown, with many longitudinal black lines, one between each series of scales; hind toes unequal, rather short.

f. Toes 4, 4, *or* 3, 3; *body very long and slender; feet far apart: ears distinct.* Seps.

Four-toed Seps. *Seps Peronii,* Fitz.

Toes 4, 4, hinder unequal; golden brown, with streaks; sides brown and white spotted; beneath white, with cross black spots.

Peron's Seps. *Tridactylus Decresiensis,* Peron.

Toes, 3. 3; hinder unequal, one short and two long, subequal; pale brown, with long dark lines; beneath netted.

Common Seps. *Seps Chalcides Zygnis,* Fitz.

Grey, with four longitudinal lines on the back. Brit. Mus.

Striated Seps. *Seps Striata.* *Zygnis Striata,* Fitz.

Grey, with eight or nine equal spaced dorsal lines.

g. Toes 3, 3; *body long; feet far apart, slender; ears* 0; *femoral pores* 0. Saiphos, *Gray.* Cophias, *Fitz.*

Lacertine Saiphos. *Saiphos Equalis,* Gray. *Seps Equalis,* Gray Ann. Phil. *Anguis Lacertina,* E. W. Gray, MSS.

Grey; scales of the head equal; toes short, subequal. Brit. Mus.

h. Feet four, undivided, oblong, scaly; scales of body and tail keeled, pointed subverticulate. Monodactylus, *Merrem.* Chamæ Saura, *Fitzinger.*

Anguine Monodactyle. Monodactylus Anguinus. Lacerta Anguinus, Linn. *Lac. Monodactyle*, Lacep. Ann. Mus. ij. t. 49. f. 1. The scales of the back in straight, and those of the belly in oblique whorles.

Gen. VI. Bipes.

Feet two, posterior imperfect; scales uniform, imbricate, the ventral rather the largest; head long; tongue short, contractile, two-cut.

a. Feet undivided, oblong; scales smooth; femoral pores none; ears o; Ophoides, *Wagler.* Pygopus, *Spix.*

Brazilian Bipes. Ophiodes Striatus, Wagler. *Pygopus Cariococca*, Spix, t. 28. f. 1., adult. *Pygopus Striatus*, Spix, t. 28. f. 2. Young. *Seps fragilis*, Raddi. *Pseudopus Olfersii*, Licht. Brazils. Mus. Brit. *Bipes Lineata*, Cuv. Mus. Paris. The lines become double near the head.

Wagler cites another species, *O. Gronovii*, Wagler. *Pygopus Striatus*, Fitzinger. *Pygodactylus Gronovii*, Merrem.; but I have not seen it.

b. Feet undivided, oblong; scales smooth; femoral pores none; ears distinct. Delma, *Gray.*

Frasers Bipes. Bipes Fraseri, Gray. Brit. Mus.

Tail one-third longer than the body; black; beneath paler; head with four narrow pale cross bands. New Holland.

c. Feet undivided, oblong; scales of back keeled; femoral pores distinct. Pygopus, *Merrem.*

New Holland Bipes. Pygopus Lepidopus, Merrem. *Bipes*

Lepidopus, Lacep. Ann. Mus. iv. t. 65. Tail when perfect, longer than the body.

d. Feet with two unequal toes ; scales keeled ; femoral pores none. Bipes, Merrem. *Scelotes*, Fitz. *Pygodactylus*, Merrem. *Zignis*, Wagler.

Cape Bipes. Bipes Anguinus, Merrem. *Anguis Bipes*, Linn. *Lacerta Bipes*, Gmel. Seba, j. t. 85. f. 3.

Golden brown ; back with ten longitudinal black lines, of which the four lateral are the most distinct. The feet are considered as the generative organs by Seba. *Seps. Sex lineata*. Harlan. Jour. Acad. N. S. Philad. iv. t. 10. f. 1., according to Cuvier is a variety.

Gen. VII. ANGUIS.

Legs o. ; body and tail covered with smooth imbricate scales ; femoral pores none.

a. Ears distinct. Siguana.

Otto's Blind-worm. Siguana Ottonis, Gray.

Brown, with darker lines ; beneath paler ; tail rather longer than the body. Breslau. Discovered by Dr. Otto.

b. Ears hidden by the skin. Anguis.

Common Blind-worm. Anguis Fragilis, Lin. *Anguis Eryx*, Daud., the young. *Anguis clivica*, Wolf. *A. Lineata*, Laur.

Silvery grey, when young with a black dorsal line and black sides. Europe.

American Blind-worm. Anguis Eryx, Linn., said to have four nostrils, I have not been able to see them.

Gen. VIII. TORTRIX.

Body long, cylindrical ; back covered with imbricate scales ;

beneath with a series of larger scales; tongue short, contractile, two-cut; head long.

a. Tail blunt, with a single series of larger scales beneath; Anilius *Oken.* Tortrix *Oppel.*, not *Linn.* Ilysia, *Hemprick.* Torquatrix, *Haworth.* Cylindrophis, *Wagler.*

Scytale Coral Snake. Anilius Scytale. Anguis, Linn. Mus. Adolph. t. 6. Daud. vii. t. 87. Seba, ji. t. 20.

With irregular black and white rings. S. America.

Common Coral Snake. Anilius Corallinus. Anguis, *Laur.* Seba, ij. t. 73. f. 1. 2, 3. S. America.

Black Coral Snake. Anilius Rufus. Anguis Ater, Cuv. Seba, ij. t. 25. f. 1. t. 7. f. 3. *Anguis Rufa,* Laur. *A. Striatus,* Gmel. *Eryx Rufus,* Daud. *Tortrix Rufa,* Merr. *Illysia Rufa,* Mempr. *Cylindrophis Resplendens,* Wagler. Russel Ind. Serp. ij. t. 27. t. 28.

Copper colour, with white spot on tail.

Spotted Coral Snake. Anilius Maculatus. Anguis, Linn. Mus. Adolph. t. 21. f. 3. and *A. Tessellata,* Laur. Seba, j. t. 53. f. 7. ? ij. t. 100. f. 2 Russell, ij. t. 89.

Yellow, with black dorsal band.

Dotted Coral Snake. An. Melanostictus. Anguis, Schn. *Eryx,* Daud. *Tortrix Punctatus,* Cuv., from Seba, ij. t. 2. f. 1, 4. Ind. Serp j. t. 42.

Pale, yellowish punctated, with different coloured spots.

Black-eared Coral Snake. Anilius Melanotis. Cylindrophis, Wagler. *Tortrix,* Remw. MSS.

Like L. Resplendens; but tip of tail black; occipital band red; forehead black, and bands of body nearer together.

Indian Coral Snake, Anilius Latta; Tortrix, Cuv. Seba, ij. t. 30. f. 3. Russel. t. 46.

b. Tail obliquely truncated, with a double series of larger scales beneath; muzzle pointed. Uropeltis, *Cuv.*

Cuvier has indicated, but not described. two species of this genus.

Gen. IX. Acontias.

Body sub-cylindrical, covered with uniform imbricate scales; tongue ——— ? head short, blunt, masked; ears hid; tail short.

Cape Pintado Snake. Acontias Meleagris, Merrem. *Anguis,* Linn. *Eryx.* Daud.

Eyes distinct; back with eight lines of black spots. Cape.

Netted Pintado Snake. Acontias Reticulata, Merrem. Eyes distinct; scales rough.

Blind Pintado Snake. Acontias Cæcus, Cuv.

Eyes covered. Africa.

Gen. X. Typhlops.

Body sub-cylindrical, covered with uniform imbricate scales; tongue long, forked; muzzle produced; vent subterminal.

a. Head same size as body, blunt.

Bramin Typhlops. Typhlops Braminus, Cuv. from Russel. t. 43. *Eryx,* Daud., and *Tortrix Russelli,* Merrem.

b. Muzzle depressed and blunt, with many plates in front, Stenostoma, *Spix.*

Netted Typhlops. Typhlops Reticulatus, Merrem. *Anguis.* Cuv. Sch. Phys. Scer. t. 757. f. 4.

Ash Typhlops. Ty. Cinereus, Merrem.

Ash coloured.

Banded Typhlops. Ty. Fasciatus, Merrem.

Body equal, tail ending in a spine.

Yellow Typhlops. Ty. Crocotatus. Anguis, Schn. *T. Rostralis*, Merrem.

Body depressed, and then compressed, tail acute.

White-bellied Typhlops. Typhlops Leucogaster, Prin. Max. Beytr. j. 495.

Olive brown, beneath white.

Cuvier here also indicates, by name, two undescribed species. *T. Undecimstriatus* and *T. Leucorrhous*.

c. Muzzle covered with a single sharp edged plate in front. Typhlops.

Worm-like Typhlops. Ty. Lumbricalis. T. Vermicularis, Merrem. *Le Lombric. Anguis*, Lacep. ij. t. 20. f. 1. Seba, t. 86. f. 2.

Tail conical, blunt.

White-fronted Typhlops. Typhlops Albifrons. Stenosoma Albifrons, Spix, Braz. t. 25. f. 3.

Forehead white.

d. Muzzle ending in a conical point ; eyes quite hid ; tip of tail enveloped in an oval horny shield. Rhinophis, *Wagler.*

Phillipine Typhlops. Typhlops Phillipinus, Cuv.

Blackish ; length eight inches. The Phillipines.

Sharp-nosed Typhlops. Ty. Oxyrhynchus, Merrem.

e. Muzzle covered with a single large convex shield ; eyes none. Typhlina, *Wagler.*

Seven-streaked Typhlops. Typhlops Septemstriatus, Merr.

Tail ending in a spine.

Lined Typhlops. Typhlops Lineatus Acontias, Reimw.

Order OPHIDII. *Serpents.**

The jaws very dilatile; the mastoïd bones free from the skull, and the branches of the lower jaws free in front; legs o, body with a row of larger scales beneath.

I. *Upper jaw toothless, fangs large, distinct vertebral plates broad; tail short, conical.* Venemous.

Gen. 1. CROTALUS. Rattle-snakes.

Head broadly triangular, with a large pit before the eyes.

A. Tail with a rattle, head covered with scales like the back. Crotalus. New World.

Common Rattle Snake. Crotalus Horridus, Lin. Catesby, ij. t. 41. Crot. Catesbæi. Hempr.——Lozenge spotted Rattle-snake. Crotalus Durissus. Cr. Horridus. Pr. Max. Ch. Rhombifer, Boie. Cr.· Cascavella, Spix, t. 24.——add Cr. Confluentus, as North America, and Boie has indicated, in the Leyden Mus. C. Atricaudatus, and C. Drijinus; and Humboldt in his Zoological observation describes Crotalus Loeflingii. Crot. Strepitans, Daud. Seba, ij. t. 96. f. 2, is said to be Boa Canina; and Crot. triseriatus, Weigmann. Is referred to Mexico.

B. Tail with a rattle, head with shields. Crotalophorus, *Gray.* Caudisona, *Fitz.*

Miliar Rattle-snake. Crotalophorus Miliaris. Crotalus, Lin. Catesby, ij. t. 42. Mus Brit. See also Crot. Tergeminus, Say. North America.

* It has been found absolutely necessary to omit the specific characters of the ensuing divisions of this class ED.

The Catalogue on the Species, is greatly dependant on the one published by Boie, in the Iris, Vol. xx. The species are separated by a short dash, thus ——. J. E. G.

C. Tail ending in a spine; subcaudal plates simple, one rowed, head with shields behind the eyes. Tisiphone *Fitz.* Cenchris. *Daud.*—Ancistodan, *Pallas.*

Catesby's Tisiphone. Tisiphone Shausii, Gray. Coluber Tisiphone, Shaw. Catesby Car. ij. t. 45. Peleas Niger fusens, Merrem. Cenchris Mockeson, Daud, t. 60, and t. 70. f. 3, 4.—See also Tisiphone Cupoea, Fitz. South America.

D. Tail ending in a spine; subcaudal plates double and single; head with scales. Lachesis.

L. Rhombeata. Pr. Max. Crotalus Mutus, Lin. Boa Mutus, Lacep. Scytale Catenata, Licht. Lachesis Mutus, Daud. Cophias Crotalinus, Merrem. Curuacea, Margr. Bothrops Suruacea, Wagler. Braz. t. 23.

E. Tail simple; sub-caudal plates double; head scaly. Cophias, *Merrem.* Crassedocephalus, *Fitzenger.* Bothrops, *Spix.* Trigenocephalus, *Oppel.* Alecto, Megæra, Atropos, and Tropidolæmus, *Wagler.*

American.

Cophias Lanceolatus. Coluber, Lacep. ij. t. 5. f. 1. Vipera, Daud. Col. Megæra, Shaw. Brazils.——Cophias Mægera. Bothrops, Spix, t. 19.——Cophias Furia, Boie. Bothrops, Spix, t. 20.——Cophias Leucostigma. Bothrops, Spix, t. 21. f. 1.——Cophias Tessalatus. Bothrops, Spix, t. 21. f. 2.——Cophias Tæniatus. Bothrops, Spix, t. 21. f. 3.——Cophias Triangulum, Boie. Mus. Leyd. Vip. Brasiliana, Lacep.——Cophias Jacaraca, Pr. Max. adult and jun. Col. Atrox, Pr. Max. not Merrem.——Cophias Atrox, Merrem. Col. Atrox, Linn. Mus. Adolph. j. t. 23. f. 9. Vip. Atrox, Laur. Col. Ambiguum, Weigl. Vip. Weigelii, Daud. Vip. Tigrina, Daud.——Vipera Tigrina, Licht. Cophias Neuweidii. C. Bothrops, Spix, t. 21. f. 1.——

Cophias Leucurus, Boie, Bothrops, Spix, t. 22. f. 2. appears to be the young. Cophias Bilineatus Pr. Max. Vipera Chloris, Gravenhorst.

Asiatic,

Cophias Trigonocephalus, Merrem. Col. Capite Triangularis, Lacep. t. 5. f. 2. Col. Trigonocephalus, Dau. Vip. Trigonocephalus, Daud. Sch. P. S. t. 749. f. 11. Inhab. Isle of St. Eustach. Asia. Boie ——Cophias Viridis, Merrem. Trimesurus Viridis, Lacep. Ann. Mus. iv. t. 56. f. 2. Coluber gramineus, Shaw. Brodro. Pam. Russel, t. 9 Green, with a narrow yellowish line on each side. Coph. Punicea, Boie. Atropos, Wagler. Java.——Cophias Sumatrana, n. Col. Sumatranous, Raffles. Coph. Wagleri, Boie. Seba, ij. t. 68. f. 4. Tropidolæmus, Wagleri.—— Cophias Russelii, n. Russel, ij. t. 22. Inhab. Phillipine Islands. Mus. Paris.

F. Tail, end simple ; subcaudal plates double ; head shielded Trigonocephalus, *Fitz.*

Old World.

Trigonocephalus Rhodostoma, Reinw. Vipera Pretextata, Gray. India.——Trigon. Hypnale, Boie. Cophias Hypnale, Merrem. Col. Lebetinus, Linn. Vipera Lebetina, Daud. Inhab. North Africa and Southern Europe and Asia.—— Trigonocephalus Haly., Boie. Col. Halys, Pall. Vipera Halys, Daud. Vipera Aspis, B. Pallasii, Merrem. Lichten, Reese. Siberia.——Trigonocephalus Blomhoffi, Boie. Iris. 1826, 214. Bul. Sci. Nat. x. 151. Japan.——Trigonocephalus Orophyas, Oppel. Boie.——Trigonocephalus Boeii, Boie. Iris. xx. 561. Inhab. Levant. Olivier.

American.

Trigonocephalus Cacodæmon, Boie. Col. Cocadæmon.

80

Shaw. Pelais Niger, Merrem. 2 Catesby, Car. t. 44. Scytale Niger, Daud.——Trigonocephalus Boiei. Iris, xx. 561. Inhab. Carolina.

Gen. II. VIPERA.

Head broad, without any pit; covered with scales or small shields.

A. Head covered with scales like the back; subcaudal plates double; nostrils large. Vipera. Echidna, *Merrem.* (not Geoffrey.) Cerastes, *Wagler!*

Vipera Nasicornis, Daud. Coluber, Shaw. Nat. Misc. t. 94. Zool. t. 104. Brit. Mus.——Vipera Russelii. Coluber, Shaw. Zool. t. 108. Russel Ind. Serp t. 7. Col. Trinoculus, Schn. Vipera Elegans, Daud. Col. Triseriatus, Herm. Bengal. Brit. Mus.——Vipera Cerastes; male with horn before the eyes. Col. Cerastes, Hasselq. Ellis Phil Trans 4. vi. t. 14. Shaw Zool. t. 103. Lacep. ij. t. 1. f. 9. Daud, vi. t. 47. f. 1. Col. Cornutus, Hasselq; female hornless. Col. Vipera, Hasselq. Aspis Cleopatra, Laur. Col. Egypticus, Lacep. Vipera Ægytia, Latr. Vipera Ægyptica, Daud. North Africa. Brit. Mus.——Vipera Lophophris, Cuv. Vipera Flava, Merrem.? Paterson, Africa. t. 15.——Vipera Arietans, Echidna, Merrem. Vipera Inflata, Burchel. Seba, ij. t. 30. f. 1. Col. Intumescens, Donnd. Vipera Severa, Daud. Col. Dubius, Gmel. Col. Hebriacus, Lacep. Col. Bitis, Bonnat. Vip. Brachyura, Cuv.——Vipera Daboia, Daud. Daboia Lacep. t. 13. f. 2. Col. Braziliensis, Lacep. t. 4. f. 2.——Vipera Ocellata, Latr. Aspic, Lacep. ij. t. 2. f. 1.——Vipera Atropos. Coluber, Linn. Mus. Adolph. t. 13. f. 1.

B. Head covered with granular scales; sub-caudal plates double; nostrils moderate, Berus.

Berus Vulgaris. Col. Berus, Lin. Var. Col. Aspis, Linn. Sturm Fauna, t.——Berus Ammodytes. Col. Ammodytes, Jacq. Coll. iv. t. 24, 25. Vip. Illyrica, Ald. Sturm Fauna, t.

C. Head covered with scales, with three larger scales on the forehead; sub-caudal plates double; nostrils moderate. Pelias, *Merrem.*

Pelias Chersea. Col. Berus, Laur. and Daud. Col. Chersia, Linn. Sturm. Fauna, Pelias Berus, Merrem. Vipera Berus, Fitz. Var. black. Col. Prester, Linn. Sturm Fauna.

D. Head covered with small scales; sub-caudal plates simple. Echis, *Merrem.* Scytale, *Daud.* Pseudoboa, *Schn.*

Echis Carinata, Merrem. Pseudoboa Carinata, Schn. Scytale Bizonata, Daud. t. 70. Boa Horatta, Shaw. Russel, t. 2. India.——Echis Krait, Merrem. Boa Krait, Williams. Pseudoboa, Schn. Scytale, Daud. India.——Echis Arenicola, Boie. Geoff. Rept. Egypt, t. 7. f. 1. Savigny, t. 4. f. 1—4. North Africa.

E. Head covered with scales in front; tail ending in a hook; sub-caudal plates simple and double. Acanthopis, *Daud.* Ophrias, *Merrem.*

Acanthopis Cerastinus, Daud. v. t. 77. Merrem. Beytr. ij. t. 9. Boa Palpebrosa, Shaw. Ophria Acanthropis, Merrem. Boa Aculeata, E. W. Gray MSS. 1796. Mus. Brit. New Holland.——Acanthopis Brownii, Leach Zool. Misc. j. t. 3. Boa Antartica, Shaw. New Holland.

Gen. 3. NAIA.

Head moderate, without any pit, covered with large regular shields.

A. Head broad; neck not dilatile; sub-caudal plates double. Sepedon, *Merrem.* Ophis, *Spix.* Naia, *Boie.* Hæmacathus, *Flem.* Causus, *Wagler.*

Sepedon Rhombeata, Licht. Col. Vip. Nigrum, Licht. Scheush. Ph. Sac. iv. t. 717. Naia Rhombeata, Boie. Mus. Leyd.——Sepedon Hœmachates, Merrem. Hœmachate, Lacep. L. O. ij. t. 3. f. 2. Vipera Hœmachates, Seba, ij. t. 58. f. 1, 3. Naie Hœmachate, Boie.——Sepedon Porphyraceus. Coluber Shaw, Zool. iij. 423.——Sepedon Merremii. Ophis Merremi Spix, Braz. t. 17.

B. Head narrow; neck dilatile; dorsal scales linear; tail conical; sub-caudal plates two-rowed. Naja, *Daud.* Uræus and Aspis, *Wagler.*

Naia Tricuspidans, Merrem. Col. Naja. Latr. Russel, I. S. j. t. 5 and 6. ij. t. 1. Daud iv. t. 60. Lacep. ij. t. 3. f. 1. Has been divided into two several species by Laurente.——Naja Haje, Merrem. Col. Haje, Hasselq. Vipera Haje. Daud. Geoff. Rept. Egypt, t. 7. Savigny, t. 3. Africa.——Naja Nivea, Cuv. Boïe. Isis, xx. 557. Vipera Flava, Merrem. Cape of Good Hope.——Naja Spectatrix. Reinw. Boie. Iris, xx. 557. Elaps Fuscus, Merrem. Col. Castaneus, Oppel. Russel. ij. t. 36.——Naie Latratus. Boie. Col. Latratus, Kuhl. Beytr. 69.

C. Head narrow; neck not dilatile; dorsal scales equal; tail conical, moderate; sub-caudal plate two-rowed Elaps.

Elaps Anguiformis, Schn.——Elaps Lacteus. Col. Domicella, Lin. Col. Domicellarum, Lac. Col. Lacteus, Lin. Mus. Adolph. t. 17. f. 1. Seba, ij. t. 35. f. 2. 54. f. 1.——Elaps Lemniscatus, Cuv. Col. Lemniscatus, Lin. Seba, j. t. 10 ij. t. 76. f. 2.——Elaps Surinamensis, Cuv.

Seba, ij. t. 6. f. 2, and t. 86. f. 1.——Col. Surinamensis, Merrem, n. 184. Inhab. Surinam. Mus. Leyd. (Perhaps var. of former.)——Elaps Latonius. Col. Latonius, Merr. j. t. 2. Seba, 3. ij. t. 34. f. 4. t. 43. f. 3. Col. Lubricus, Merrem. Cape of Good Hope.——Elaps Flavius. Col. Flavius.——Elaps Micurus. Micrurus Spixii, Wagler, Spix, t. 18.——Elaps Ibibaboca, Merr. Inhab. Brazils.——Elaps Corallinus, Merrem, 10. Pr. Max. ——Elaps Corallinus Merrem, 61. Vipera Psyches, Daud. t. 100. f. 1. Inhab. Brazils. Mus. Leyd.——Elaps Coccineus, Merrem.——E. Collaris, Boie. n. 5.——Elaps Furcatus, Schneider. Russel, ij. t. 19. Java——Elaps Bivirgatus, Kuhl. Boie. Isis, xx. 556. Java.——Elaps Boie nob. Elaphoides Fusca, Boie. Isis, xx. Java.—Elaps Chalybeum, Latostoma, Wagler.

D. Head indistinct; body cylindrical; dorsal scales equal; tail compressed, two-edged; caudal plates two-rowed. Platurus, *Latr.*

Platurus Laticaudatus. Col. Laticaudatus, Lin. Hydrus Colubrinus, Shaw. Platurus Faciatus, Boie. Daud. viii. t. 85. Lin. Mus. Alolph. j. t. 16. f. 1.——Platurus Semifaciatus, Reinw. Boie. (Mus. Leyden,) adult. Laticauda Scutale, Laur. Bechst. ij. t. 20. f. 1. Shaw, Zool. ij. t. 233.

E. Head distinct; body fusiform; dorsal scales ——; tail conical; caudal plates partly double, partly simple. Trimesurus, *Lacep.*

Trimesurus Leptocephalus, Lacep. Ann. Mus. iv. t. 56. f. 1.

F. Head distinct; body fusiform; dorsal scales ——; tail conical; caudal plates all simple. Oplocephalus, *Cuvier.*

Oplocephalus Cuvieri, nob. Mus. Paris.

II. Upper jaw toothed, fangs none, or small.

Gen. 3. COLUBER.

Ventral shields broad, anal spurs none, tail, usually conical, and elongate.

A. Body subcylindrical, scales equal, subcaudal plates, double or single, head moderate, above shielded.

a. Head regularly shielded, nose acute, recurved. Heterodon.

Coluber constrictor, Lin. (Ed. 10.) Boa contortrix, Lin. (Ed. 12.) Col. Heterodon, Daud. Heterodon platyrhincus, Latr. Boa porcaria, Lacep. Catesby, 7. t. 56.——Heterodon simus. Col. simus, Lin. Merrem. 96. Col. borealis, Schoeff. ——Heterodon nasua, Vipera nasua, Oppel. Rhinostoma nasua, Wagler. Rhinostoma proboscideum, Fitz.

b. Head regularly shielded, nose rounded, Coluber.

A. Head distinct, oblong, ovate before, slightly truncated, gape, very wide, body long, cylindrical, tail half as long as the body, continued, acute, scales imbricate lanceolate, mostly keeled, placed in longitudinal series, abdominal shields, simple, arched. Tropidonotus, *Kuhl.*

European.

Trepidomurus natrix, Col. natrix, Lin. Col. murorum. Col. mur. Fitz.——Col. Oppelii, Dumeril. C. Murorum, Mus.——Col. Viperinus, Daud. Latr. Col. pseudo echidna, Herm.——Col. hybridus, Merrem.——Tr. scaber, Boie. Col. scaber, Lin. Mus. Adolph. t. 10, f. 1. Dacypeltis scaber, Wagler.

Asiatic.

Tr. melanozostus, Gravenh. Iris, 1826, 206.——Tr. funebris, Oppel.——Tr. chrysargus, Kuhl.——Trop. sub-

85

miniatus. Col. subminiatûs, Reinw. n. 5.——Col. stolatus, Lin. Mus. Adolph. t. 22. f. 1. Coronella cervina, Laur. Col. malpolon, Lacep. C. cervinus. Gm. Vipera stolata, Lat. Col. sibitans, Lat.——Tr. cyanocephalus, Boie, n. 5.——Col. palustris, Mer. 121. Hydrus palustris, Schn. Col. Braminus, Daud. Russel, t. 20.——Tr. spilogaster, Boie, n. 5.—— Tr. rhodomelas, Kuhl. and V. H. n. 5.——Tr. trianguliferus. Col. trianguligerus, Reinw. ——Tr. Mortuarius, Kuhl. Russel, t. 28.——Tropidonotus hypostictus, Boie, n. 5.——Tropidonotus piscator, Mer. Hydrus piscator, Seb. Enchydus piscator, Latr. Col. anastomosatus, Daud. Russel, t. 33. Tropidonotus dora. Col. dora, Merrem, 104. Russel.——Tropidonotus, t. 5, lugubris, Oppel. ——Tropidonotus lacrimans, Oppel.——Tropidonotus vibikari, Boie, Iris, 1826. 207.——Tropidonotus tigrinus, Boie, Do. 206.

American.

Col. variabilis, Pr. Max.——Tropodontus tesselata. Coronella tesselata. Laur. Col. viperinus, Daud. Col. Hydrus, Pallas. Natrix cherscoides. Var. bivittatus and Natrix occellata, Wag.——Trop. æstirus, Boie. Col. æstirus, Lin. Catesb. Col. subrirides, Lac.——Tr. poecilostoma. Col. poecilostoma, Pr. Max. Natrix sculptura, Wag.——Col. saurita, Lin. Ribon Snake, Cates. Daud. vi. t. 81. f. 2.——Tr. vittatus. Col. vittatus, Lin. Mus. Adolph. t. 18. f. 2. Scheuchz, t. 66. f. 8. Seba, t. 15. f. 3.——Col. fasciatus, Lin. Wampum Snake, Cates. t. 58. Dirty yellow with two white lines down the back, and cross black bars.——Tr. Nattereri, Pr. Max.——Tr. parietalis. Col. parietalis, Say, Isis, 1824, 255. ——Tr. proximus. Col. proximus, Say, Isis, 1824.——Tr. ordinatus. Col. ordinatus, Lin.——Tr. dimidiatus, Boie. ——Tr. porcatus, Boie. Col. porcatus, Latr. Cates. Col. erythrogaster, Sh.

B. Head distinct; gape very small; tail half the length of the body; scales placed in longitudinal series, mostly smooth, abdominal shields angularly recurved. Coluber.

American.

Col. punctatus, Lin.——Col. cyaneus, Lin. Col. viridicæruleus, Lac. Seba, 17. t. 48. f. 2.——Col. azurus, Lac. ——Col. caninana, Merr. Seba, ij. t. 20. f. 1. &c. C. pullatus, Lin. Mus. Adolph. Col. peruvianus. Sh. Col. plutonicus, Daud. C. humanus and C. coronatus, Gm. Spiletes pullatus, Wag. Col. Nova Hispaniæ, Gm. Col. leucomelas, Gm. Col. variabilis, Pr. Max.——Col. Lichtenstenii, Pr. Max. Wag. Col. capistratus, Lich.——Col. bifossatus, Radd.——Col. pantherinus, Merr.——Col. scurrula, Wag. Natrix scurrula, Wag. Braz. t. 8.——Col. pileatus, Pr. Max. i. 344. Col. olfersii, Hemp. Col. olivaceus, Olfers. Philodryas olfersii, Wag.——Col. plumbeus, Pr. Max.—— Col. triangulum, Lac.——Col. corais, Daud. Boie. Isis, 1826, 538.——Col. getulus, Daud.——Col. bahaiensis, Spix. Braz. t. 10. f. 1.——Col. variabilis, Kuhl.

Africa.

Col. Hippocrepis, Lin. Col. canus, Lin. Periops hippocrepis, Wag. Lin. Mus. Adolph. —— Col. barbarus, Boie. ——Col. Geoffroyii. Coulêvre aux raies parelleles, Geoff. Rept. Egypt. Periops, Wag.——Col. trabalis, Pall.—— Var. *a.* fuliginosus, Oppel. *b.* minochronis, Cuv.

European.

Col. Elaphis.——Col. scalaris, Sh. Col. Meifrenii, Oppel. Col. bilineatus, Dumeril.——Col. flavescens, Scop.——Col. viridiflavus, Scop. Col. personatus, Daud. young, Col. atrovirens, Metaxa. Zamerinus viridiflavus, Wag.——Col. Æsculapii, Lacep. not Lin. Metaxa. Zamenis Æsculapii, Wag.

Asiatic.

Col. Oppelii, Wag. Russel, ii. t. 30.——Col. melanopis, Oppel. Col. flavolineatus, Rienw.——Col. trivergatus, Boie. Iris, 1826, 209.—— Col. clamacophorus, Do. 210. ——Col. vulneratus, Do. 212.——Col, conspicillatus, Do. 211. Col. canus, Lin. Merrem. i. t. 7. C. cinarascens, Lacep. C. ammobates, Shaw, Seba, ii. t. 3. f. 2. t. 78. f. 2. young C. margaritaceus, Daud. B. ij. t. 9. C. crucifer, Sh. C. Parias, Hem.——Col. radiatus, Russel, t. 42.——Col. mucosus, Lin. ——Col. melanurus, Oppel.—— Col. cancellatus, Oppel. Col. korros, Rein. Col. Saturninus, Merr. 84. ——Col. tricolor, Boie.——Col. geminiatus, Oppel. Isis, 1826.——Col. oxycephalus, Rein. Gonyosoma viride, Wag. t. 9. ——Col. trabalis, Pallas. ——Col. Helena, Daud. Rus. t. 32.——Col. obscurus, Daud. Rus. t. 18.

C. Head large, distinct, depressed ; gape very wide ; nostrils large, open ; body fuciform, thick ; tail shorter than half the length of the body ; scales equal, placed in oblique transverse series. Xenodon, *Boie.*

American, Ophis, *Wagler.*

Col. severus, Lin. Col. versicolor, Lin. Sch. Phy. Sac. t. 661. f. 7. Seba, j. t. 85. f. 1.——X. Merremii, Ophis Merremii, Wag. Spix, Braz. t. 17.——Col. rhapdocephalus, Pr. Max. Braz. i. 351.——Dipsas scholti, Fitz.—— X. aneus, Boie, n. 5.——X. ocellatus, Boie.——X. saurocephalus, Boie. Col. saurocephalus, Boie, Pr. Max.

D. Asiatic, Xenodon, *Wagler.*

Xenodon inornatus, Boie. Col. inornatus, Kuhl.

Head short, blunt, and narrow : tail short, continued, Oligodon, *Boie.*

O. bitorquatus, Boie.——O. atriventus, Boie. Col. atri-

88

ventus, Daud. Merr. Seba, ij. t. 86. f. 5. Col. lutrix, Lin. ? Homalosoma arctiventris, Wag. Elaps Lutonia and E. Duberria, Sch. Duberria arctiventris, Fitz.——O. punctatus. Col. punctatus, Latr. Homolosoma punctatum, Wagler.

E. Head, very long, very distinct ; gape very wide ; tail, very long ; scales small and oblique series, partly keeled. Herpotodryas, *Wag.* Erpetodryas, *Boie.* Inhab. South America.

C. carinatus, Boie. Col. carinatus, Lin. Merr. Seba, ii· t. 54, t. 56, f. 3, t. 71, f. 12. C. fuscus, Lin. Col. subfuscus, Lacep. Natrix bicarinatus, Wag.——E. lævicollis, de Col. lævicollis, Pr. Max.——E. exoletus, Lin. Mus. Adolph. Col. bicarinatus, Pr. Max.——Col. carinatus, var. Merr. Natrix exoleta, Lau.——Col. quadricarinatus, Fitz. Erpet, Boie. Col. pyrrhopogon. Pr. Max.——Col. sexcarinatus, Pr. Max. Natrix cinnamomea, Wag. Without epidermis, back with six series of keeled scales.

F. Tail conical, sub-caudal plates first simple, then two rowed, dorsal scales uniform. Hurria, *Daud.*

Hurria bilineatus, Daud, Rup.——Hurria Nympha, Mer. Col. Nympha, Daud. from Russel, t. 36, 37.

G. Head scarcely distinct from the body, gape small or moderate, teeth colubrine, the hinder rather larger, scales imbricate, very smooth, abdomen arcuate, Coronella *Laur.* Erythrolampus Liophis *and* Cloelia, *Wag.*

American.

Cor. venustissima. Col. venustissima, Pr. Max. Col. binatus, Lich.——Col. agilis, Lin. Col. Æsculapii, Lin. Mus Adolph. t. 21.——Col. poecelogyrus, Pr. Max. Liophis poe Wag. Body marbled the whole length.——Col. formosus Pr. Max. Ery. formosus, Wag.——Col. clœlia, Daud

t. 78.——Col. melanocephalus, Lin. Mus. Adolph. t. 15, f. 2. Col. capite niger, Lacep. Clœlia melanocephalus, Wag.——Col. cobella, Lin. Elaps Cob. Sch. Cerastes Cob. Laur. Merr. Col. serpentinus, Daud. Cates. Leophis Cobella, Wag.——Col. miliaris, Lin. Col. Merremii, Pr. Max. Col. Dictyodes, Id. Natrix chiametla, Wag. t. 2, f. 2. Col. ammobates, Shaw. Liophis miliaris, Wag.——Natrix Forsteri Wag. t. 4, f. 1. Liophis Forsteri, Id.——Col. typhlus, Lin. Lac.——Col. viridissimus, Lin. Mer. 1, t. 12. Col. Ianthinius, Daud. Chlorosoma viridissima, Wag. Col. crassicaudatus, Mer. C. Africanus, Bonat. Seba, ij. 35, f. 4. Col. Reginæ, Lin. Liophis Reginæ, Wag. Natrix semilinea, Wag. t. 11, f. 2. Natrix alamensis, jun. Wag. t. 10, f. 3.——Col. Boddartii, Scitz.——Col. doliatus, Daud. Pr. Max. 368.——Col. meleagris, Sh. from Seba, ij. 32, f. 1, 56.——Col. orythrogastra, Pr. Max.——Natrix occipitalis, Wag. Spix, Braz. t. 6. Cloelia occipitalis, Wag. Col. melanocephalus, Boie. Elaps melanocephalus, Wag. t. 2. f 1. Cloelia dorsata, Wagler.——Col. lineatus, Lin.——Col. Nicandri, Mer.——Col. bicinctus, Herm.——Col. raninus, Boie, from Seba, ij. t. 9.——Elaps triscalis, Mer.

Africa.

Col. annulatus, Lin. Mus. Adol. t. 8. Col. ignobilis. Laur. Col. atrofuscus, Lacep. C. caudiolus, Lac. C. orientalis, Gm. C. epidaurius, Herm. Seba, Thes. i. t. 72. ij. t. 9, t. 13, t. 52, 57, 82, f. 2.——Col. rufulus, Lich. Col. 1823.——Col. rufescens, Gm. Coron. Kotamboija, Laur. C. Hitambocia, Gm. Seba, t. 33, f. 6.——Col. Aurora, Lin. Cloelia aurora, Wag. Lin. Merr. Adol. 19. Seb. ij. 22.——Col. annulatus, Boie. Lin. Mus. Adolph. i. t. 8. Seba, i. t. 54. Merrem, t. 11. Oxyrhopus annulatus, Wagler.——Cor. Petolarius. Coluber pet. Lin. Ib. t. 6. Col. pethola, Bech.

Lacep. t. 3. Oxyrhopus pet. Wagler.——Col. rhombeatus, Boie. Col. rhom. Lin. Ib. t. 24. f. 1.

Asiatic.

Col. Baliedeira, Boie. Col. bal., Kuhl.——Col. octolineatus, Boie. Elaps oct., Schneid. Russel, 4, 38.——Col. Bilineatus, Boie. Col. bilineatus, Lac.——Col. tæniolatus, Boie. Col. tæn. Russel, t. 19.

European.

Col. meridionalis, Daud.——Col. lævis, Lacep.——Col. tessalata, Lat.

H. Head not distinct from body ; no hind frontal scale ; body cylindrical, tail very short, blunt abdominal shields, entire caudal shields, in two rows, scales smooth. Calamaria, *Boie.*

Col. Linnæ, Boie. Col. calamarius, Lin. Mus. Adol. 1. t. 6. f. 1. Anguis calamaria, Laur.——Col. lumbricoides, Boie. ——Col. tessalata, Boie.——Col. maculosa, Boie.——Col. multipunctata, Boie.——Col. virgulata, Boie. ——Col. reticulata, Boie.

I. Head not distinct ; eyes small ; tail short, acute. Brachyorrhos, *Kuhl.* Atractos, *Wagler.*

Br. Albus, Kuhl. Col. albus, Lin. Mus. Adol. t. 14, f. 2. Anguis alba, Laur. Mer. t. 7. Col. brachyurus, Shaw.—— Br. dimidiatus, Kuhl. Col. dimidiatus, Oppel.——Br. Kuhlii, Boie. Br. brachyurus, Kuhl. ——Br. Decussata, Kuhl.——Br. torquatus.——Br. badius, Boie.——Br. flammigerus, Boie.——Br. trilineatus, Atractus trilineatus, *Wagler. Isis,* 1828.——Br. Schach, Boie.

K. Body, long ; sub-compressed ; tail continued short, abdominal shields convex ; scales rhomboidal, nearly square, imbricate. Lycodon, *Boie.*

Col. Maximiliana, Merr. Col. audax, Daud. Col. cancellatus, Merr. Col. cattenularis, Daud. Tar tutta, Russel, 1, t. 15.——Lycedon Hebe, Boie. Col. Hebe, Daud. Col. fasciolatus, Sh. Russel, 1, t. 21.——Lycodon aulicus, Boie. Col. aulicus, Lin. Mus. Adol. t. 12, f. 2.; Seba, t. 91. f. 5. Nat. aulica, Laur. Lappiata, Lacep.——Lycodon subcinctus, Boie. Col. sub. Rein; Seba, t. 109, f. 7; Russel, ij. t. 41. Ophites, sub. Wagler.——Lyc. capucinus, Boie; Russel, 11, t. 37. Col. aulicus, Kuhl.——Lyc. fuliginosus, Boie.—— Lyc. pethola. Col. pethola, Lin.; Seba, t. 28, f. 2, t. 54; f. 4, t. 110; f. 3.: ii. t. 38, f. 2. Cor. pethola, Laur.—— Lyc. leucocephalus. Col. leucocephalus, Merr.——Lyc. Russelii. Col. Russelius, Daud. Katla tutta, Russel, 1, t, 38.——Lyc. unicolor, Boie; Russel, ij. 39.——Lyc. malignus, Boie. Col. Malignus, Merr.——Lyc. macrochinus. Col. macro, Boie.——Lyc. galathea, Boie. Col. galathea, Daud. Russel, 1, t. 26.

L. Nose rounded; head shielded; shield triangular, larger than the dorsal scales; tail conical. Xenopeltis.

Xen. Concolor Rein. Boie. Isis. 564.——Xen. Unicolor. Rein. Boie. Ib. 564. Col. alvearius Oppel.——Xen. leucocephalus, Rien. Boie. Isis, 564.

M. Head ovate indistinct, shielded; body cylindrical, scales equal, tail conical, with a single series of subcaudal scales. Scytale. *Merr.* not *Daud.* Pseudoboa. *Schn.*

Scytale coronata. Mer. Seba, ij. t. 41. f. 1. Pseudoboa coronata. Schn.——Scy. coronata. Pr. Max.——Hurria carinata. Kuhl. Ceraspis carinata. Wag.——Scy. Brachyorrhos, Boie. Isis, 20. Aspidura Brachyorrhos. Wag.

N. Head regularly shielded; body long; slender scales, equal; head long; muzzle acute; tail very long; subcau-

dal plates in two rows. Dryophis. *Fitz.* Dryinus, *Mer.* Oxybelis, *Wag.*

Col. fulgidus, Daud., t. 80. Seba, ij. t. 53. Col. nasutus, Shaw, and Col. purpurascens, Shaw. Dryophis fulgidus, Wag.——Col. acuminatus. Pr. Max. Dryinus Æneus. Spix. Braz. Oxybelis Œneus, Wag. Dryinus auratus. Bell.——Dryophis Panthoraria, Kuhl.

O. Head very long, regularly shielded; body very long; dorsal scales triangular, lateral scales linear.

Muzzle acute appendaged. Passerita.

Dryinus oxyrhynchus, Bell. Dryophis nasutus, Boie. Col. nasutus, Russel, 1. t. 12. Tragops, Wagler. Dryinus nasutus, Merr. Dryophis Russelii, Bottla. Passericki, Russel, t. 13.——Dry. pavoninus. Cuv. Boie. Tragops, Wag. ——Dry. xanthozonica, Kuhl. Tragops, Wag.——Dry. prasina, Rein. Russel, ij. t. 25. Seba, ij. t. 53. f. 4. Dry. nasutus, Bell. Tragops, Wag.——D. rostratus, Reinw.

Muzzle blunt, rounded: ventral plates angularly keeled on the sides. Ahœtula. Leptophis, *Wagler.*

Ahæ Linnei. Col. ahætuta, Lin. Natrix ahætula, Laur. Boiga. Lacep. ij. t. 2. Dendrophis ahætula, Boie.——Ahæ. Richardi. Col. Richardi, Bory.——Ahæ. liocercus. Col. liocercus, Pr. Max. Dendrophis lio, Boie. These two are considered by Wagler as varieties of the first.——Col. caracarus, Gm. Seb. ij. Bungarus filiformis, Oppel.——Daud. periophthalmica, Boie.

Psamnophis, *Boie.*

Psam. Girondicus, Boie. Col. Girond, Daud. Col. Gallicus, Hem. Zacholus Girondicus, Wagler.——Coronella Austriaca, Laur. Zacholus Austriacus, Wagler.——Col. Crucifer, Daud. Mer. Seba, j. t. 109. f. 8.——Col. moniliger, Lac. Col. sibelans, Seba, ij. t. 56.——Macrosoma elegans,

Leach. Col. elegans, Sch.—Psam. pulverulenta, Kuhl, Boie. Isis. xx. 547.——Psam. schokari, Boie. Col. schokari, Forsh.——Psam. lacertina, Boie. Col. lac. Spix, t. 5. Coclopeltes, Lac. Wag. Geoff. Egypt, t. 5. f. 2.——Psam. rhombeatus. Col. Rhom. Lin. Mus. Ad. t. 24. f. 2. Col. Tyria, Kuhl. Coclopeltis, Wagler.——Col. condanarus, Mer. Russel, t. 27. j. t. 27.——Psam. bicolor, Wag.

Dendrophis, *Boie.*

Den. chaireacos, Boie. Russel, ij. t. 26. Elaps bilineatus, Sch.——Den. maniar, Russel, ij. 25. Leptophis mancas, Bell. Zool. Jour.——Den. tristis. Col. tristis, Daud. Col. scandens, Mer. Russel, t. 31.——Den. formosa, Rien. Boie. ——Den. picta, Boie. Col. pictus, Gm. Col. decorus, Sh. Col. cœruleus. Bonnat. Dips. Schokari, Kuhl. Bungarus filum, Oppel. Seba, j. t. 99.——Den polychroa, Boie.—— Den scandens, Col. scandens, Mer. Russel, j. t. 31.

Ventral plates two keeled. Chysopelea, *Boie.*

Col. ornatus, Mer. Col. Ibibcea, Daud. Russel, ij. f. 2. Seba, j. t. 94. ij. t. 7. f. 1. t. 56. f. 1. t. 61. f. 2.——Chrys. rhodopleura, Rern.——Chup. pardalis, Boie.——Chry. paradisi, Boie. Isis, 20. 547. Seba, i. 94. f. 7. i. j. 61. f. 2.

P. Head very high, above flat, shielded before, truncate ; body compressed, with one or three series of longer vertebral scales, lateral scales imbricate. Amblicephalus, *Kuhl.* Pareas, *Wag.*

A. carinata, Boie. Dipsas carinata, Kuhl. Pareas car. Wag.——A. mikani. Col. mikani, Fitz.——A. coccineus. Col. coccineus, Blum.——A. lœvis, Khul.——A. Nattereri. Col. Natereri, Mikan. Dryophyllax Nattereri, Wag.—— Natrix punctatissima, Wag. t. 14. Col. lineolatus, Oppel. Thammodynastes punctatissima, Wag.——Rhinostomus probiscidens, Fitz.

Q. Head short and broad, above shielded; body long, compressed; vertebral scales square; lateral scales linear. Dipsas.

a. Subcaudal scales double. Dipsas, *Laur.* Bungarus, *Oppel.*

Dipsas Indica, Laur. Col. bucephalus, Shaw, Seba, i. t. 43. Col. atrox, Gmel. Dipsas bucephalus, Boie.——Dipsas dendrophila, Rein. Wagler, t. 8. Col. peruvianus, Sh. ——Dip. Drapiezii, Boie, Isis, xx. 549.——Dipsas multomaculata, Rein. Russ. t. 23.——Dipsas cynodon, Cuv. Boie, Isis.——Dipsas trigonatus, Boie.——Dipsas irregularis. Col. irreg, Mer. t. 4. Hurria pseudotriga, Daud.——Dip. cenchoa, Boie. Col. cenchoa, Lin. Seba, ij. t. 16. f. 2.——Dip. Catesbœi. Col. Cates. Weig. Bungarus, Oppel.——Dip. nebulatus. Col. nebulatus, Lin.——Dip. compressus, Boie. Col. compressus, Daud.——Dip. Savignii, Boie.

b. Sub-caudal plates simple, fangs, with the teeth. Bungarus, *Daud.*

Bungarus cæruleus, Daud, Boie, Russel, t. 1. Boa latotecta Herm. —— Bungarus annularis, Daud. Russel, t. 3. —— Bung. semifasciatus, Oppel. Seba. Col. candidus, Lin.

R. Head, scales, with small plates over the face and between the eyes, with plates, simple, double, a simple and double. Cerebus, Cuv. Homolopsis, *Boie.*

Col. cerberus, Daud. Hydrus rhynchops, Schn. Eubydris rhynchops, Latr. Hydrus cinereus, Sh. Python rhynchops, Mer. Homalophis rhinchops, Boie.——Homolopsis obtusatus, Boie.——Hern. erythrogammus, Boie. Col. erythrogammus, Daud.——Hern. carinicaudatus, Boie. Col. carinicaudatus, Pr. Max.——Col. monilis, Lin. Col. buccatus, Lin. Mus. t. 19. Vipera buccata, Daud. Vipera semifasciata, Merr. Col. varius, Merr. Seba.——Col. angulatus, Lin. Merr. t. 9. Natrix aspera, Spix. Braz.——Homolophis aer, Boie. Col. aer, Oppel.——Col. plicatus, Lin. Mus. t. 6. f. 1. Bali. Lacep. Elaps plicatilis, Schn. Cerastes plicatilis, Laur.——Col. Æneus, Spix.——Hom. moluroides, Boie. Col. molurus, Lac. t. 10.—— Hom. plumbea, Kuhl, Boie.——Hom. Jara, Boie. Col. Jara, Shaw, Russel, t. 14. Col. Linnæi, Merrem.

Genus. Boa.

Ventral shields narrow; body thick; anal spurs distinct; nostrils opical; subcaudal plates simple, a double tail conical.

Body compressed fusiform; head depressed; tail prehensile; back of the head scaly; subcaudal plates simple. America. Boa.

Muzzle covered with scales; labial scales flat. Boa, Wagler. Boa constrictor, Lin. Devin or Boa, Empereur, Daud.

Muzzle covered with scales; labial scales pitted. Epicrates, *Wagler.*

Epicrates Cenechria. Boa cenchris, Lin. Daud. Seba. Boa cenchrya, Pr. Max.

Muzzle covered with plates; labial scales flat. Eunectes, *Wag.*

Eunectes murina. Boa murina. Boa scytale, Linn. Seba. Boa anaçondo, Daud. Boa aquatica, Pr. Max.——Eunectus laterisţrigota. Bòa lateristriga, Bœie.

Muzzle covered with plates; labial scales pitted. Xiphosoma, Wag.

Xiphosoma canina. Boa canina. Lin. Boa hypnale, Lin. Xip. araramboya, Wag.——Xiph. hortulana. Boa hortulana.

Lin. Sel. Vipera et Echidna cærulescens. Mer. Vip. moderans and Boa elegans, Daud. Coluber glaucus, Gm. Xiphes dorsale, and X. ornatum, Wag.——Xip. Merrem. Corallus obtusirostris, Daud. accidental deformity.

Body much compressed, keeled; head obliquely cut behind and truncated in front, covered with small shields in front; sub-caudal plates simple. Cenchris, *Gray*, Engyrus, *Wagler*. India.

Cenchris regia, Boa regia, Shaw. Boa carinata, Merrem. Wetter, Amer. t. 9. —— Cenchris ocellata, Boa ocellata, Oppel. Boa carinata, Var. C. Mer.

Body fusiform; head covered with scales; muzzle with small plates; sub-caudal plates, two rowed. Python. Asia.

a. Labial scales with deep pits, dorsal scales rhombic. Constrictor, Wagler.

Python Schneideri, Mer. Boa reticulata and Boa rhombeata, Schn. Python Javanicus, Kuhl. Seba, t. 62. Col. Javanicus, Shaw.——Python Poda, Boie, Pedda Poda. Russel. t. 23. Col. boaformis, Sh. Python Tigris, Daud. ——Python bivittatus, Kuhl. Col. Sebœ, Lin.——Python amethystina, Schn.——Boa orbiculata, Schn. Bora, Russel, t. 39.

a. Labial scales equal, flat. Python, *Wagler*.

Python Peronii, Cuv. Wag. Python punctatus, Mer. K. H.

Body fusiform; head covered with keeled scales like the back; tail short; sub-caudal plates simple; spurs distinct. Gongylophis, *Wagler*, Boa, part *Cuv*. Asia.

Boa conica, Schn. Boa ornata, Daud. Boa viperina, Shaw. Padani Cootoo, Russel, t. 4. Mus. Brit.

Body cylindrical; head covered like the back with small

scales ; muzzle shielded, tail short. Eryx and Clothonia, *Daud.*

Eryx jaculus and E. turcica, Daud. Boa Tartarica, Lich. but Cuvier states that this is without spurs.——Eryx anguiformis. Boa anguiformis, Schn. Eryx Indica, Cuv.

Hydrus, Water or Sea Snakes.

Ventral shields narrow, formed of two united scales; body compressed; anal spurs none; tail compressed, except in Achrocordus; nostrils vertical operculated.

A. Head small, shielded ; body covered with scales. Hydrophis, Hydrus, *Wagler.*

a. Abdomen with two rows of larger scales.

Hydrus major, Sh.——Hyd. nigrocinctus, Merr. from Russel, t. 6. Anguis Xiphura, Herm.——Hyd. chloris, Mer. Russel, t. 7.——Hyd. obscurus, Merr. Russel, t. 8. Leioseloma obscura. Var. a Fitz.——Hyd. schistosus, Mer. from Russel, t. 10.——Hyd. cyanocinatus, Mer. from Russel.

b. Abdomen with two rows of larger scales united into a band.

Hydrus Shawii, Hyd. major, var. Sh. MSS.——Hyd. cœrulescens, Sh. Mus. Brit.—— Hyd. leiolepis, Gray.——Hyd. lanceolatus, Gray. Mus. Brit.——Hyd. spiralis, Shaw.——Hyd. doliatus, Mer. Distera, Lacep. Pelamis, Wag.——Hyd. striatus, Mer.—— Hyd. lævis, Mer. —— Hyd. Brugmansii, Boie.——Hyd. atricapillus, Rein.

B. Head large, depressed, dilated behind, shielded ; body with square plates. Pelamis, Hydrophis, *Wag.*

a. The central abdominal line with a series of scales on each side.

Pelamis bicolor, Daud. Hyd. platura, Lin.——Pelamis ornatus, Gray, Mus. Brit.——Pelamis gracilis, Hydrus gracilis, Sh. Pel. Valakadin, Boie, Russ. t. 11. Disteria, Fitz.——Pelamis Rupelii, Hyd. Schiddil, Boie. Russ. t. 12.——Pel. melanurus, Hyd. melanurus. Wag. t. 3.

b. The central abdominal line formed of a series of long plates.

Pelamis fasciatus, Daud. Hydrus, Sh. Russ. t. 44. Anguis laticauda, Lin.——Pel. Kadll, Russ. t. 13.——Pelamis curtus, Hyd. curtus, Shaw.——Pelamis carinatus, Hyd. carinatus, Cuv. Pel. Lindsayi. Gray. Mus. Brit.

c. Head and body covered with small keeled scales ; tail compressed. Chersydrus, *Cuv.*

Ch. granulatus, Mer. Achrocordus fasciatus. Sh. Hydrus granulatus, Schn.

d. Head and body covered with small keeled scales ; tail conical. Achrocordus, *Horst.*

Ach. dubius, Sh. Ach. Javanicus, Horst.

e. Head shielded, with two soft scaly tentacula, ventral plates narrow ; tail long. Herpeton, *Wagler.*

Herpeton tentaculatus, Daud. Erpeton, Lacep. Merrem.

AMPHIBIA.

Body with a naked skin.

Order 1. MUTABILIA.

Undergoing a transformation; gills deciduous, covered with a deciduous operculum.

Gen. 1. RANA.

Body thick, tailless; feet four, long; sternum and clavicles distinct. Larva elongate, fishlike, tailed and without legs; gills four on each side.

A. Tongue and tympanum distinct ; mouth toothed ; skin smooth ; toes clawless. Rana.

a. Toes simple, eyebrows rounded.

Green Frog. Rana esculenta, Lin.——Common Frog. Rana temporaria, Lin.——Rana cultripes, Cuv.——Rana punctata, Daud. R. Daudini, Mer. Hypscaphelia, Boie. ——R. plicata, Daud. R. Daudini, β Mer.——R. paradoxa, Lin.—— R. tigrina, Daud. R. pipens, Daud. t. 18 ——R. cyanophlyctis, Schn. Bufo cyanophlyctis, Latr. ——R. rudibunda, Pallas. Bufo, Sch.——R. vespertina, Pal. Bufo, Schn.——R pipiens, Lin. not Daud. R. maxima, Cates. R. Catesbiana, Shaw. R. mugiens, Mer.——R. palmipes, Spix.——R. Virginica, Gm. R. Halecina, Daud. R. pipiens, Mer. Shad Frog, Bartram.——R. clamitans, Latr. Daud. t. 16. Grunting Frog, Bartram. Argus Frog, Shaw. R. ocellata, Lin. R. pentadactyla, Laur. R. sonans, Bonnat. Cystygnathus, Wag. R. gruniens, Daud. is thought by Harlan to be the same.——R. melanota, Raffinesque.——R. cutricularis, Harlan.——R. scapularis, Harlan.——R. flaviventris, Har. R. fontinalis, Leconte. Spring Frog, Bartram.——R. sylvatica, Leconte.—— R. palustris, Leconte. R. pardalis, Harl.——R. pumilla, Leconte.——R. Gryllus, Id.——R. dorsalis, Harlan. Savannah Cricket of the Americans.——R. nigrita, Leconte.—— R. typhonia, Daud. R. Virginica, Mer. Cystignathus, Wagler.——R. Gigas, Spix. R. coriacea, Id. R. pachypus, Spix Cystignathus pachypus, Wag.——R. sibilatrix, Pr. Max. R. pygmœa, Spix. Cystignathus, Wagler.—— R. rubella, Latr. R. Daudini, Mer.——R. mystacea, Spix. Cystignathus, Wagler.——R. miliaris, Spix.——R. labyrinthica, Spix. Cystignathus, Wagler.——R. maculata, Latr. Daud.

b. Toes simple ; eyebrows horned. Ceratophys.

America. Ceratophys cornuta, Pr. Max. C. varius, Boie. C. cornuta, Lin. Seba, j. t. 72. f. 1. 2. Ceratophys dorsata, Pr. Max. Back with separate shield-like plates, the front one the largest.——Ceratophys Spixii, Cuv. R. megastoma, Spix, perhaps the same species.——Cer. clypeata, Cuv. Mus. Par. ——Cer. Boiei, Pr. Max. Cer. granosus, Cuv.——Cer. scutata. R. scutata, Spix. Hemiphractus Spixii, Wagler.—— Asia. Cer. montanus, Megalophrys Montana, Kuhl. Wagler.

c. End of toes dilated. Hyla, *Laurent.* Calamita, *Schn.* Tree Frogs.

Hyla arborea. R. arborea, Lin. Hyas, Wagler.—— Hyla ranœformis, Lau. Calamita, Mer. La Bossue, Lacep. Seba.——Hyla Indica, Gray.

Hy. bilineata Daud.——Hy. Blochii, Daud. Calamita cinereus, Schn. Cal. bilineatus, β Mer.——Hyl. Scychelerii, Peron.——Hy. vermiculata, Peron.——Hy. Leschenaultii, Gray.——Hy. Peronii, Gray.——H. pulchra, Gray. ——Hy. bicolor, Daud. Phyllomedusa, Wagler.——Hy. boans, R. Boans, Lin. R. maxima, Lau. Calamita maximus, Schn. Hyla palmata, Latr. Rana Zebra, Shaw. Auletris, Wagler.——Hy. venulosa, Latr. Rana ven., Laur. Hy. viridi-fusca, Laur. Calamita Boans, Schn. Hyla Boans, Mer. Rana Meriana, Shaw. Rana arborea, Var. Gm. R. squamigera, Walb.——Hy. marmorata, Latr. Rana gibbosa, β Gm.——Hy. tinctoria, Latr. Daud. Rana, Shaw. Rametta a tapiner, Lacep.

* Some of the Indian species deposit their eggs on the under side of leaves, hanging over water. General Hardwicke has observed them place the eggs on a leaf which stood over a pail of water, so that the young dropt into the water.

Hy. Surinamensis, Daud. Calamita, Mer.——Hy. hypochondrialis, Latr. Daud. Calamita, Mer.——Hy. intermixta.——Hy. ruber, Laur. La Rouge, Lacep. Hy. sceleton, Laur. —— Hy. aurantiaca, Daud. Auletris, Wagler.—— Hy. lactea, Laur. Rana lactea, Lin. Calam. melanorabdotus, Sch. Hy. Boans, Daud.——Hy. tibiatrix, Laur. H. aurantiaca, Laur.——Hy. lactea, Latr. Daud. Hy. verrucosa, Daud. Hy. variegata, Daud. Hy. lecophyllatus, Beirus. Hy. frontalis, Latr.——Hy. bufonia, Spix.——Hy. geographica, Id.——Hy. albomarginata, Id ——Hy. papillaris, Id.——Hy. pardalii, Id. ——Hy. cinerascens. Id——Hy. affinis, Id.——Hy. trivitata, Id.——Hy. abbreviata, Id.——Hy. lateralis, Daud. Rana arborea, β Gm. Hyla viridis, Laur. Calamita Carolinensis, Penn. C. cinerea, Schn. Rana bilineata, Shaw.—— Hy. femoralis, Daud.— Hy. syuirella, Daud. H. occularis, var Leconte.——Hy. delitescens, Leconte.——Hy. versicolor, Id.—Hy. occularis, Latr.——Hy. Quoyii, Bory.——Hy. Gaimardii, Id.——Hy. Seuerii, Desm. Hy. bifasciata, Gray, Mus. Paris.——Blue tree-frog, White's Journal, Hy. Cyanea, Daud. R. austraciæ, Schn. R. cærulea, Daud.

White's specimen of the last mentioned seems to have wanted a toe on its hind feet, and from this Fitzinger has made of it a genus, under the name Calamita. Hyla violacepoda, and Hyla erythropoda, of the French museum, appear to be nearly applied to, and probably the young of this species. Hyla bufonoides of the Paris Museum, of which H. centripoda of the same Museum may be a variety. Hy. ocellata, and Hy. rubeola, all also of the same Museum, are from New Holland.

d Tongue and tympanum distinct; mouth toothless. Nose rounded. Bufo.

Common Toad, Rana Bufo, Lin. Bufo cinereus, Schn.

Bufo vulgaris, Laur. Bufo Roeselii, Latr. Daud. t. 27?—— Bufo roseus, Wasserkrote, Meyer.——Bufo fuscus, Laur. Rana bombina, Gm. Roesel, Fro. t. 15. Pelobates fuscus, Wagler.——Bufo variabilis, Mer. R. variabilis, Pallas. t. 6, f. 12. Bufo viridis, Laur. Rana viridis, Shaw. Bufo Schreberianus, Laur. Rana sitibunda, Pallas. Bufo sitibundus, Lepeche. See also Geoffroy Reptiles of Egypt, t. 4. Bufo obstetricans, Laur. R. obstetricans, Sturm. Bombina obstetricans, Mer. Alytes, Wagler.——Bufo calamita, Laur. R. protentosa, Blumen. R. Bufo, var. Gm. B. cruciatus, Schn. R. fætidissima, Hern. R. mephitica, Shaw. Bufo salsus Schrank, Daud. Common on heaths near London.—— Bufo marina, Daud. t. 37. B. maculiventris, Spix.——Rana marina, Lin. Bufo marinus, Sch., is from Seba; t. 76, f. 7, it has toes like an Hyla. Copied R. maxima, Shaw.——Bufo ictericus, Spix, Brazils. ——B. lazarus, Id.——B. stellateus, Id. ——B. scaber, Id.——Bufo cinctus, Pr. Max.——B. Agua, Id.——Bufo scaber, Daud.——Bufo pustulosus, Laur. Bufo melanostrictus, Schn. Seba.——Bufo cognatus, Say, Brickcoloured toad, Bartram.——Bufo musicus, Latr. Daud. ——Rana musica, Lin? Bufo clamosus, Schn. Land-frog, Bartram.——Bufo Bengalensis, Daud. Bufo marinus, Mer. R. dubia, Shaw.——Bufo flaviventris, Daud. B. chlorogaster, Daud.——Bufo prætextatus, Boie, MSS.

e. Nose acute, head rounded behind. Rhinella, *Fitz.* Oxyrhynchus, *Spix.*

Rhinella proboscideus, Bufo, Spix. ——Rhi. semilineatus, Bufo, Spix.——Rhi. granulosus.——Rhi. acutirostris—— Rhi. nasicus——all Bufo, Spix, t. 21, and 24.

f. Nose acute; head with a crest extending on each side, to the paratoids. Otilopha, Les Otilophes, *Cuv.*

Otilopha typhonia, R. typhonia, Lin. Seba. R. margiratifera, Laur. Bufo margaritifera, Daud.— Oli. ocellata, Tympanum, obscurely seen through the skin. Dark brown, with white edged brown spots.

B. Tongue distinct; tympanum hidden; toes 3, innerclawed; mouth toothed. Dactylethra, Cuv. Engystoma, *Fitz.* not *Merrem.* Xenopus, *Wagler.*

Xenopus Boiei, Wag. Pipra Buffonia, Mer. Bufo lævis, Daud. Pipa lævis, Mer. Engystoma ovalis, Fitz. ?—Rana fasciata of Burchel's Travels, may belong to this genus.—— Xenopus ovalis, Rana ovalis, Schn. Microps unicolor, Wag.

C. Tongue distinct, tympanum hidden; toes clawless; mouth large, eyebrows rounded. Bombinator.

B. igneus. Rana bombina, Gm. variegata, Lin.——Bombinator Daudini, Bufo ventricosus, Daud. Engystoma ventricosa, Fitz.

D. Mouth large; eyebrows acute. Strombus, *Gray.*

Ceratophys granulata, Cuv. C. Boiei, Pr. Max. Physalamis Cuvieri, Fitz.

E. Mouth and head small, paratoides none; toes, 5, 4. Breviceps. Engystoma, part Fitz.

Breviceps gibbosus, Bufo gibbosus, Daud. Engystoma dorsatum, Cuv. Seba. R. systoma, Schn.——Breviceps marmoratum, Engyst. marmoratum, Cuv. not described.—-Bre. granosus, Engyst. granosus, Cuv.—— Bre. Surinamensis, R. Suninamensis, Daud. R. Bufonia, Mer.——Bre. globulosus, Bufo, glo. Spix, t. 19, f. 1. Chaunus marmoratus, Wagler. ——Bre. albifrons, Bufo albifrons, Spix, t. 19, f. 2. Paludicola albifrons, Wagler.——Bufo ephippium, Spix. This species has only three toes on all the feet, and Fitzinger has made a genus of it, therefore, under the name of Brachycephalus.

F. Tongue indistinct, tympanum, hidden, end of toes, four, cut. Pipa, *Daud.* Asterodactylus, *Wagler.*

Common Pipa, R. Pipa, Lin. Seba, j. t. 77. The females of this species hatch their eggs on the back.——Pipa Curucura, Spix, t. 22. The female of this species do not carry their eggs on the back. Inhabits the bottom of the lakes in Brazil.

Gen. 2. SALAMANDRA.

Body subcylindrical, tailed, feet four, short; sternum clavicules none. Larva, with four feet, gills, three on each side.

* *Toes* 4, 4. *Salamandrina.*

Salamandrina perspicilleta. Savi; les trois doights, Lacep. ij. t. 36. Molge tridactylus, Merrem. Salam. tridactylus, Daud. See also, Salam. Savi, Gosse, with the toes half webbed. Mus. Paris.

** *Toes* 4, 5. *Salamandra.*

a. Skin smooth, paratoids glandular, with group of tubercles on the sides, in two series down the back and tail, each ending in a large pore, lateral line, simple.

Common Salamandra. Salamandra maculosa, Laur. Lac. Salamandra. Lin. Black, yellow spotted.——Black Salamandra. Salamandra atra, Laur. Sal. fusca Gesner, Sturm fauna. Black may be only a variety.

b. Skin granular, granicles ending in a group of minute pores, paratoids porous, with a series of pores on each side, forming a line between the legs.

Common Salamander. Salamandra palustris. Salam. cristata, Latr.——The Marbled Salamander. Salem marmorata, Latr. Triton Gesneri, Laur., and the Alpine Salamander.

Salam. alpestris, Bechst. Lacep. t. 20, perhaps belong to this section.

c. Skin and paratoids granular, each granule ending in a large pore, lateral line porus. Pleurodeles ?

Spanish Salamander. Salamandra major, nob, Mus. Brit. Salam. fenestrata, Mus. Par. Inhab. Gibraltar, Mus. Col. Surg. and Brit. length ten, or twelve inches. This species is very like Pleurodeles Waltl. Michaellis, Isis, 1830, t. 2, said to be found in the South of Spain, and said to differ from the other Salamanders, by having fourteen perfect ribs.

d. Skin and paratoids smooth, with scattered granules, each ending in a pore, place of lateral line occupied by a large blood vessel.

Common Eft. Salamandra vulgaris. Sal. punctata, Latr. ——The Webfooted Eft. Salam. palmata, Latr., scarcely appears to be distinct.

e. Skin and paratoids smooth, minutely punctulated, the pustules along the upper surface of the tail rather larger, and more distinct.

Violet Salamander. Salamandra subviolocea, Barton. Scargus Valeno Mus. Par. Sal. Venenosa, Barton, Amer. Phil. Trans. vi. t. 1. Brit. Mus.

f. Skin and paratoids smooth, minutely punctated, lateral line none, toes long and free, Molge. nob.

Glutinous Salamander. M. glutinosa. Salamandra glutinosa, Green. Mus. Brit.——Red backed Salamander. Salamandra Erythronota, Green. Brit. Mus.—— Cuvier says that Salam. Japonica, Bechst. Lacep. ij. t. 10, f. 1, is allied to this species.——Ashy Salamander. Salamandra cinerea, Green.——Banded Salamander. Salamandra fasciata, Green, J. A. N. S. Phil. Mus. Brit.

Probably the following American species belong to this section.

Punctate Salamander. Salam. symetrica. Sal. punctata, Bechst. Lacep. ij. t. 10, f. 2.——Tigerine Salamander. Salam. tigrina, Green. N. America.——Two lined Salamander. Salam. bislineata, Green, New Jersey.——Red Salamander. Salem. rubra, Daud. Salam. rubiventris, Green. North America——Variolated Salamander. Salam. variolata, Gilliams. Jour. Sic. Nat. Philad. 1, t. 18, f. 1, Pensylvania.——Cylindrical Salamander. Salam. cylindracea, Harlan. South Carolina.——White faced Salamander. Salamandra frontalis. Sal. sinciput-alba, Green.——Brown Salamander. Salamandra fusca, Green, Jour. Acad. Philad. New Jersey.——Dorsal Salamander. Salamandra dorsalis, Harlan. South Carolina.——Painted Salamander. Salamandra picta, Harlan. Pensylvania.——Yellow Salamander. Salamandra flavissima, Harlan. Pensylvania.——Capt. Beechy's Salamander. Salamandra Beechii, Gray.——Spotted Salamander. Salamandra maculata, Green. New Jersey. Mus. Brit.——Brownish Salamander, Salamandra subfusca, Green. New Jersey.——Long Tail Salamander. Salamandra longicaudata, Green.——Black Salamander. Salamandra nigra, Green. Pennsylvania.——Green's Salamander. Salamandra Greeni, Gray.

g. Skin and paratoids smooth, minutely punctuated; lateral line none; toes very short webbed.

Salamandra variegata. Gray. Sal. platydactylus. Cuv. Mus. Paris. Brown, with irregular dorsal bands. Mexico.

Sect. II. Amphipneusta. *Not undergoing transformation.*

Gen. I. Proteus.

Gills free, persistent; skull formed of several distinct bones;

body rather depressed; tail compressed; legs four; muzzle depressed.

a. Toes 3 before, 2 behind. Hydochthon.

Common Proteus. Proteus anguinus, Laur. Siren anguinus, Schn. See Configliachi, Monografra and Schrieber's, Monograph, Carniola.

b. Toes 4 before, 4 behind. Menobranchus, *Harlan.* Necturus, *Raffin.* Phanerobranchus, *Fitz.*

Say's Menobranchus. Menobranchus lateralis, Harlan. Triton lateralis. Acad. Nat. Sci. Philad. iv. t. 21. Ann. Lyc. N. ij. 1. 16. Proteus of the Lakes, Mitchell, Silliman Jour, iv. and vii.——Salamandra alleghaniensis, Jun. Say. ——Lacepedes Menobranchus. Menobranchus Lacepedii. Proteus tetradactylus. Lacep. Ann. Mus. x. t.

c. Toes 4 before, 5 behind. Phyllhydrus, *Brooks*, Gyrinus. *Shaw* not *Lin.* Siredon, *Wagler.*

Common Axolotl. Phyllhydrus pisceformormis. Siren pisciformis. Shaw Zool. iij. t. 140. Misc. t. Axolotl Humboldt. Obs. Zool. t. 12. Home Phil. Trans. and Lacep. Anat. Siredon Axolotl, Wagler. Mexico. Consult also Proteus Nove Cæsarcenris, Green. Jour Acad. Phil. and Siren operculata, Pal de Beauv. Amer. Phil. Trans. iv.

Gen. II. Siren.

Gills free, persistent; skull of distinct bones: body subcylindrical; tail compressed; legs two, anterior.

a. Toes 4, 4. Siren, *Linn.* Muræna, *Gmel.*

Common Siren. Siren lacertina, Linn. Mud Iguana. Ellis Phil. Trans.——Intermediate Siren. Siren intermedia, Leconte Ann. Lyc. New York ij. t. 1.

b. Toes 3. 3. Pseudobranchus, *Gray.*

Striated Siren. Pseudobranchus striatus. Siren striata, Leconte Ann. Lyc. New York, j. t. 4. Captain Leconte has unjustly and unscientifically attacked this genus. See Ann. Lyc. New York, ij. 133, though it was established originally on the authority of his erroneous account of the Opercula.

Gen. III. Amphiuma.

Gills none; skull formed of a solid bony substance; body subcylindrical; tail compressed; legs four.

a. Legs rather strong; toes 4, 5, *outer edge of the feet fringed.* Abranchus.

Large Hellbender. Abranchus horridus. Salamandra gigantea. Salamandra horrida, and Protonopsis horrida, Barton. Menopoma Alleghaniensis, Harlan. Triton Alleghaniensis, Daud. Molge gigantea, Merrem. Salamandra Alleghaniensis, Michaux. Abranchus alleghaniensis, Leconte, Ann. Lyc. j. t. 17.

b. Legs weak, boneless. Amphiuma.

Common Congo Snake. Amphiuma means. Garden. Florida and Georgia. Mus. Brit. Chrysodonta larvaformis, Mitchel.——Three Toes Congo Snake. Amphiuma tridactylum. Cuv. Mem. Mus. xiv. t. 1. Mus. Col. Surg. Four footed. Siren. Barton.

Gen. IV. Cæcilia. Linn.

Gills none, head depressed, formed of a solid bony substance; legs none; body cylindrical; tail short, blunt; vent roundish, nearly terminal—perhaps this should be placed near Amphisbœna.

a. Muzzle blunt ; nostrils bearded ; body distinctly ringed. Cæcilia. Siphonops, *Wagler*.

Ringed Cœcilia. Cæcilia annulata. Spix, t. 27. f. 1.—— Interrupted Ringed Cæcilia. Cæcilia interrupta, Cuv.—— Bearded Cæcilia. Cæcilia tentaculata, Linn. Amon. Acad. j. t. 17. f. 1. Cæcilia albiventris, Daud. vii. t. 122. f. 2.—— Glutinous Cæcilia. Cæcilia glutinosa, Lin. Mus. Adolph, t. 4. f. 1. Seba, t. 25. f. 2. Ceylon.——Two Banded Cæcilia. Cæcilia bivittata, Cuv. America.——Sharp-nosed Cæcilia. Cæcilia rostrata, Cuv.

b. Muzzle prominent, body slender, very long, smooth, tentacula distinct. Ichthyophis, *Fitz*. Cæcilia, *Wagler*.

Wormlike Cæcilia. Cæcilia lumbricoides, Cuv. Daud. t. 92, f. 2, Lin. Mus. Adolph. t. 5, f. 2.

c. Muzzle depressed, bearded; body, rather fusiform, very closely ringed, eyes small. Epicrium, *Wagler*.

Javanesse Cæcilia. Cæcilia hypocyana, Van. Hasselt, Isis. 1827. Epicrium Hasseltii, Wagler, Java. Cæcilia nasuta, is said to be a fish, the Sphagebranchus Coromandelicus of Russel, t. 37.